Digital Electronics 3

Series Editor
Robert Baptist

Digital Electronics 3

Finite-state Machines

Tertulien Ndjountche

iSTE

WILEY

First published 2016 in Great Britain and the United States by ISTE Ltd and John Wiley & Sons, Inc.

ISTE Ltd
27-37 St George's Road
London SW19 4EU
UK

www.iste.co.uk

John Wiley & Sons, Inc.
111 River Street
Hoboken, NJ 07030
USA

www.wiley.com

Library of Congress Control Number: 2016950312

British Library Cataloguing-in-Publication Data
A CIP record for this book is available from the British Library
ISBN 978-1-84821-986-1

Contents

Preface

The omnipresence of electronic devices in everyday life is accompanied by the size reduction and the ever-increasing complexity of digital circuits. This comprehensive and easy-to-understand book deals with the basic principles of digital electronics and allows the reader to grasp the subtleties of digital circuits, from logic gates to finite state machines. It presents all the aspects related to combinational logic and sequential logic. It introduces techniques to establish logic equations in a simple and concise manner, as well as methods for the analysis and design of digital circuits. Emphasis has been especially laid on design approaches that can be used to ensure a reliable operation of finite state machines. Various programmable logic circuit structures by practical examples and well-designed exercises with worked solutions.

This series of books discusses all the different aspects of digital electronics, following a descriptive approach combined with a gradual, detailed and comprehensive presentation of basic concepts. The principles of combinational and sequential logic are presented, as well as the underlying techniques for the analysis and design of digital circuits. The analysis and design of digital circuits with increasing complexity is facilitated by the use of abstractions at the circuit and architecture levels. The series is divided into three volumes devoted to the following subjects:

1) combinational logic circuits;

2) sequential and arithmetic logic circuits;

3) finite state machines.

A progressive approach has been chosen and the chapters are relatively independent of each other. To help master the subject matter and put the different concepts and techniques into practice, the book is complemented by a selection of exercises with solutions.

Summary

This volume deals with finite state machines. These machines are characterized by a behavior that is determined by a limited and defined number of states, and the holding conditions for each state and the branching conditions from one state to another. They only allow one transition at a time and can be divided into two components: a combinational logic circuit and a sequential logic circuit. This third volume contains the following three chapters.

1) Synchronous Finite State Machines;

2) Algorithmic State Machines;

3) Asynchronous Finite State Machines.

The reader

This book is an indispensable tool for all engineering students in a bachelors or masters course who wish to acquire detailed and practical knowledge of digital electronics. It is detailed enough to serve as a reference for electronic, automation engineers and computer engineers.

Tertulien NDJOUNTCHE
August 2016

1

Synchronous Finite State Machines

1.1. Introduction

Digital circuits composed of combinational and sequential logic sections are generally described as finite state machines.

A machine is synchronous when the state transitions are controlled or synchronized by a clock signal.

A machine whose operation is not dependent on a clock signal is said to be asynchronous.

The present state (PS) of a state machine is determined by the variables stored in the flip-flops of the sequential section. The next state (NS) of the state machine is defined by the circuit of the combinational logic section.

Among finite state machines, we can differentiate between the *Moore* model and the *Mealy* model:

– Moore state machine: the state machine output depends entirely on the PS;

– Mealy state machine: the state machine output depends on the inputs and PS.

It must be noted that there are also hybrid machines with some outputs being of Moore type and others of Mealy type.

A machine always has a finite number of states. For N variables, the machine must have between 2 and 2^N states.

A machine is defined by specifying the number of inputs and outputs, the initial state, the PS and the NS.

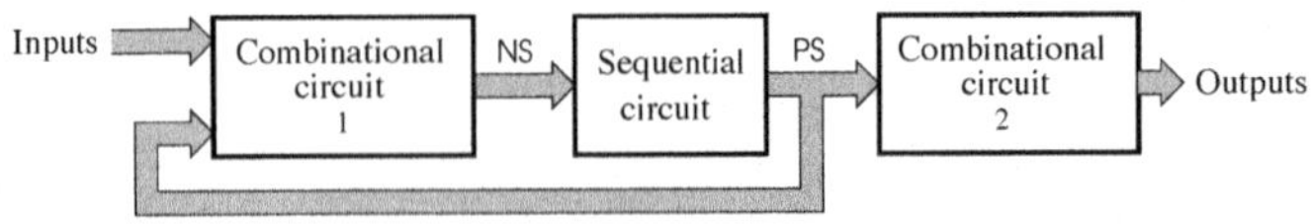

Figure 1.1. *Finite state machine: Moore model (NS: next state; PS: present state)*

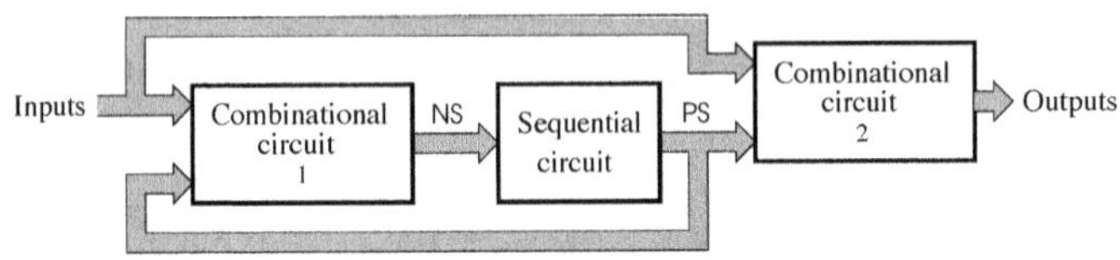

Figure 1.2. *Finite state machine: Mealy model (NS: next state; PS: present state)*

1.2. State diagram

Consider the state diagram for the Moore state machine shown in Figure 1.3. Starting from the initial state S_0, the machine goes to the state S_1 regardless of the logic state of the input X. Assuming that the PS corresponds to S_2 and that the output is set to 1, the NS will be either S_1, with the output remaining at 1 if the logic level of the input X becomes 0, or S_3, with the output being set to 0 if the input X takes the logic level 1.

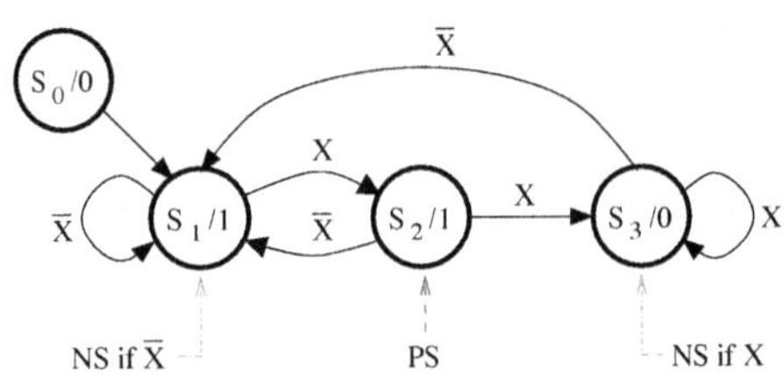

Figure 1.3. *Moore state machine: state diagram with present state and next state*

Figure 1.4(a) shows a section of the state diagram for a Mealy state machine. The states whose binary codes are 000, 010, 001 and 011 are denoted by A, B, C and D, respectively, and the outputs are S_1 and S_2.

We assume that B is the PS. The holding condition in state B is XY and the outputs S_1 and S_2 take the logic state 1. The input condition $\overline{X}$ causes the machine to enter the state D and the output, S_2, is set to 0. Once in this state, the $\overline{X}$ condition allows the machine to remain in this state. When the logic condition $X\,\overline{Y}$ is true, the machine goes to the state C, where there is no holding condition and the output S_1 is set to 0.

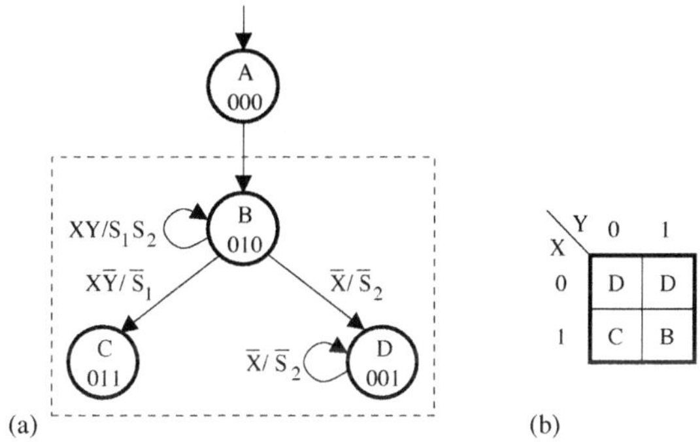

Figure 1.4. *Mealy state machine: a) state diagram; b) map showing input/next state from state B*

Figure 1.4(b) shows what state the machine may move to once in the state B based on the logic levels of inputs X and Y.

A state diagram is constructed according to certain rules. For a section of the state diagram, such as the one illustrated in Figure 1.5, where the conditions that cause the machine to remain in state S_j and to move from S_j to S_k ($k = 1, 2, \cdots, n-1$) are represented by F_0 and F_k, respectively, the following logic equations must be verified:

– *Sum rule*: the Boolean sum of all conditions under which a transition from a given state occurs must be equal to 1:

$$F_0 + F_1 + \cdots + F_{n-1} = 1 \quad [1.1]$$

– *Mutual-exclusion requirement*: each condition under which a transition from a given state occurs cannot be associated with more than one transition path:

$$F_0 = \overline{F_1 + F_2 + \cdots + F_{n-1}} \quad [1.2]$$

$$F_1 = \overline{F_0 + F_2 + \cdots + F_{n-1}} \quad [1.3]$$

$$\vdots$$

$$F_{n-1} = \overline{F_0 + F_1 + \cdots + F_{n-2}} \quad [1.4]$$

As a result, the Boolean product of both state transition conditions, $F_l \cdot F_k$ $(l, k = 0, 1, 2, \cdots, n-1$ and $l \neq k)$, is equal to 0.

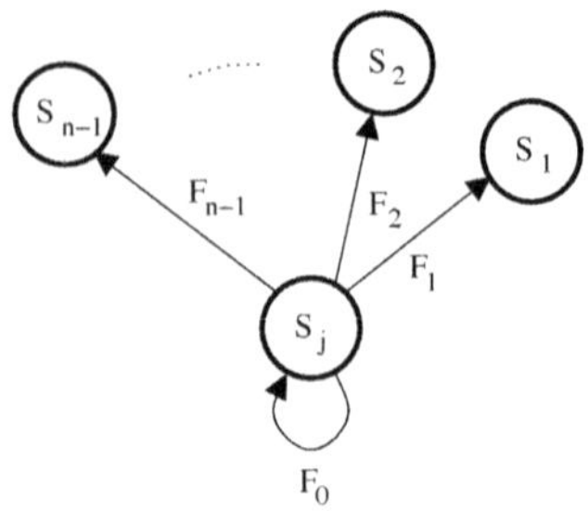

Figure 1.5. *Section of a state diagram*

However, these relationships need not be verified for applications where certain conditions will never happen or are not allowed (don't-care conditions).

EXAMPLE 1.1.– Let us consider the section of the state diagram illustrated in Figure 1.6(a). Using the Boolean transformation, we have:

$$\overline{X} + X \cdot Y + X \cdot \overline{Y} = \overline{X} + X(Y + \overline{Y}) = \overline{X} + X = 1 \quad [1.5]$$

and

$$\overline{X} = \overline{X \cdot Y + X \cdot \overline{Y}} = (\overline{X} + \overline{Y})(\overline{X} + Y) = \overline{X}(1 + Y + \overline{Y}) = \overline{X} \quad [1.6]$$

$$X \cdot Y = \overline{\overline{X} + X \cdot \overline{Y}} = X(\overline{X} + Y) = X \cdot Y \quad [1.7]$$

$$X \cdot \overline{Y} = \overline{\overline{X} + X \cdot Y} = X(\overline{X} + \overline{Y}) = X \cdot \overline{Y} \quad [1.8]$$

Thus, the sum rule and the mutual-exclusion requirement are both satisfied. Figure 1.6(b) depicts the map showing the input/NS from state A.

EXAMPLE 1.2.– Analyzing the state diagram shown in Figure 1.6(c), we can see that the sum rule is verified while the mutual-exclusion requirement is not fulfilled because the product of the terms X and $X \cdot Y$ is not equal to 0. Figure 1.6(d) shows the map for the input/NS from state A. For the branching condition $XY = 11$, the NS can be either B or C, while only one transition at a time can be carried out from a given state. Thus, when the mutual-exclusion requirement is satisfied for a given state, no cell in the input/NS map should contain more than one state symbol.

EXAMPLE 1.3.– A section of the state diagram of a finite state machine is depicted in Figure 1.6(e). We can verify that the sum rule is satisfied, but not the mutual-exclusion requirement. This is because the product of the terms $\overline{X}$ and $\overline{Y}$ is not equal to 0. As shown in Figure 1.6(f), that illustrates the input/NS starting from the state A, the $\overline{X} \cdot \overline{Y}$ condition causes the state machine either to remain in state A or to advance to state C. However, this ambiguity can be ignored if it is assumed that the condition $\overline{X} \cdot \overline{Y}$ will never occur.

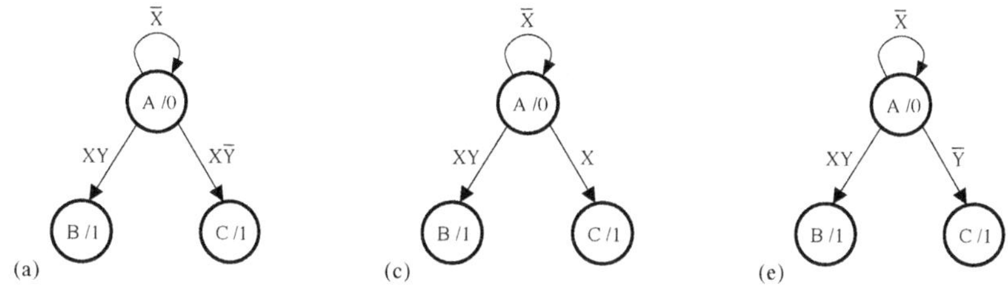

(b)

X \ Y	0	1
0	A	A
1	C	B

(d)

X \ Y	0	1
0	A	A
1	C	B+C

(f)

X \ Y	0	1
0	A+C	A
1	C	B

Figure 1.6. *Examples of state diagram sections and maps showing the input/next state from state A*

In a state diagram, we can differentiate between conditional and unconditional transitions:

– conditional transitions are only carried out on the edge of a clock signal when a certain condition, relating to the inputs, is verified. There are always at least two conditional transitions from the same state;

– unconditional transitions are automatically carried out on the occurrence of a clock signal edge. Only one unconditional transition is possible from a given state.

Let us consider an example: the state diagram for an incompletely specified Moore state machine shown in Figure 1.7. There are two inputs and the output can take either 0 or 1 or a don't-care state, represented by (–). The only unconditional transition exists between the states S_3 and S_0.

The operation of this machine can also be described based on the state table shown in Table 1.1. Starting from the state S_3, where the output is in the don't-care state, the machine goes to S_0 regardless of the logic combination of the inputs.

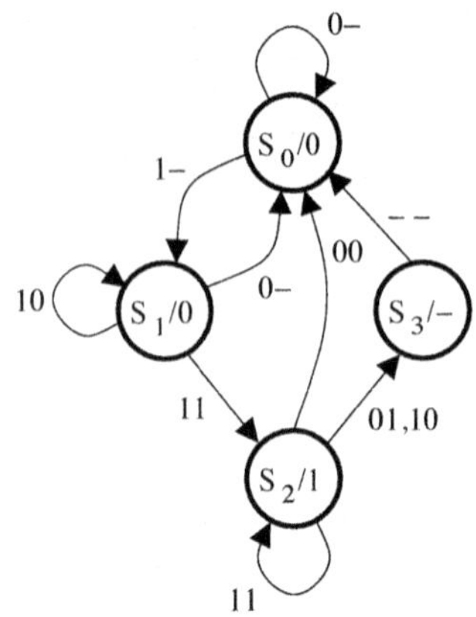

Figure 1.7. *State diagram for an incompletely specified state machine based on Moore model*

PS	NS				Output
	$XY = 00$	01	10	11	
S_0	S_0	S_0	S_1	S_1	0
S_1	S_0	S_0	S_1	S_2	0
S_2	S_0	S_3	S_3	S_2	1
S_3	S_0	S_0	S_0	S_0	–

Table 1.1. *State table of an incompletely specified state machine based on Moore model*

1.3. Design of synchronous finite state machines

The procedure for designing synchronous finite state machines may include the following steps:

1) derive the state diagram;

2) draw up the state table;

3) assign bit combinations to the variables in order to represent the different states (encoding the different states) and draw up the corresponding state table;

4) choose the flip-flop type;

5) derive the input equations based on Karnaugh maps;

6) represent the resulting logic circuit.

There are several possibilities for encoding the states, leading to different logic equations. However, it is preferable to assign bits to states such that the simplest logic expression can be obtained. The minimal logic equation is obtained if the 1s in the Karnaugh map are adjacent to each other. In general, this can be achieved by using one of the following rules (see Figure 1.8):

– adjacent codes, or codes differing by one bit, are assigned to states that lead to the same NS for a combination of inputs;

– adjacent codes, or codes differing by one bit, are assigned to the NSs from the same state.

– adjacent codes, or codes differing by one bit, are assigned to states that produce the same output for a combination of inputs.

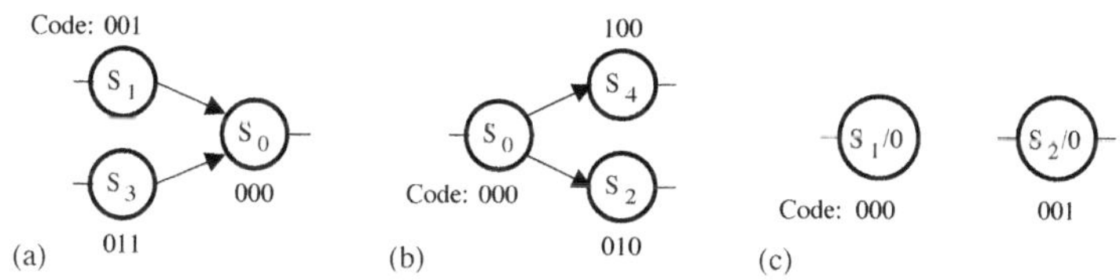

Figure 1.8. *State encoding*

However, it should be noted that the only viable method to obtain optimum encoding of the states is to test all possibilities.

For a state machine, the number of bits and flip-flops will be N if the number of states is between $2^{N-1}+1$ and 2^N.

When designing a finite state machine, the excitation table of each flip-flop can be used to derive the logic equations for the inputs (D, J and K, T, or S and R) based on the output's PS (Q) and NS (Q^+). Tables 1.2–1.5 give the excitation tables of the D, JK, T and SR flip-flops, respectively.

In the case of synchronous state machines, it must be noted that the T and SR flip-flops are implemented on programmable logic circuits using a D or JK flip-flop.

1.4. Examples

1.4.1. *Flip-flops*

The operation of the SR (without forbidden states), D, JK and T flip-flop can be described using a finite state machine model.

Q	$\to$	Q^+	D
0	$\to$	0	0
0	$\to$	1	1
1	$\to$	0	0
1	$\to$	1	1

Table 1.2. *Excitation table of the D flip-flop*

Q	$\to$	Q^+	J	K
0	$\to$	0	0	x
0	$\to$	1	1	x
1	$\to$	0	x	1
1	$\to$	1	x	0

Table 1.3. *Excitation table of the JK flip-flop*

Q	$\to$	Q^+	T
0	$\to$	0	0
0	$\to$	1	1
1	$\to$	0	1
1	$\to$	1	0

Table 1.4. *Excitation table of the T flip-flop*

Q	$\to$	Q^+	S	R
0	$\to$	0	0	x
0	$\to$	1	1	0
1	$\to$	0	0	1
1	$\to$	1	x	0

Table 1.5. *Excitation table of the SR flip-flop*

Only the inputs can be used to annotate the transitions in the state diagram, because each flip-flop can have only two states, which are the same as the output states.

Figure 1.9(a) depicts the logic circuit and symbol of the SR flip-flop; Figure 1.9(b) shows the state diagram, and the transition table is given in Table 1.6.

Figure 1.10(a) presents the symbol of the D flip-flop; Figure 1.10(b) depicts the state diagram, and the transition table is shown in Table 1.7.

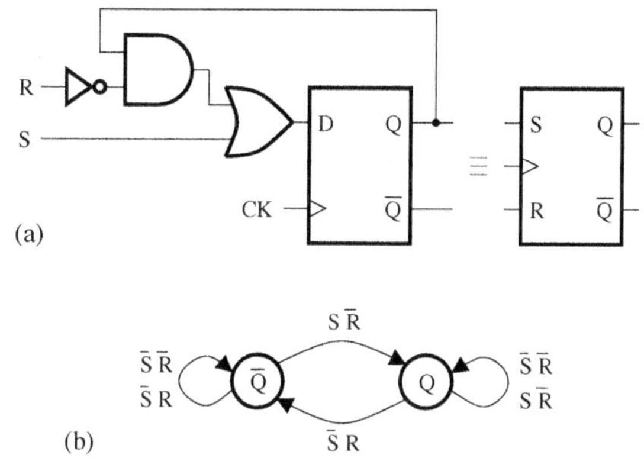

Figure 1.9. *SR flip-flop: a) circuit and symbol; b) state diagram*

R	S	Q	Q^+
0	0	0	0
0	0	1	1
0	1	0	1
0	1	1	1
1	0	0	0
1	0	1	0
1	1	0	x
1	1	1	x

Table 1.6. *Transition table of the SR flip-flop*

Figure 1.11(a) depicts the symbol of the JK flip-flop; Figure 1.11(b) depicts the state diagram, and the transition table is represented in Table 1.8.

Figure 1.12(a) shows the logic circuit and symbol of the T flip-flop; Figure 1.12(b) depicts the state diagram, and the transition table is represented in Table 1.9.

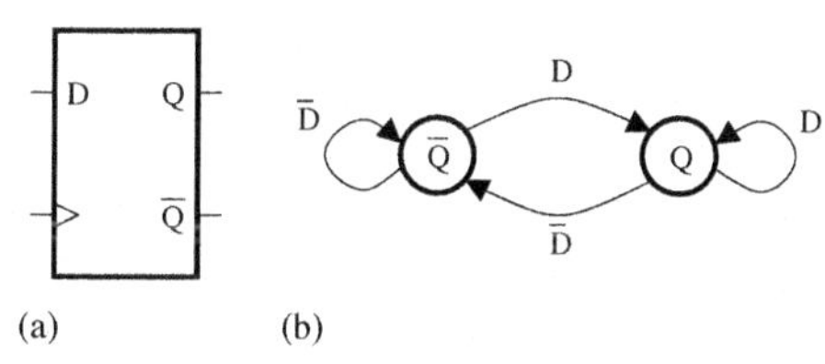

Figure 1.10. *D flip-flop: a) symbol; b) state diagram*

D	Q	Q^+
0	0	0
0	1	0
1	0	1
1	1	1

Table 1.7. *Transition table of the D flip-flop*

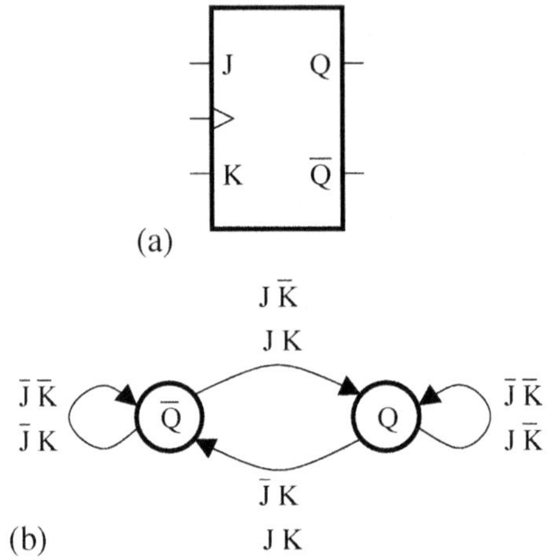

Figure 1.11. *JK flip-flop: a) symbol; b) state diagram*

J	K	Q	Q^+
0	0	0	0
0	0	1	1
0	1	0	0
0	1	1	0
1	0	0	1
1	0	1	1
1	1	0	1
1	1	1	0

Table 1.8. *Transition table of the JK flip-flop*

The characteristic equations can be written as follows:

– SR flip-flop: $Q^+ = S + \overline{R} \cdot Q$ (without a forbidden state);

– D flip-flop: $Q^+ = D$;

– JK flip-flop: $Q^+ = J \cdot \overline{Q} + \overline{K} \cdot Q$;

– T flip-flop: $Q^+ = T \oplus Q$.

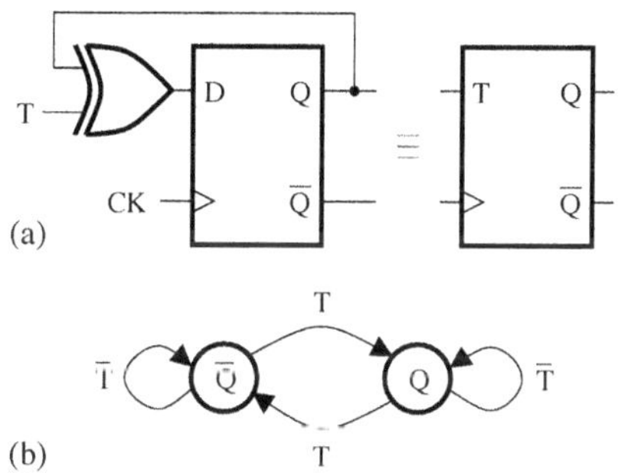

Figure 1.12. *T flip-flop: a) circuit and symbol; b) state diagram*

T	Q	Q^+
0	0	0
0	1	1
1	0	1
1	1	0

Table 1.9. *Transition table of the T flip-flop*

Some applications require the use of flip-flops with a synchronous reset input. The logic circuit of a D flip-flop with a synchronous reset input and enable signal is shown in Figure 1.13(a).

By analyzing the circuit, the timing diagram can be represented as shown in Figure 1.13(b) and the following characteristic equation can be obtained:

$$Q^+ = D = (D_i \cdot \mathrm{EN} + Q \cdot \overline{\mathrm{EN}})\mathrm{RST} \qquad [1.9]$$

where EN is the enable signal and RST designates the reset signal. When the reset is only taken into account at the active edge of the clock signal, it is possible to eliminate the effect of parasitic disturbances that may affect RST.

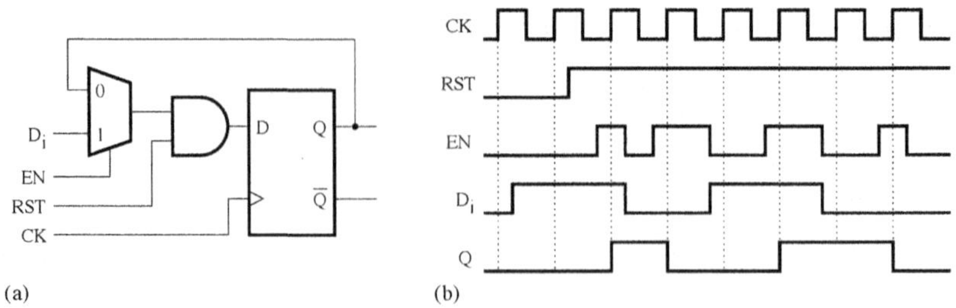

Figure 1.13. *D flip-flop with a synchronous reset input and enable signal: a) logic circuit; b) timing diagram*

1.4.2. *Binary sequence detector*

A logic circuit that can recognize the binary sequence 101 is to be designed. The output, Y, will be set to 1 immediately after the last bit of this sequence is applied at the input, X, of the circuit.

The initial step consists in representing the state diagram describing the state changes and the output in response to each of the input bits. This diagram is then converted into a state table that, in conjunction with the excitation table of a given flip-flop, helps to fill in the Karnaugh maps used for the determination of the logic equations of the NSs and output.

1.4.2.1. *Mealy model*

Based on the Mealy model of the sequence detector for the sequence 101, the state diagram can be represented as shown in Figure 1.14, where the initial state is designated by S_0. With a single input variable, each state must have two exit paths (one for each logic state of the input variable). However, depending on the specifications, it can have any number of incoming paths. Tables 1.10 and 1.11 depict possible representations of the state tables. Starting from a given state, the NS depends on the value of the input X. As each flip-flop can only have two states, two flip-flops are required to implement the 101 sequence detector.

The Karnaugh maps for the NSs A^+ and , and for the output Y are represented in Figures 1.15(a)–(c), respectively. Choosing to use the *D* flip-flop, whose characteristic equation is of the form, $Q^+ = D$, the logic equations for the NSs are given by:

$$A^+ = D_1 = B \cdot \overline{X} \qquad [1.10]$$

and:

$$B^+ = D_2 = X \qquad [1.11]$$

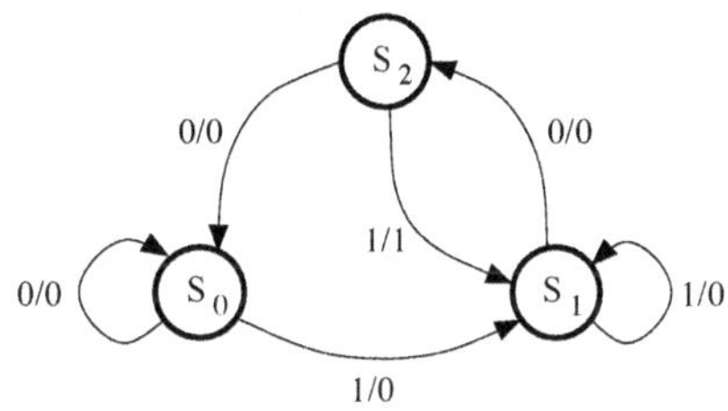

Figure 1.14. *State diagram (Mealy model)*

PS	NS		Output Y	
	$X = 0$	1	$X = 0$	1
S_0	S_0	S_1	0	0
S_1	S_2	S_1	0	0
S_2	S_0	S_1	0	1

Table 1.10. *State table (Mealy model)*

PS AB	NS A^+B^+		Output Y	
	$X = 0$	1	$X = 0$	1
00	00	01	0	0
01	10	01	0	0
10	00	01	0	1

Table 1.11. *Transition table (Mealy model)*

The logic equations for the output is written as:

$$Y = A \cdot X \quad [1.12]$$

Figure 1.16 shows the logic circuit for the 101 sequence detector based on the Mealy state machine. The timing diagrams represented in Figures 1.17(a) and (b) show the case where the input signal is ideal and when it is affected by an undesirable transient disturbance (or *glitch*). In general, the output of a machine based on a Mealy model can be sensitive to a transient disturbance affecting the input signal.

NOTE 1.1.– It is possible to represent four states with two flip-flops. There is, therefore, a state that is unused by the 101 sequence detector based on the Mealy model. In general, when there are unused states their effect on the state machine operation must be analyzed.

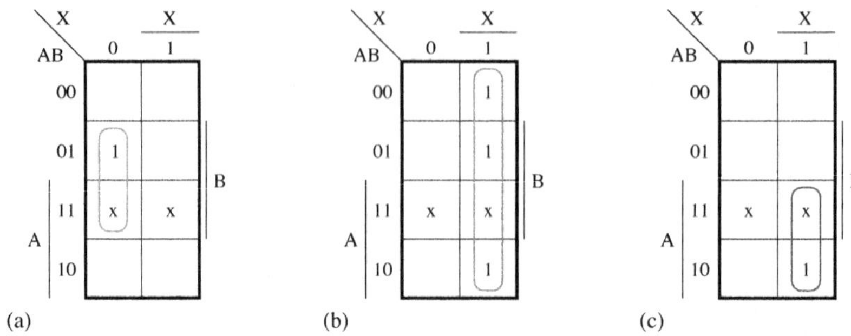

Figure 1.15. *Mealy state machine: Karnaugh maps for a) A^+; b) B^+; c) Y*

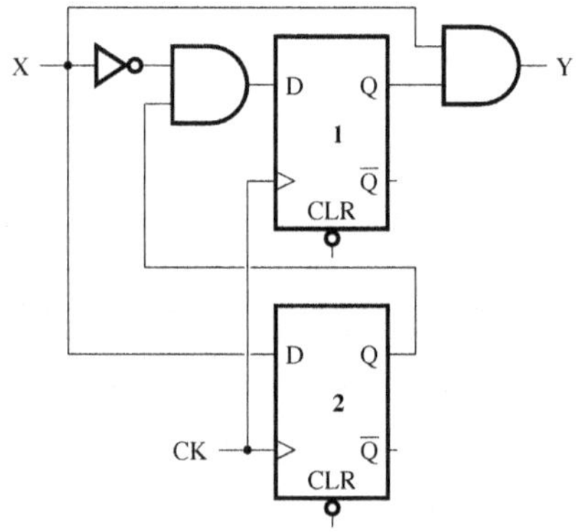

Figure 1.16. *Logic circuit (Mealy model)*

1.4.2.2. *Moore model*

For implementation of the detector based on the Moore model, we proceed as previously described, but by associating possible outputs with the states rather than with the transitions between states. The state diagram of the Moore state machine is represented in Figure 1.18. The output will be set to 1 only if the sequence 101 is detected. With the initial conditions, which are defined by choosing S_0 as the current state and by setting the output to 0, three conditions, S_1, S_2 and S_3, are required for the recognition of the sequence 101. Tables 1.12 and 1.13 give the possible representations of the state table. The output is determined by the PS and is not dependent on the input. The Karnaugh maps for the NSs, A^+ and B^+, and for the output Y are shown in

Figures 1.19(a)–(c), respectively. When a D bascule is used to represent each variable, we have $Q^+ = D$, and the logic equations for the NSs are given by:

$$A^+ = D_1 = B \cdot \overline{X} + A \cdot \overline{B} \cdot X \qquad [1.13]$$

and:

$$B^+ = D_2 = X \qquad [1.14]$$

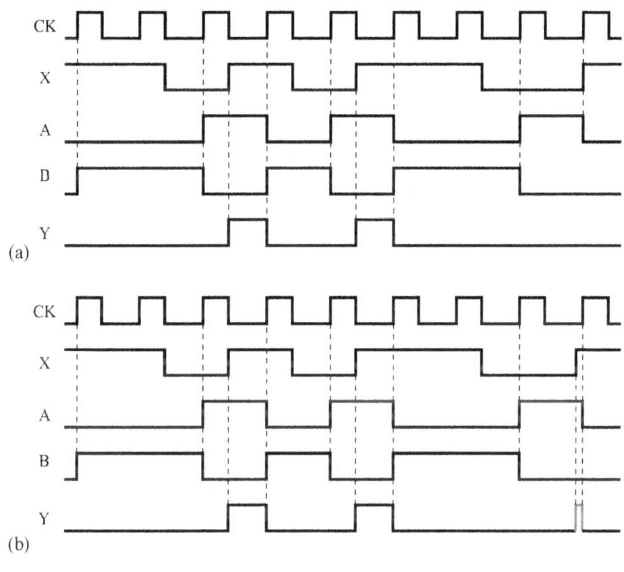

Figure 1.17. *Mealy state machine: a) ideal timing diagram; b) non-ideal timing diagram*

The logic equation for the output can be reduced to:

$$Y = A \cdot B \qquad [1.15]$$

The logic circuit for the 101 sequence detector based on the Moore model is represented in Figure 1.20. The timing diagram shown in Figure 1.21 illustrates the working of this circuit, where X designates the input signal, and A and B are the outputs of the flip-flop 1 and 2, respectively. The output of the machine, Y, can change only at the rising edge of the clock signal, CK, and is memorized for at least one clock cycle.

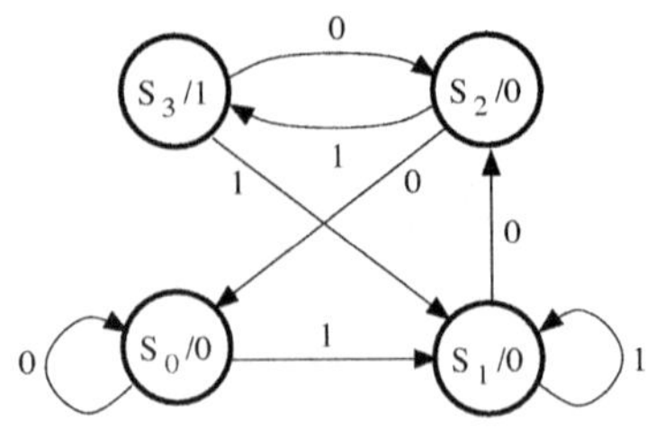

Figure 1.18. *State diagram (Moore model)*

PS	NS $X=0$	NS 1	Output Y
S_0	S_0	S_1	0
S_1	S_2	S_1	0
S_2	S_0	S_3	0
S_3	S_2	S_1	1

Table 1.12. *State table (Moore model)*

PS AB	NS A^+B^+ $X=0$	NS A^+B^+ 1	Output Y
00	00	01	0
01	10	01	0
10	00	11	0
11	10	01	1

Table 1.13. *Transition table (Moore model)*

The state encoding of a finite state machine determines the complexity of the logic equations that can be obtained for the NSs and for the output and is not unique. Tables 1.14 and 1.15 show the correspondence between the state tables when the states are represented using Gray code (or reflected binary code). The Karnaugh maps shown in Figures 1.22(a)–(c), respectively, can be used to deduce the following equations:

$$A^+ = D_1 = A \cdot \overline{B} \cdot \overline{X} + \overline{A} \cdot B \cdot \overline{X} + A \cdot B \cdot X \qquad [1.16]$$

$$B^+ = D_2 = A \cdot \overline{B} + \overline{A} \cdot B + \overline{A} \cdot X \qquad [1.17]$$

and:

$$Y = A \cdot \overline{B} \quad [1.18]$$

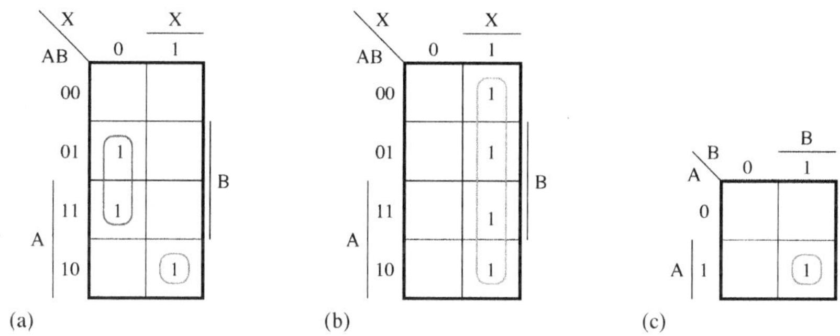

Figure 1.19. *Karnaugh maps (Moore model)*

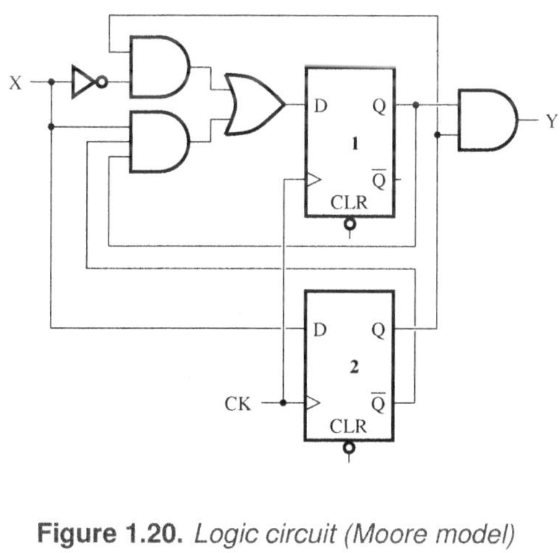

Figure 1.20. *Logic circuit (Moore model)*

A heuristic method to encode states in an optimal manner consists in assigning codes with the highest number of possible zeros to the states with the most incoming transition arrows.

NOTE 1.2.– (Comparison of the Mealy and Moore Machines). In general, a finite state machine based on the Mealy model uses fewer states than one based on the

Moore model. This is because the dependence of the output to the inputs is exploited to reduce the number of states required in order to satisfy the specifications of a given application. A state machine based on the Mealy model is faster than one based on the Moore model. The output of the Moore model is generally obtained one clock period later. The output of a state machine based on the Mealy model can be affected by transient disturbances superposed on the input signal. This is not the case for state machines based on the Moore model, which are, therefore, preferable for applications that require level triggering or control.

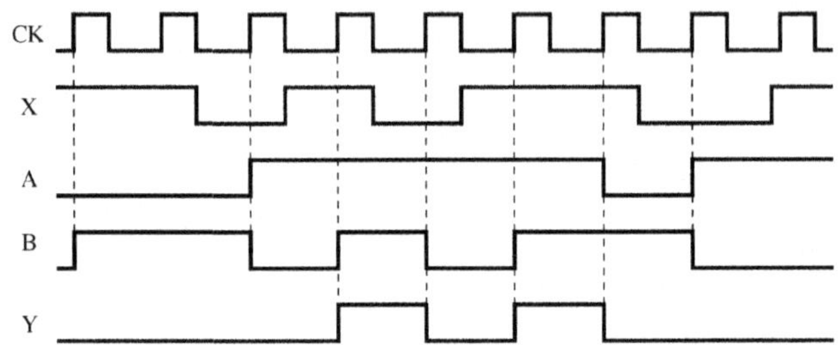

Figure 1.21. *Timing diagram (Moore model)*

PS	NS $X=0$	NS $X=1$	Output Y
S_0	S_0	S_1	0
S_1	S_2	S_1	0
S_2	S_0	S_3	0
S_3	S_2	S_1	1

Table 1.14. *State table of the Moore state machine*

PS AB	NS A^+B^+ $X=0$	NS A^+B^+ $X=1$	Output Y
00	00	01	0
01	11	01	0
11	00	10	0
10	11	01	1

Table 1.15. *Transition table with Gray encoding*

NOTE 1.3.– By choosing a code to represent the states of a machine, a tradeoff is made between the size and electrical performance (power consumption, response time) of the circuit. At least two bits are required to represent four states, and various possible codes are given in Table 1.16. Gray code is used when the states are decoded asynchronously. For instance, if a machine must proceed from 01 to 10, as is the case with natural binary code, and the flip-flops do not switch exactly at the same instant, transient states may appear, taking the form of 11 or 00. This can affect the normal operation of the state machine.

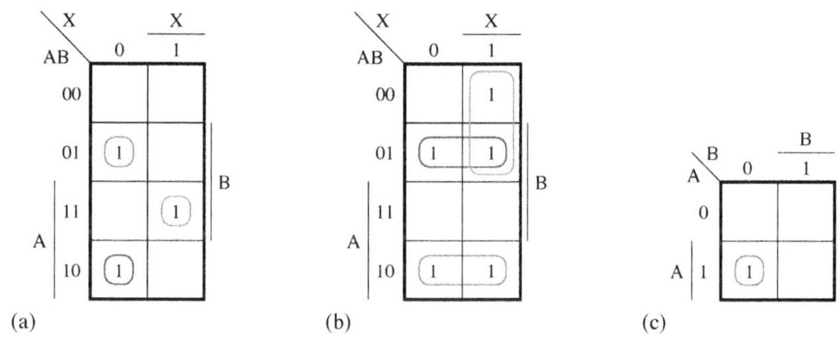

Figure 1.22. *Karnaugh maps for the Moore state machine with the states represented according to Gray code*

	Codes		
PS	Natural binary	Gray	Random binary
S_0	00	00	00
S_1	01	01	11
S_2	10	11	01
S_3	11	10	10

Table 1.16. *Example of two-bit codes*

NOTE 1.4.– A sequence detector accepts a bit string as its input and its output takes the logic state 1 only if a given sequence is detected. When this sequence is recognized, the detection can continue by taking into account any overlapping that may occur between the input bits, as illustrated in the following example:

Input X	1010110110111001010101
Output Y (with overlapping)	0010100100100000010101
Output Y (without overlapping)	0010000100100000010001

Figure 1.23 depicts state diagrams that correspond to detection without overlapping.

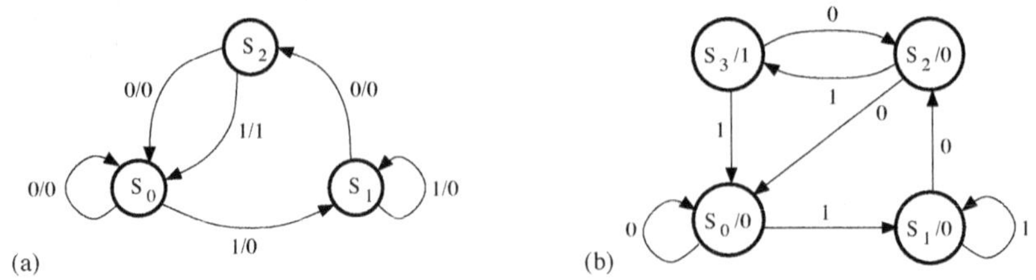

Figure 1.23. *State diagram when the overlapping of bits is not taken into account: a) Mealy model; b) Moore model*

NOTE 1.5.– (Detector for the Sequences 110 and 101). A finite state machine can also be implemented to detect more than one binary sequence. This is, for instance, the case with the 110 and 101 sequence detector, whose output is set to 1 to indicate the detection of either sequence. After each detection, it is assumed that the next detection is realized by taking into account any overlapping of the input bits, as shown below:

Input X	0110100101010
Output Y (with overlapping)	0001100001010

The state diagrams for the 110 and 101 sequence detector is given in Figures 1.24(a) and 1.24(b) for the Moore and Mealy machines, respectively.

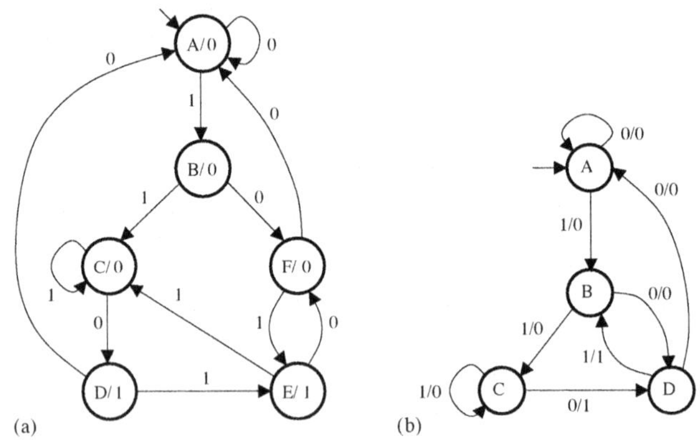

Figure 1.24. *Detector for the sequences* 110 *and* 101*: a) Moore model; b) Mealy model*

Starting from the initial state A, the state machine based on the Moore model changes to the state D or E if either of the sequences, 110 or 101, is recognized. The output of the state machine based on the Mealy model is set to 1, indicating the recognition of one of the sequences, 110 or 101, when there is a transition from the state C to D or from D to B. Transitions involving several states (C, D, E, and F for the Moore state machine; B, C and D for the Mealy state machine) are added to take into account any eventual overlapping of input bits.

1.4.3. *State machine implementation based on a state table*

Implement a synchronous finite state machine whose operation is described by the state table given in Table 1.17 by using, respectively:

– D flip-flops;

– JK flip-flops.

PS $A\ B$	NS A^+B^+ $X = 0$	 1	Output Y
0 0	0 0	0 1	0
0 0	0 0	1 0	0
0 0	0 0	1 1	0
1 1	0 0	1 1	1

Table 1.17. *State table*

1.4.3.1. *D flip-flop*

By referring to the state table and the D flip-flop excitation table, we can construct the transition table (see Table 1.18), which yields the data needed to fill the Karnaugh maps associated with the D inputs of flip-flops, as shown in Figures 1.25 and 1.26.

The logic equations obtained for the D inputs can be used to construct the logic circuit shown in Figure 1.27.

1.4.3.2. *JK flip-flop*

As before, in order to determine the logic equations for the inputs of each JK flip-flop, we construct the transition table as shown in Table 1.19, and the Karnaugh maps as shown in Figures 1.28–1.31.

The logic circuit using JK flip-flops is represented in Figure 1.32. It should be noted that the use of JK flip-flops results in a reduction of the number of logic gates.

X	$A = Q_1$	$B = Q_2$	$A^+ = D_1$	$B^+ = D_2$
0	0	0	0	0
0	0	1	0	0
0	1	0	0	0
0	1	1	0	0
1	0	0	0	1
1	0	1	1	0
1	1	0	1	1
1	1	1	1	1

Table 1.18. *Transition table that can be used to derive the logic expressions for the D inputs*

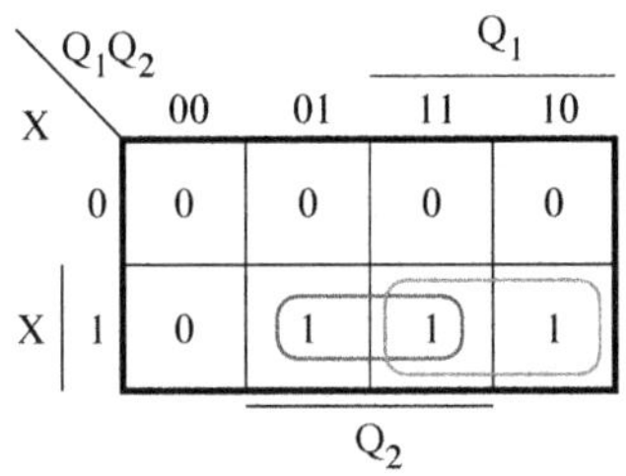

Figure 1.25. *Input* D_1
$D_1 = Q_1 \cdot X + Q_2 \cdot X$

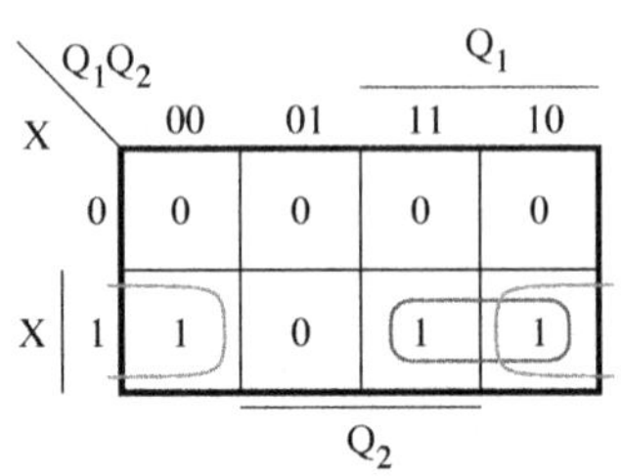

Figure 1.26. *Input* D_2
$D_2 = Q_1 \cdot X + \overline{Q_2} \cdot X$

1.4.4. *Variable width pulse generator*

The generator to be implemented has two inputs, X_1 and X_2, and the output Y. The output is set to logic level 1 during a certain number of clock signal cycles; this number is specified by the bits applied to both inputs. Thus:

– if $X_2X_1 = 01$, $Y = 1$ for one cycle;

– if $X_2X_1 = 10$, $Y = 1$ for two cycles;

– if $X_2X_1 = 11$, $Y = 1$ for three cycles.

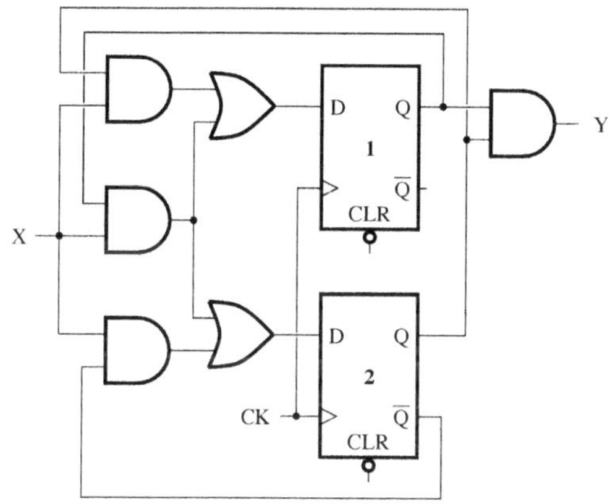

Figure 1.27. *Logic circuit using D flip-flops*

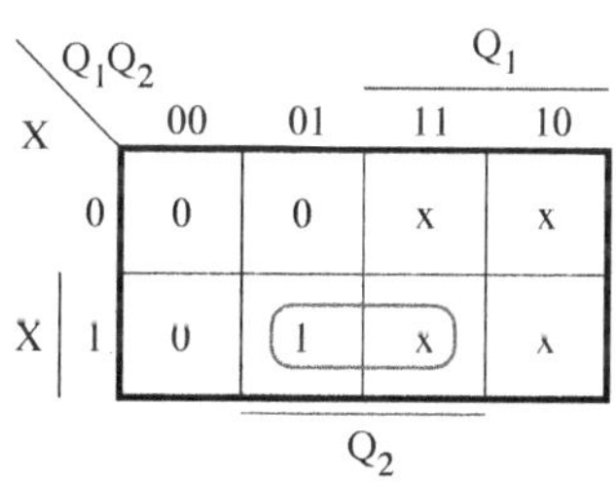

Figure 1.28. *Input J_1*
$J_1 = Q_2 \cdot X$

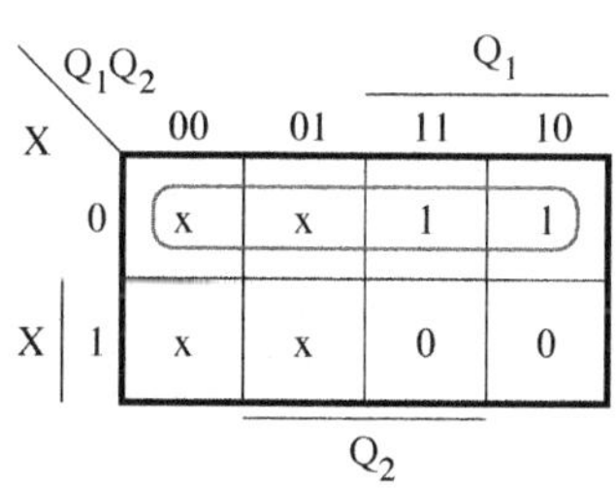

Figure 1.29. *Input K_1*
$K_1 = \overline{X}$

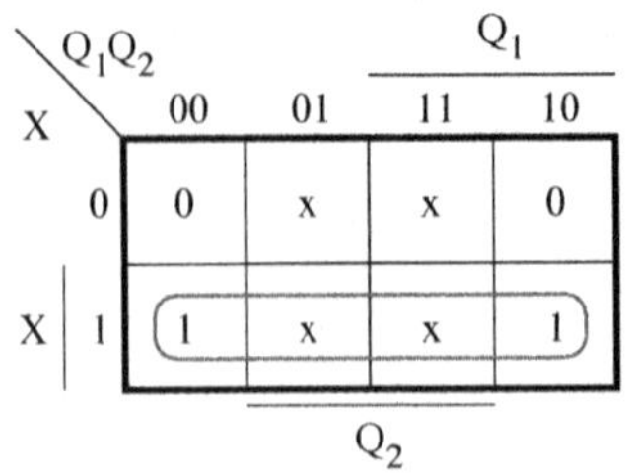

Figure 1.30. *Input J_2*

$J_2 = X$

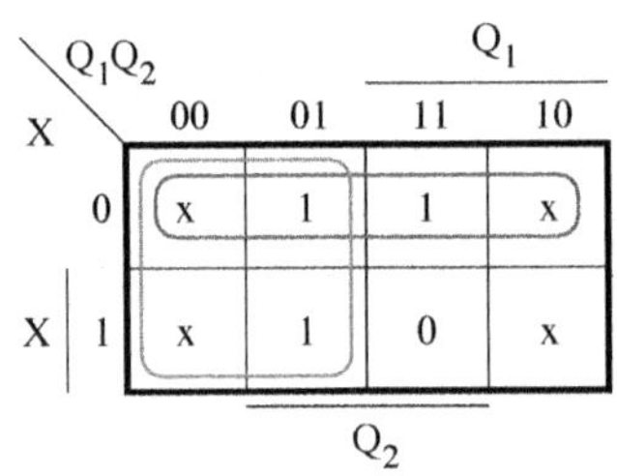

Figure 1.31. *Input K_2*

$K_2 = \overline{X} + \overline{Q_1}$

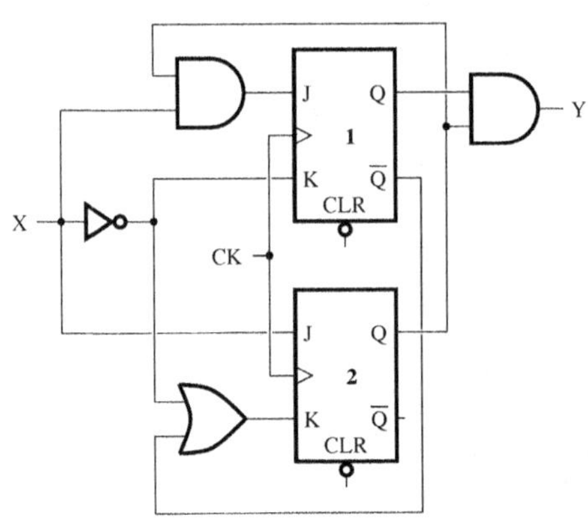

Figure 1.32. *Logic circuit using JK flip-flops*

The Moore model can be used to describe a finite state machine based on the working described above.

There are four states (S_1, S_2, S_3 and S_4) and S_1 is considered as the initial state.

X	$A = Q_1$	$B = Q_2$	$A^+ = Q_1^+$	$B^+ = Q_2^+$	J_1	K_1	J_2	K_2
0	0	0	0	0	0	x	0	x
0	0	1	0	0	0	x	x	1
0	1	0	0	0	x	1	0	x
0	1	1	0	0	x	1	x	1
1	0	0	0	1	0	x	1	x
1	0	1	1	0	1	x	x	1
1	1	0	1	1	x	0	1	x
1	1	1	1	1	x	0	x	0

Table 1.19. *Transition table that can be used to derive the logic expressions for the J and K inputs*

The state table of the generator is represented in Table 1.20.

PS	NS				Output
	$X_2X_1 = 00$	01	11	10	Y
S_1	S_1	S_2	S_2	S_2	0
S_2	S_1	S_1	S_3	S_3	1
S_3	S_1	S_1	S_4	S_1	1
S_4	S_1	S_1	S_1	S_1	1

Table 1.20. *State table*

Transitions that depend on several input variables are difficult to represent when bits (0 and 1) are used. In this case, it is more convenient to express the transition conditions as Boolean expressions.

Figure 1.33 shows the state diagram of the generator. After having reached the state S_4, the machine returns to the state S_1, regardless of the state of the inputs. The transition table may be obtained by adopting Gray code to represent these states, as shown in Table 1.21.

Using the excitation table for the D flip-flop, Karnaugh maps can be constructed as shown in Figures 1.34 and 1.35.

The logic equations for the flip-flop inputs are thus given by:

$$D_1 = A^+ = \overline{A} \cdot B \cdot X_2 + B \cdot X_1 \cdot X_2 \quad [1.19]$$

$$D_2 = B^+ = \overline{A} \cdot \overline{B} \cdot X_1 + \overline{A} \cdot X_2 \quad [1.20]$$

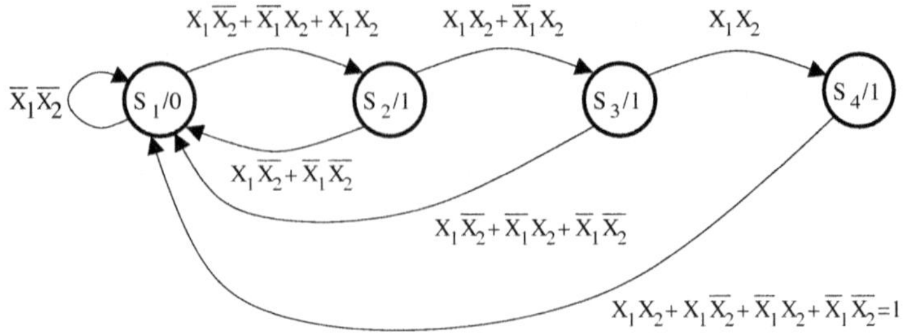

Figure 1.33. *State diagram of the pulse generator*

PS AB	NS A^+B^+ $X_2X_1 = 00$	01	11	10	Output Y
00	00	01	01	01	0
01	00	00	11	11	1
11	00	00	10	00	1
10	00	00	00	00	1

Table 1.21. *Transition table*

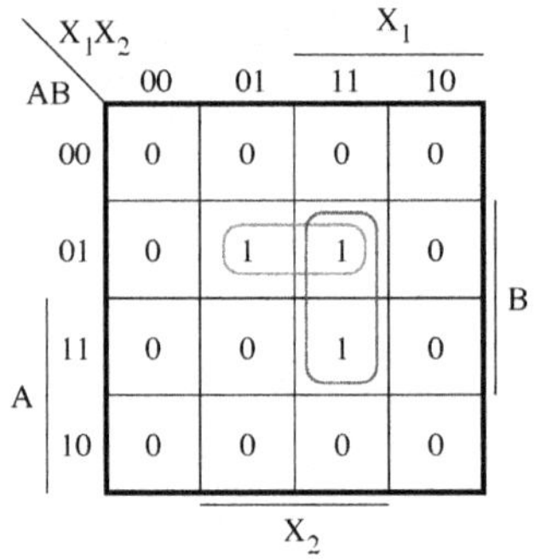

Figure 1.34. *Function A^+*

The logic equation for the output is written as follows:

$$Y = A + B \quad [1.21]$$

The logic circuit of the pulse generator is represented in Figure 1.36.

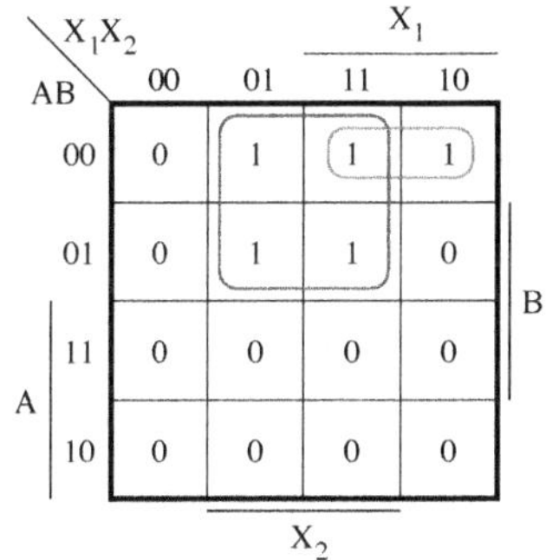

Figure 1.35. *Function* B^+

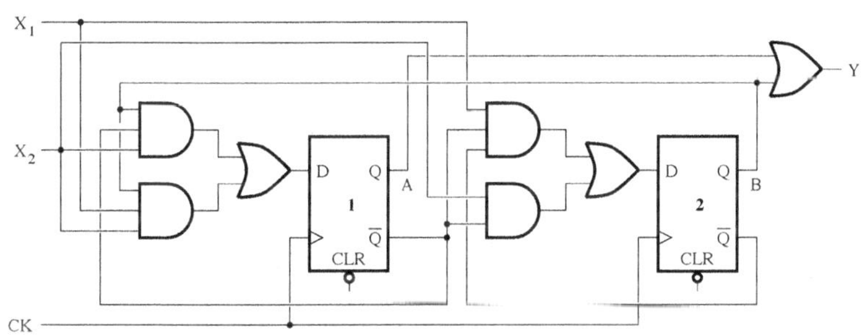

Figure 1.36. *Logic circuit for the pulse generator*

1.5. Equivalent states and minimization of the number of states

The process of designing finite state machines can, in general, be optimized for certain applications that require minimizing the area occupied by the circuit or the number of components. This can be achieved by reducing the number of states.

DEFINITION 1.1.–

Two states, S_1 and S_2, are said to be equivalent and are referred to as $S_1 \equiv S_2$ if, from each of these states, a finite state machines generates the same output bit sequence in response to any input bit sequence.

In practice, two states, S_1 and S_2 can be considered to be equivalent if for any input bit sequence the corresponding outputs, Y_1 and Y_2, are identical. That is, $Y_1 = Y_2$, and the NSs, S_1^+ and S_2^+, are equivalent, or to put it another way, $S_1^+ \equiv S_2^+$.

A state that is the equivalent of another state can be considered redundant.

As an example, let us consider the finite state machine whose state table is represented in Table 1.22. Regardless of whether it starts from state S_2 or S_4, the machine goes through identical NSs and yields the same output. The states S_2 and S_4 are, therefore, equivalent, and state S_4 can be eliminated from the state table as shown in Table 1.23.

PS	NS		Output Y
	$X = 0$	1	
S_0	S_0	S_1	0
S_1	S_0	S_2	0
S_2	S_0	S_3	0
S_3	S_0	S_3	1
S_4	S_0	S_3	0

Table 1.22. *State table*

PS	NS		Output Y
	$X = 0$	1	
S_0	S_0	S_1	0
S_1	S_0	S_2	0
S_2	S_0	S_3	0
S_3	S_0	S_3	1

Table 1.23. *Reduced state table*

For a completely specified finite state machine, the equivalence relationship between states is symmetrical and transitive.

1.5.1. *Implication table method*

Let us consider the state table shown in Table 1.24, where the NSs and the output are dependent on the logic level, X_k, at the input. Table 1.25 gives the implication table, which has as many cells as there are possible pairs of states. However, as the cells on the diagonal show the pairs formed by identical states and there is symmetry between the cells on either side of the diagonal, only the cells in the lower triangle are necessary. By eliminating redundant cells, the implication table is reduced as shown in Table 1.26. In general, for a machine with N states, the implication table must have $(N^2 - N)/2$ cells.

PS	NS			Output Y		
	$\cdots$	X_k	$\cdots$	$\cdots$	X_k	$\cdots$
$\vdots$	$\vdots$	$\vdots$	$\vdots$	$\vdots$	$\vdots$	$\vdots$
S_i	$\cdots$	S_i^+	$\cdots$	$\cdots$	Y_i	$\cdots$
S_j	$\cdots$	S_j^+	$\cdots$	$\cdots$	Y_j	$\cdots$
$\vdots$	$\vdots$	$\vdots$	$\vdots$	$\vdots$	$\vdots$	$\vdots$

Table 1.24. *State table*

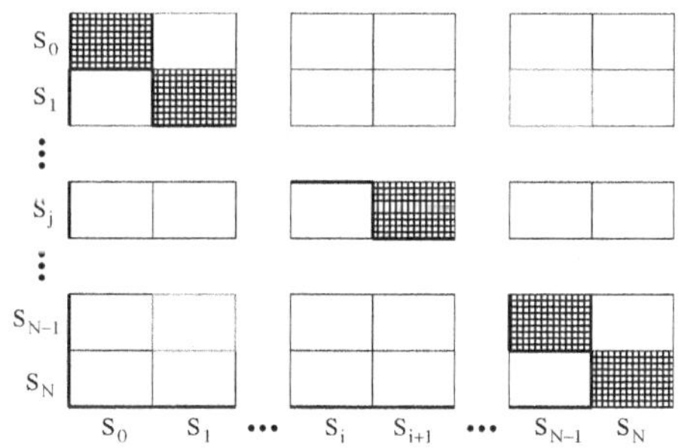

Table 1.25. *Implication table*

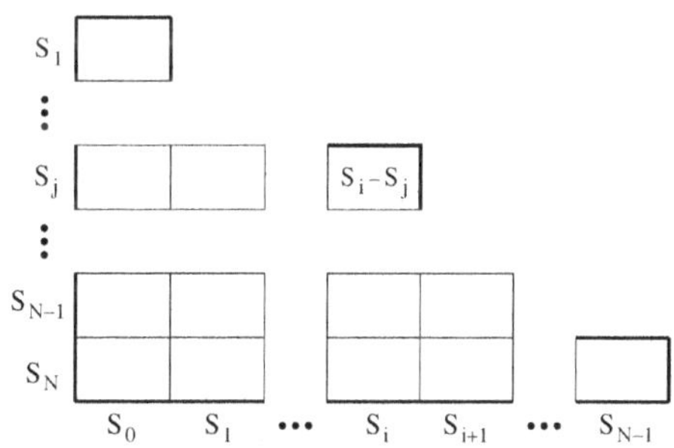

Table 1.26. *Implication table*

Referring to the section of the state table for the logic level X_k at input, the pair of states, S_i and S_j, may be inserted in the implication table as shown in Table 1.26. If the corresponding outputs are different, the two states are not equivalent and a cross can be entered into cell (i, j), as shown in Table 1.27. On the other hand, if the outputs are identical, the NSs, S_i^+ and S_j^+, must be entered into cell (i, j). This results in the implication table given in Table 1.27, where S_i^+ and S_j^+ are assumed to be different.

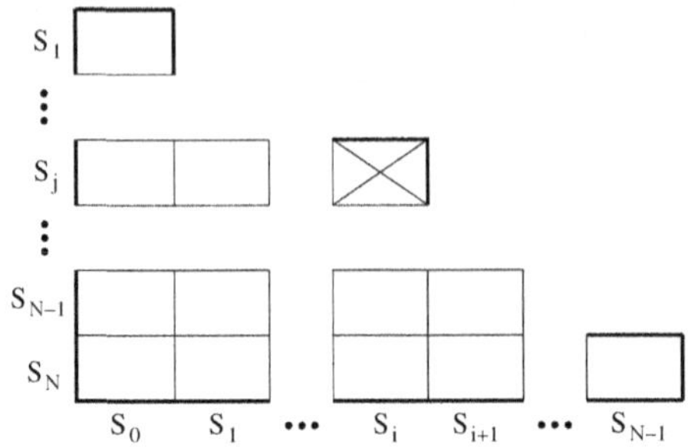

Table 1.27. *Implication table when $Y_i \neq Y_j$*

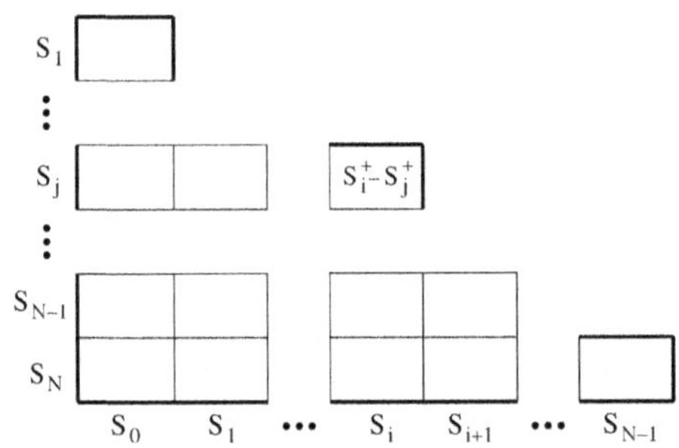

Table 1.28. *Implication table when $Y_i = Y_j$*

The following operations must be carried out in order to determine the NSs:

1) Construct an implication table with a cell (i, j) for each pair of states, S_i and S_j.

2) Identify the outputs, Y_i and Y_j, and the NSs,S_i^+ and S_j^+, for each combination of input bits, X_k, and for each pair of states, S_i and S_j:

- if $Y_i \neq Y_j$, insert a cross (X) in the cell (i, j) to indicate that $S_i \not\equiv S_j$;

- if $Y_i = Y_j$, $(S_i^+, S_j^+) \neq (S_i, S_j)$ and $S_i^+ \neq S_j^+$, enter $S_i^+ - S_j^+$ in the cell (i, j), if not, leave the cell (i, j) empty.

3) Examine each cell that is not yet marked by a cross. Insert a cross in each of these unmarked cells that contain a pair of states associated with a cell that already contains a cross.

4) Repeat step 3 until there is no more cell that can be marked with a cross.

Two states, S_i and S_j, are, thus, said to be equivalent if the corresponding cell (i, j) of the implication table does not contain a cross.

The reduction of the number of states consists in identifying and eliminating the redundant states.

1.5.1.1. *Example 1*

Let us consider a finite state machine with four states, whose operation is described by the state table shown in Table 1.29, where X is the input.

PS	NS		Output Y	
	$X = 0$	1	$X = 0$	1
S_0	S_3	S_3	1	1
S_1	S_2	S_2	1	0
S_2	S_1	S_1	1	1
S_3	S_0	S_2	1	0

Table 1.29. *State table*

Reduce the number of states using the implication table method.

For a state machine with four states, the implication table can be represented as shown in Table 1.30. There are six distinct pairs of states: S_0 and S_1, S_0 and S_2, S_0 and S_3, S_1 and S_2, S_1 and S_3 and S_2 and S_3.

S_1			
S_2			
S_3			
	S_0	S_1	S_2

Table 1.30. *Implication table*

Table 1.31 presents the implication table after a single pass. For the pair of states, S_0 and S_1, the outputs are different when $X = 1$. Thus, states S_0 and S_1 are not equivalent and a cross cannot be inserted in the cell $(0, 1)$. By examining the pair of states, S_0 and S_2, we can see that the outputs are identical and that the NSs are S_3

and S_1, regardless of the value of X. The cell $(0, 2)$ now contains the pair of states, S_3 and S_1. A cross can also be inserted in cells $(0, 3)$, $(1, 2)$ and $(2, 3)$ because the outputs are different in each of these cases when $X = 1$. The pair of NSs, S_2 and S_0 when $X = 0$, is entered in the cell $(1, 3)$ that contained the pair of states S_1 and S_2 because the corresponding outputs are identical, while the pair of NSs when $X = 1$ is not considered as it consists of the same state, S_2.

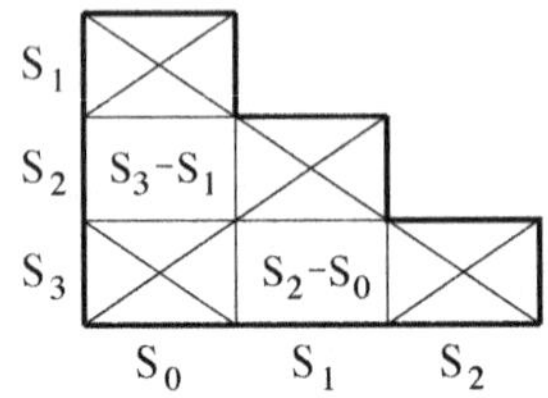

Table 1.31. *Implication table after a marking transition*

Given that the states S_0 and S_2, and S_1 and S_3, are equivalent, the state table can be reduced to the form given in Table 1.32.

PS	NS		Output Y	
	$X = 0$	1	$X = 0$	1
S_0	S_1	S_1	1	1
S_1	S_0	S_0	1	0

Table 1.32. *Reduced state table of the state machine*

1.5.1.2. *Example 2*

The operation of a finite state machine is described by the state table shown in Table 1.33.

Reduce the number of states using the implication table method.

As there are seven states, the implication table has $(7^2 - 7)/2$ or 21 cells. Table 1.34 shows the implication table that can be constructed initially. Based on the state table, the implication table can be filled in as shown in Table 1.35. The cell $(0, 1)$ contains $S_1 - S_3$ and $S_2 - S_5$ as the same output is obtained starting from the states S_0 and S_1. The NS starting from the state S_0 is S_1 when $X = 0$ or S_2 when $X = 1$, while starting from the state S_1, the NS is S_3 when $X = 0$ or S_5 when $X = 1$. The cell $(0, 2)$ contains a cross to indicate that starting from the states S_0 and S_2, we obtain different outputs and that these two states cannot be equivalent. The cell $(0, 3)$ only contains $S_2 - S_6$ because the same output is obtained when starting from the states

S_0 and S_3 and the machine moves to the same NS S_1 if $X = 0$, or to the NSs S_2 and S_6, respectively, if $X = 1$. The filling in of the other cells is carried out in a similar manner. The inputs $S_1 - S_3$, for the cell $(1, 3)$, and $S_2 - S_4$, for the cell $(2, 4)$, must be eliminated as they are identical to the initial states.

PS	NS		Output Y
	$X = 0$	1	
S_0	S_1	S_2	1
S_1	S_3	S_5	1
S_2	S_5	S_4	0
S_3	S_1	S_6	1
S_4	S_5	S_2	0
S_5	S_4	S_3	0
S_6	S_5	S_6	0

Table 1.33. *State table*

S_1						
S_2						
S_3						
S_4						
S_5						
S_6						
	S_0	S_1	S_2	S_3	S_4	S_5

Table 1.34. *Implication table*

Table 1.36 shows the implication table after the first marking pass. Each of the cells $(4, 5)$, $(2, 5)$ and $(5, 6)$ must be marked by a cross as each of the cells $(2, 3)$, $(3, 4)$ and $(3, 6)$ corresponding, respectively, to the implied states, already contain a cross.

Table 1.37 shows the implication table after the second marking pass. Each of the cells $(2, 5)$ and $(5, 6)$ must be marked by a cross as each of the cells $(0, 1)$ and $(1, 3)$ corresponding, respectively, to the implied states, already contain a cross. As it is no longer possible to add a cross to the implication table, it can be deduced that the states S_2, S_4 and S_6 are equivalent, ($S_2 \equiv S_4 \equiv S_6$), as are the states S_0 and S_3 ($S_0 \equiv S_3$).

S_1	S_1-S_3 S_2-S_5					
S_2	X	X				
S_3	S_2-S_6	S_1-S_3 S_5-S_6	X			
S_4	X	X	S_2-S_4	X		
S_5	X	X	S_4-S_5 S_3-S_4	X	S_4-S_5 S_2-S_3	
S_6	X	X	S_4-S_6	X	S_2-S_6	S_4-S_5 S_3-S_6
	S_0	S_1	S_2	S_3	S_4	S_5

Table 1.35. *Implication table based on the state table*

S_1	S_1-S_3 S_2-S_5					
S_2	X	X				
S_3	S_2-S_6	S_5-S_6	X			
S_4	X	X	X	X		
S_5	X	X	S_4-S_5 S_3-S_4	X	S_4-S_5 S_2-S_3	
S_6	X	X	S_4-S_6	X	S_2-S_6	S_4-S_5 S_3-S_6
	S_0	S_1	S_2	S_3	S_4	S_5

Table 1.36. *Implication table after the first marking pass*

By eliminating redundant states, the state table can be reduced to the form shown in Table 1.38.

1.5.1.3. *Example 3*

Construct implication tables to minimize the number of states for the finite state machine whose state table is illustrated in Table 1.39, where X and Y represent the inputs.

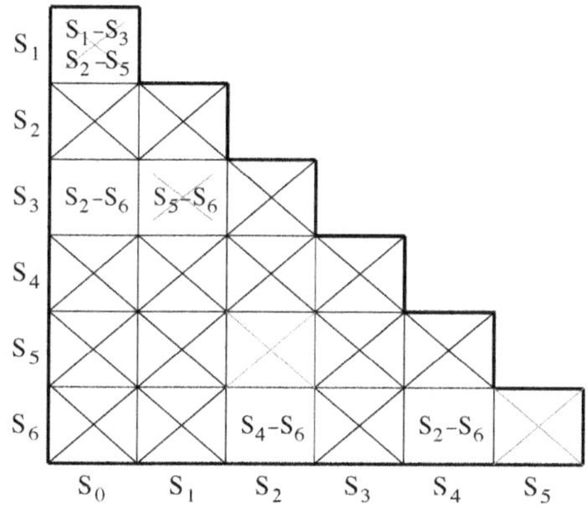

Table 1.37. *Implication table after the second marking pass*

PS	NS $X = 0$	NS 1	Output Y
S_0	S_1	S_2	1
S_1	S_0	S_5	1
S_2	S_5	S_2	0
S_5	S_2	S_0	0

Table 1.38. *Reduced state table of the state machine*

PS	NS $XY = 00$	01	10	11	Output Y
S_0	S_0	S_1	S_2	S_3	1
S_1	S_0	S_3	S_1	S_5	0
S_2	S_1	S_3	S_2	S_4	1
S_3	S_1	S_0	S_4	S_5	0
S_4	S_0	S_1	S_2	S_5	1
S_5	S_1	S_4	S_0	S_5	0
S_6	S_4	S_1	S_2	S_3	1

Table 1.39. *State table*

	S_0	S_1	S_2	S_3	S_4	S_5
S_1	X					
S_2	S_0-S_1 S_1-S_3 S_2-S_2 S_3-S_4	X				
S_3	X	S_0-S_1 S_3-S_0 S_1-S_4 S_5-S_5	X			
S_4	S_0-S_0 S_1-S_1 S_2-S_2 S_3-S_5	X	S_1-S_0 S_3-S_1 S_2-S_2 S_4-S_5	X		
S_5	X	S_0-S_1 S_3-S_4 S_1-S_0 S_5-S_5	X	S_1-S_1 S_0-S_4 S_4-S_0 S_5-S_5	X	
S_6	S_0-S_4 S_1-S_1 S_2-S_2 S_3-S_3	X	S_1-S_4 S_3-S_1 S_2-S_2 S_4-S_3	X	S_0-S_4 S_1-S_1 S_2-S_2 S_5-S_3	X

Table 1.40. *Implication table based on the state table*

	S_0	S_1	S_2	S_3	S_4	S_5
S_1	X					
S_2	X S_0-S_1 S_1-S_3 S_2-S_2 S_3-S_4	X				
S_3	X	X S_0-S_1 S_3-S_0 S_1-S_4 S_5-S_5	X			
S_4	S_3-S_5	X	X S_1-S_0 S_3-S_1 S_2-S_2 S_4-S_5	X		
S_5	X	X S_0-S_1 S_3-S_4 S_1-S_0 S_5-S_5	X	S_0-S_4	X	
S_6	S_0-S_4	X	X S_1-S_4 S_3-S_1 S_2-S_2 S_4-S_3	X	S_0-S_4 S_5-S_3	X

Table 1.41. *Implication state after the first marking pass*

The implication table of a state machine with seven states must have $(7^2 - 7)/2$ or 21 cells. Based on the state table, the implication table can be drawn up as shown in Table 1.40. A cross is inserted in each cell, related to states with different outputs; and the NSs for the four different input combinations are inserted in each cell for pairs of states yielding the same output.

By eliminating the pairs made up of identical states, $S_0 - S_0$, $S_1 - S_1$, $S_2 - S_2$, $S_3 - S_3$ or $S_5 - S_5$, the implication table after the first marking pass can be constructed as shown in Table 1.41. As the cell (0,1) is already marked with a cross, a cross must be inserted into each of the cells (0,2), (1,3), (1,5) and (2,4), where we have the term $S_0 - S_1$. Similarly, we also add a cross in the cell (2,6), where we have the term $S_1 - S_4$, because the cell (1,4) is already marked. As there are no more marking possibilities, the states S_0, S_4 and S_6 are equivalent: $(S_0 \equiv S_4 \equiv S_6)$, as are the states S_3 and S_5 $(S_3 \equiv S_5)$.

After the elimination of redundant states, the reduced state table is obtained as shown in Table 1.42.

PS	NS				Output Y
	$XY = 00$	01	10	11	
S_0	S_0	S_1	S_2	S_3	1
S_1	S_0	S_3	S_1	S_3	0
S_2	S_1	S_3	S_2	S_0	1
S_3	S_1	S_0	S_0	S_3	0

Table 1.42. *Reduced state table of the state machine*

1.5.2. *Partitioning method*

The partitioning method is used to determine the equivalent states of a finite state machine and can also be considered as an approach to reduce the number of states. It consists of successively forming partitions, P_k, $k = 1, 2, 3, \cdots$, which are composed of a certain number of blocks that group one or more states.

The PSs are placed in the same partition block P_1 if and only if the corresponding outputs of the state machine are identical for each input combination.

For other partitions, P_k, $k = 2, 3, \cdots$, the PSs are placed in the same block if and only if the NSs of the state machine for each input combination are in the same block of the partition, P_{k-1}.

The states belonging to different blocks of the partition P_k cannot be equivalent, while the states that are in the same block are called k-equivalent. The partitioning process continues until it is no longer possible to obtain a partition with smaller blocks. All the states that are in the same block of the last partition are, thus, equivalent.

1.5.2.1. *Example 1*

Use the partitioning method to reduce the number of states for the finite state machine whose state table is shown in Table 1.43.

PS	NS $X = 0$	NS 1	Output Y
S_0	S_3	S_6	0
S_1	S_2	S_4	1
S_2	S_1	S_6	0
S_3	S_0	S_1	1
S_4	S_5	S_4	0
S_5	S_6	S_1	1
S_6	S_5	S_0	0

Table 1.43. *State table*

We begin by grouping all the states into a single block. That is:

$$P_0 = (S_0 S_1 S_2 S_3 S_4 S_5 S_6) \quad [1.22]$$

In order to form the first partition, the states are distributed across two blocks depending on whether the state machine generates an output value of 0 or 1. As the same output value, 0, is obtained when starting from each of the states S_0, S_2, S_4 and S_6, and the same output value, 1, is obtained when starting from the states S_1, S_3 and S_5, we have:

$$P_1 = (S_0 S_2 S_4 S_6)(S_1 S_3 S_5) \quad [1.23]$$

To form each of the remaining partitions, we must verify whether the NSs for states in the same block are in the same or in different blocks.

The machine goes from the states S_0, S_2, S_4 and S_6 to the states S_3, S_1, S_5 and S_5, respectively, when $X = 0$, or to S_6, S_6, S_4 and S_0 when $X = 1$. States S_3, S_1 and S_5, like the states S_6, S_4 and S_0, belong to the same block of the partition P_1.

That is, the states S_0, S_2, S_4 and S_6 must remain grouped in the same block. Starting from the states S_1, S_3 and S_5, the state machine advances to the states S_2, S_0 and S_6, respectively, when $X = 0$, or S_4, S_1 and S_1 when $X = 1$. The states S_2, S_0 and S_6 are in the same block of the partition P_1, while the states S_4 and S_1 belong to different blocks of the partition P_1. State S_1 differs, therefore, from states S_3 and S_5, and the block containing these states must be split into two. Thus:

$$P_2 = (S_0S_2S_4S_6)(S_3S_5)(S_1) \quad [1.24]$$

The state machine goes from the states S_0, S_2, S_4 and S_6 to the states S_3, S_1, S_5 and S_5, respectively, when $X = 0$, or to S_6, S_6, S_4 and S_0 when $X = 1$. The states S_3 and S_5 are in a block of the partition P_2, that is different from the block where S_1 is placed, while the states S_6, S_4 and S_0 belong to the same block of P_2. It thus appears that the block grouping the states S_0, S_2, S_4 and S_6 must be split into two. Starting from the states S_3 and S_5, the state machine proceeds to the states S_0 and S_6, when $X = 0$, or S_1 and S_1, when $X = 1$. The states S_0 and S_6 are in the same block of the partition P_2, and the state S_1 is the only element in one of the blocks of P_2. We thus have:

$$P_3 = (S_0S_4S_6)(S_2)(S_1)(S_3S_5) \quad [1.25]$$

Operating in the same manner as stated before, we obtain the following partition:

$$P_4 = (S_0S_4S_6)(S_2)(S_1)(S_3S_5) = P_3 \quad [1.26]$$

As P_4 is identical to P_3, it follows that the states in each block are equivalent: $S_0 \equiv S_4 \equiv S_6$ and $S_3 \equiv S_5$. Table1.44 lists the different steps to follow in order to determine the equivalent states. The reduced state table is shown in Table 1.45.

1.5.2.2. *Example 2*

Let us consider the finite state machine whose operation is described by the state table in Table 1.46. Reduce the number of its states using the partitioning method.

Initially, we have:

$$P_0 = (S_0S_1S_2S_3S_4) \quad [1.27]$$

		Blocks to be formed
P_0	$(S_0S_1S_2S_3S_4S_5S_6)$	
Output Y	0 1 0 1 0 1 0	$S_0S_2S_4S_6$ and $S_1S_3S_5$
P_1	$(S_0S_2S_4S_6)(S_1S_3S_5)$	
NS		
$X = 0$	$S_3S_1S_5S_5$ $S_2S_0S_6$	
$X = 1$	$S_6S_6S_4S_0$ $S_4S_1S_1$	S_1 and S_3S_5
P_2	$(S_0S_2S_4S_6)(S_1)(S_3S_5)$	
NS		
$X = 0$	$S_3S_1S_5S_5$ S_2 S_0S_6	$S_0S_4S_6$ and S_2
$X = 1$	$S_6S_6S_4S_0$ S_4 S_1S_1	
P_3	$(S_0S_4S_6)(S_2)(S_1)(S_3S_5)$	
NS		
$X = 0$	$S_3S_5S_5$ S_1 S_2 S_0S_6	
$X = 1$	$S_6S_4S_0$ S_6 S_4 S_1S_1	
$P_4 = P_3$	$(S_0S_4S_6)(S_2)(S_1)(S_3S_5)$	

Table 1.44. *Determining the equivalent states using the partitioning method (example 1)*

PS	NS $X = 0$	NS $X = 1$	Output Y
S_0	S_3	S_0	0
S_1	S_2	S_0	1
S_2	S_1	S_0	0
S_3	S_0	S_1	1

Table 1.45. *Reduced state table*

Starting from the states S_0, S_1 and S_2, and the states S_3 and S_4, the machine generates the output value 1 and 0, respectively, when $X = 0$, or the output value 0 and 1, when $X = 1$. The first partition, thus, takes the following form:

$$P_1 = (S_0S_1S_2)(S_3S_4) \quad [1.28]$$

The state machine goes from the states S_0, S_1 and S_2 to the states S_2, S_2 and S_1, respectively, when $X = 0$, or to S_1, S_4 and S_4 when $X = 1$. The states S_1 and S_4 belong to different blocks of the partition P_1 and it can be concluded that the state S_0 is different from the states S_1 and S_2. Starting from the states S_3 and S_4, the machine moves to the states S_3 and S_4, when $X = 0$, or to S_1 and S_0 when $X = 1$. The states S_3 and S_4, just like the states S_1 and S_0, are in the same block of the partition P_1. We can then conclude that the states S_3 and S_4 must remain in the same block. Thus:

$$P_2 = (S_0)(S_1S_2)(S_3S_4) \quad [1.29]$$

PS	NS		Output Y	
	$X = 0$	1	$X = 0$	1
S_0	S_2	S_1	1	0
S_1	S_2	S_4	1	0
S_2	S_1	S_4	1	0
S_3	S_3	S_1	0	1
S_4	S_4	S_0	0	1

Table 1.46. *State table*

Starting from the states S_1 and S_2, the state machine advances to the states S_2 and S_1, when $X = 0$, or to the state S_4, when $X = 1$. Thus, the block grouping the states S_1 and S_2 remains unaffected. The machine is held either in the state S_3 or in the state S_4 when $X = 0$, it moves from the states S_3 and S_4 to the states S_1 and S_0, respectively, when $X = 1$. The block formed by the states S_1 and S_0 must be split into two as the states S_1 and S_0 belong to different blocks of the partition P_2. We thus have:

$$P_3 = (S_0)(S_1S_2)(S_3)(S_4) \quad [1.30]$$

Similarly, the next partition can be obtained as follows:

$$P_4 = (S_0)(S_1S_2)(S_3)(S_4) = P_3 \quad [1.31]$$

As P_4 is identical to P_3, it follows that the states S_1 and S_2, which are in the same block, are equivalent. Table 1.47 lists out the different steps to be followed in order to determine the equivalent states. The reduced state table is represented in Table 1.48.

1.5.2.3. *Example 3*

Using the partitioning method, minimize the number of states of the finite state machine (Moore model) whose state table is illustrated in Table 1.49, where X and Y represent the inputs.

		Blocks to be formed
P_0	$(S_0S_1S_2S_3S_4)$	
Output Y		
$X = 0$	1 1 1 0 0	$S_0S_1S_2$ and S_3S_4
$X = 1$	0 0 0 1 1	$S_0S_1S_2$ and S_3S_4
P_1	$(S_0S_1S_2)(S_3S_4)$	
NS		
$X = 0$	$S_2S_2S_1 \quad S_3S_4$	
$X = 1$	$S_1S_4S_4 \quad S_1S_0$	S_0 and S_1S_2
P_2	$(S_0)(S_1S_2)(S_3S_4)$	
NS		
$X = 0$	$S_2 \quad S_2S_1 \quad S_3S_4$	
$X = 1$	$S_1 \quad S_4S_4 \quad S_1S_0$	S_3 and S_4
P_3	$(S_0)(S_1S_2)(S_3)(S_4)$	
NS		
$X = 0$	$S_2 \quad S_2S_1 \quad S_3 \quad S_4$	
$X = 1$	$S_1 \quad S_4S_4 \quad S_1 \quad S_0$	
$P_4 = P_3$	$(S_0)(S_1S_2)(S_3)(S_4)$	

Table 1.47. *Determining the equivalent states using the partitioning approach (example 2)*

Table 1.50 summarizes the different steps to be followed in order to determine the equivalent states. From the last partition, P_2, it can be deduced that the states S_0 and S_3 are equivalent, $(S_0 \equiv S_3)$, as are the states S_1 and S_4 $(S_1 \equiv S_4)$, and the states S_2, S_5 and S_7 $(S_2 \equiv S_5 \equiv S_7)$. Assuming that:

$$A = S_0 = S_3 \quad [1.32]$$

$$B = S_1 = S_4 \quad [1.33]$$

$$C = S_2 = S_5 = S_7 \quad [1.34]$$

and:

$$D = S_6 \quad [1.35]$$

we can obtain the reduced state table, as illustrated in Table 1.51. Hence, the number of states of the state machine has been reduced from eight to four.

1.5.3. *Simplification of incompletely specified machines*

Finite states machines used in some applications have incompletely specified state tables. Their behavior is dependent on don't-care states and therefore cannot be predicted uniquely.

PS	NS		Output Y	
	$X=0$	1	$X=0$	1
S_0	S_1	S_1	1	0
S_1	S_1	S_4	1	0
S_3	S_3	S_1	0	1
S_4	S_4	S_0	0	1

Table 1.48. *Reduced state table*

PS	NS				Output Y
	$XY=00$	01	10	11	
S_0	S_0	S_0	S_4	S_6	1
S_1	S_2	S_5	S_3	S_6	0
S_2	S_6	S_2	S_4	S_0	1
S_3	S_0	S_3	S_4	S_6	1
S_4	S_5	S_7	S_0	S_6	0
S_5	S_6	S_2	S_4	S_3	1
S_6	S_2	S_3	S_4	S_6	0
S_7	S_6	S_7	S_4	S_3	1

Table 1.49. *State table*

Consider the incompletely specified state machine whose state table is represented in Table 1.52.

The direct approach for the simplification of this type of machine consists in assigning all the possible values to the don't-care states, then proceeding to the reduction of the number of states and, finally, choosing only the specified machine that can be described using the smallest number of states. In the present case, this leads to the minimization of two completely specified machines.

Table 1.53 depicts the state table when the don't-care state is assumed to be 0. The states S_0 and S_1 cannot be equivalent unless the states S_1 and S_2 are equivalent. This is not the case here as the outputs corresponding to the states S_1 and S_2 are different. For the same reason, the states S_0 and S_2 are not equivalent. Therefore, no simplification is possible.

		Blocks to be formed
P_0	$(S_0S_1S_2S_3S_4S_5S_6S_7)$	
Output Y	1 0 1 1 0 1 0 1	$S_0S_2S_3S_5S_7$ and $S_1S_4S_6$
P_1	$(S_0S_2S_3S_5S_7)(S_1S_4S_6)$	
NS		
$XY = 00$	$S_0S_6S_0S_6S_6$ $S_2S_5S_2$	S_0S_3 and $S_2S_5S_7$
$XY = 01$	$S_0S_2S_3S_2S_7$ $S_5S_7S_3$	
$XY = 10$	$S_4S_4S_4S_4S_4$ $S_3S_0S_4$	S_1S_4 and S_6
$XY = 11$	$S_6S_0S_6S_3S_3$ $S_6S_6S_6$	$S_2S_5S_7$ and S_0S_3
P_2	$(S_0S_3)(S_2S_5S_7)(S_1S_4)(S_6)$	
NS		
$XY = 00$	S_0S_0 $S_6S_6S_6$ S_2S_5 S_2	
$XY = 01$	S_0S_3 $S_2S_2S_7$ S_5S_7 S_3	
$XY = 10$	S_4S_4 $S_4S_4S_4$ S_3S_0 S_4	
$XY = 11$	S_6S_6 $S_0S_3S_3$ S_6S_6 S_6	
$P_3 = P_2$	$(S_0S_3)(S_2S_5S_7)(S_1S_4)(S_6)$	

Table 1.50. *Determining the equivalent states using the partitioning approach (example 3)*

PS	NS $XY = 00$	01	10	11	Output Y
A	A	A	B	D	1
B	C	C	A	D	0
C	D	C	B	A	1
D	C	A	B	D	0

Table 1.51. *Reduced state table*

By assigning the value 1 to the don't-care state, we can obtain the state table shown in Table 1.54. The state S_0 cannot be equivalent to either of the states S_1 or S_2. On the other hand, the states S_1 and S_2 are equivalent and the number of states can, thus, be reduced to two, as shown in the reduced state table of Table 1.55.

1.5.3.1. *Definition and basic concepts*

Two states are said to have compatible outputs if and only if they are associated with outputs that are identical when specified.

It must be noted that the compatibility relationship for the outputs is reflexive and symmetrical but is not, generally, transitive.

PS	NS		Output Y	
	$X = 0$	1	$X = 0$	1
S_0	S_2	S_1	0	0
S_1	S_1	S_2	–	0
S_2	S_2	S_1	1	0

Table 1.52. *State table of an incompletely specified machine*

PS	NS		Output Y	
	$X = 0$	1	$X = 0$	1
S_0	S_2	S_1	0	0
S_1	S_1	S_2	0	0
S_2	S_2	S_1	1	0

Table 1.53. *State table in the case where – is assumed to be 0*

PS	NS		Output Y	
	$X = 0$	1	$X = 0$	1
S_0	S_2	S_1	0	0
S_1	S_1	S_2	1	0
S_2	S_2	S_1	1	0

Table 1.54. *State table when – is assumed to be 1*

PS	NS		Output Y	
	$X = 0$	1	$X = 0$	1
S_0	S_1	S_1	0	0
S_1	S_1	S_1	1	0

Table 1.55. *Reduced state table when – is assumed to be 1*

Two states are said to be compatible if and only if they have compatible outputs and involve only pairs of compatible NSs.

In general, a set of states is said to be compatible if and only if it is made up entirely of states that are pairwise compatible.

A compatibility class is a set of mutually compatible states.

A compatibility class is said to be maximal if it is strictly not a subclass of another compatibility class.

A set of compatibility classes forms a cover of the state machine if each of the states of this machine belongs to at least one of these compatibility classes.

A set of compatibility classes is said to be closed and constitutes a closed cover if any compatibility class implied by one of the compatibility classes of this set is also contained in this set.

In the case of the machine described by the state table shown in Table 1.52, the pairs of states, (S_0, S_1) and (S_1, S_2), are compatible. As the concept of compatibility can be used to merge rows in the state tables in order to simplify an incompletely specified machine, it is possible to obtain a reduced state table that is identical to the previous one, as shown in Table 1.56, by assuming that $A = (S_0, S_1)$ and $B = (S_1, S_2)$.

PS	NS		Output Y	
	$X = 0$	1	$X = 0$	1
A	B	B	0	0
B	B	B	1	0

Table 1.56. *Reduced state table*

Another example of an incompletely specified machine is characterized by the state table shown in Table 1.57. By assigning each value, 0 and 1, to the don't-care state, we obtain Tables 1.58 and 1.59, respectively. In both cases, no other simplification is possible.

PS	NS		Output Y	
	$X = 0$	1	$X = 0$	1
S_0	S_2	S_1	0	0
S_1	S_1	S_0	–	0
S_2	S_0	S_1	1	0

Table 1.57. *State table for an incompletely specified machine*

On the other hand, if we take into account the fact that the pairs of states, $A = (S_0, S_1)$ and $B = (S_1, S_2)$, are compatible, we can merge the row S_0 and the row S_1

in the state table, as well as the rows S_2 and S_1. As a result, the reduced state table can be obtained as shown in Table 1.60.

PS	NS		Output Y	
	$X = 0$	1	$X = 0$	1
S_0	S_2	S_1	0	0
S_1	S_1	S_0	0	0
S_2	S_0	S_1	1	0

Table 1.58. *State table when – is assumed to be 0*

PS	NS		Output Y	
	$X = 0$	1	$X = 0$	1
S_0	S_2	S_1	0	0
S_1	S_1	S_0	1	0
S_2	S_0	S_1	1	0

Table 1.59. *State table when – is assumed to be 1*

PS	NS		Output Y	
	$X = 0$	1	$X = 0$	1
A	B	B	0	0
B	A	B	1	0

Table 1.60. *Reduced state table*

In order to simplify incompletely specified machines, the concept of compatibility is used instead of the concept of equivalence. Two states that are equivalent to a third state are also equivalent to one another. However, two states that are compatible with a third state are not necessarily compatible with each other.

As a result of this difference, the simplification of incompletely specified machines can appear to be complicated and, in general, does not yield a unique solution.

The following steps are used for the simplification of incompletely specified machines:

1) determine the pairs of compatible states;

2) form all compatibility classes, including those said to be maximal;

3) select the smallest set of compatibility classes making up a closed cover of the state machine;

4) construct a reduced state table by replacing each compatibility class with a single state.

The rows of the state table are merged by selecting a single state to represent each set or subset of compatible states and by considering, each time, the specified state (or specified output) as representative of the combination of this state (or output) with an indefinite state (or indefinite output) when these appear in the same column.

The implementation of the simplification requires the use of tools such as a merger graph to determine the set of compatible state pairs and compatibility classes, and the compatibility graph to find the different sets of closed compatibility classes.

The merger graph is an undirected graph, where the number of nodes is equal to the number of states of the machine.

For each pair of compatible states (S_i, S_j), an uninterrupted arc can be drawn between the two nodes associated with S_i and S_j.

For each pair of states, (S_i, S_j), whose compatibility depends on different pairs of NSs, an arc interrupted by a label, indicating the conflicting pairs of states, can be drawn between the two nodes associated with S_i and S_j.

It should be noted that the conditions relating to the states S_i and S_j are ignored.

For each pair of incompatible states, (S_i, S_j), no arc is drawn between the two nodes associated with S_i and S_j.

The maximal compatibility classes can be determined from the merger graph by searching for complete polygons that are not contained in any other complete polygons of a higher order.

A polygon is said to be complete when each of its nodes is connected to all the other nodes.

A compatibility graph is a directed graph, where the number of nodes is equal to the number of compatible states. An arc goes from the compatible states (S_i, S_j) toward (S_k, S_l) if and only if the compatibility of (S_i, S_j) implies the compatibility of (S_k, S_l).

A subgraph of the compatibility graph is considered closed if, for all nodes of this subgraph, the emerging arcs are directed toward nodes that are also a part of this subgraph.

A closed subgraph with at least one node associated with each state of the state machine corresponds to a closed cover.

The minimal form of an incompletely specified finite state machine is often not unique and can be determined only after several trials. In this case, knowledge of the minimum and maximum number of states that can be used to represent the minimal form of the state machine could prove useful. The minimum number of states is given by:

$$N_m = \max(N_{I_1}, N_{I_2}, \cdots, N_{I_k}, \cdots) \qquad [1.36]$$

where N_{I_k} is the number of states of the *k*th incompatibility class, while the maximum number of states is of the form:

$$N_M = \min(N, N_C) \qquad [1.37]$$

where N is the number of states of the machine and N_C is the maximal number of compatibility classes.

1.5.3.2. *Example 1*

Consider the incompletely specified finite state machine, whose state table is represented in Table 1.61:

– determine the compatibility classes for this machine;

– construct the reduced state table.

PS	NS		Output Y	
	$X = 0$	1	$X = 0$	1
A	D	A	–	–
B	E	A	0	–
C	D	B	0	–
D	C	C	–	–
E	C	B	1	–

Table 1.61. *State table for example 1*

The merger graph is constructed as shown in Figure 1.37(a). The states B and D cannot be compatible because the implied NSs, C and E, are not compatible. The simplified merger graph for the compatible states is represented in Figure 1.37(c).

The maximal compatibility classes are as follows: (AB), (AC), (AD), (AE), (BC), (CD), (DE), (ABC), (ACD) and (ADE).

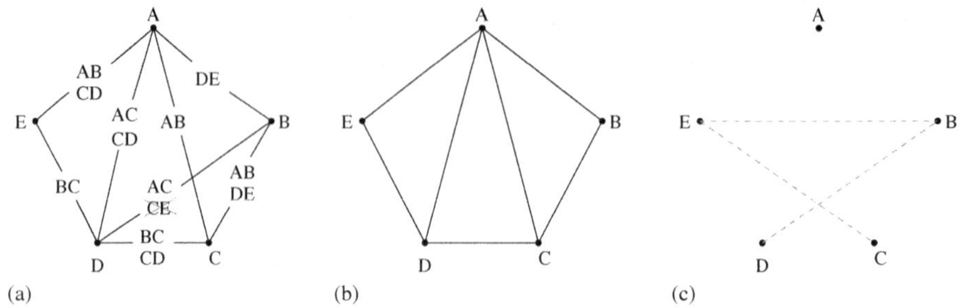

Figure 1.37. *a) Merger graph. Simplified merger graph: b) compatible states and c) incompatible states*

From the compatibility graph shown in Figure 1.38, it can be deduced that the set of states (ABC) and (DE) form a closed cover.

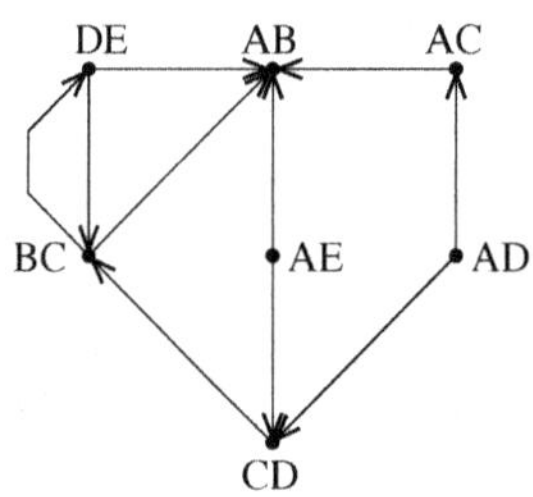

Figure 1.38. *Compatibility graph*

PS	NS		Output Y	
	$X = 0$	1	$X = 0$	1
S_0	S_1	S_0	0	–
S_1	S_0	S_0	1	–

Table 1.62. *Reduced state table for example 1*

By assuming that $S_0 = (ABC)$ and $S_1 = (DE)$, we can obtain the reduced state table as illustrated in Table 1.62.

1.5.3.3. *Example 2*

An incompletely specified finite state machine is characterized by the state table shown in Table 1.63:

– determine the compatibility classes for this machine;

– construct a reduced state table.

PS	NS				Output Z
	$XY = 00$	01	10	11	
A	A	H	B	–	0
B	F	–	B	C	0
C	–	H	–	C	1
D	A	D	–	E	1
E	–	D	G	E	1
F	F	D	–	–	0
G	F	–	G	–	0
H	–	H	–	E	0

Table 1.63. *State table for example 2*

Figures 1.39(a) and (b) show the compatible and incompatible states, respectively. The maximal compatibility classes are as follows: (AH), (BFG), (C), (DE) and (GH).

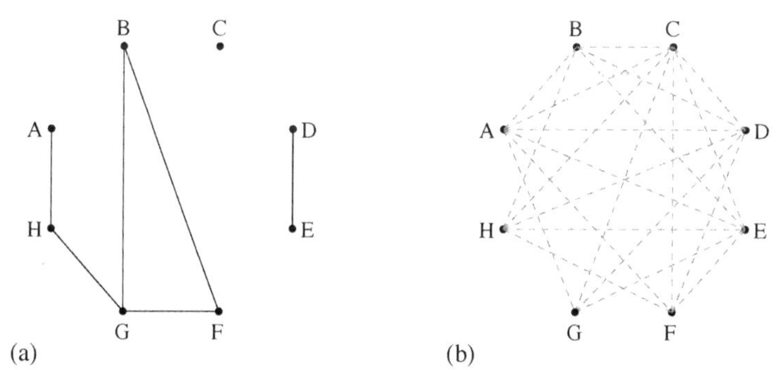

Figure 1.39. *Merger graph: a) compatible states; b) incompatible states*

The largest number of states that can be grouped in an incompatibility class is equal to the minimum number of states necessary for the representation of the state table. This value is 4, as the largest incompatibility classes are as follows: $(ABCD)$, $(ACDG)$, $(ACEG)$ and $(BCEH)$.

Considering the closed cover that consists of the compatibility classes $S_0 = (AH)$, $S_1 = (BFG)$, $S_2 = C$ and $S_3 = (DE)$, we can obtain the reduced state table shown in Table 1.64.

PS	NS				Output Z
	$XY = 00$	01	10	11	
S_0	S_0	S_0	S_1	S_3	0
S_1	S_1	S_3	S_1	S_2	0
S_2	–	S_0	–	S_2	1
S_3	S_0	S_3	S_1	S_3	1

Table 1.64. *Reduced state table for example 2*

1.5.3.4. *Example 3*

Using the merger graph and the compatibility graph, simplify the incompletely specified finite state machine whose state table is given in Table 1.65.

From the merger graph, as shown in Figure 1.40, we can obtain the pairs of compatible states (AB), (AC), (AD), (BC), (BD), (BE), (CD), (CF) and (EF). We can also deduce that the set of states $(ABCD)$ forms a compatibility class. As no closed polygon can be identified on the merger graph for incompatible states, we can conclude that there are only pairs of incompatible states.

A choice of maximal compatibility classes that covers all the states of the state machine consists of $(ABCD)$ and (EF). However, the closure conditions are not satisfied as the set $(ABCD)$ implies (CF) for $XY = 01$ and (BE) for $XY = 10$.

From the compatibility graph shown in Figure 1.41, we can identify the closed cover that consists of the pairs of states (AB), (CD) and (EF). Table 1.66 presents the reduced state table, obtained by assuming that $S_0 = (AB)$, $S_1 = (CD)$ and $S_2 = (EF)$.

NOTE 1.6.– We can also determine the pairs of compatible or incompatible states by making use of the concept of compatibility in the construction of an implication table. An implication table is filled in by inserting, in each cell, either a cross, when two states are incompatible, or closed bracket when two states or the implied NSs are compatible.

PS	NS				Output Z			
	$XY = 00$	01	10	11	$XY = 00$	01	10	11
A	–	C	E	B	–	1	1	1
B	E	–	–	–	0	–	–	–
C	F	F	–	–	0	1	–	–
D	–	–	B	–	–	–	1	–
E	–	F	A	D	–	0	0	1
F	C	–	B	C	0	–	0	1

Table 1.65. *State table for example 3*

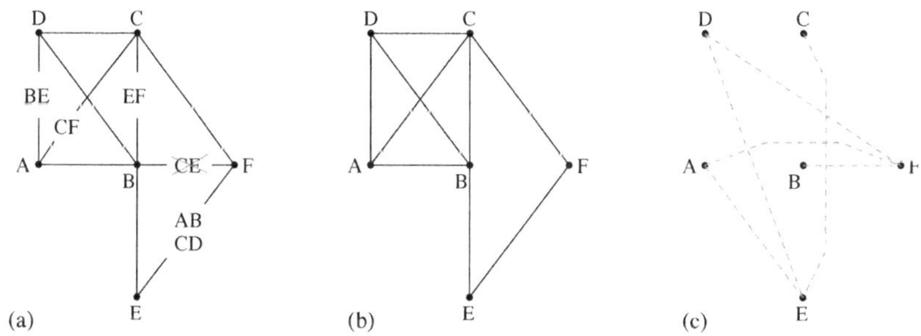

Figure 1.40. *a) Merger graph. Simplified merger graph: b) compatible states and c) incompatible states*

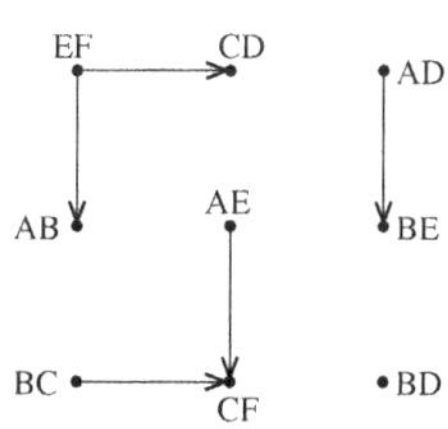

Figure 1.41. *Compatibility graph*

For example 3, Table 1.67 presents the implication tables obtained. The first table (see Table 1.67(a)) is constructed based on the state table. The second table (see Table 1.67(b)) is drawn up by taking into account the implication relationship that exists between pairs of incompatible states, or state pairs that correspond to cells that already contain a cross, and the other states. Thus, as states B and F imply the NSs

C and E, which are incompatible, a cross must be inserted in each cell corresponding to the states B and F to indicate that they are also incompatible.

PS	NS				Output Z			
	$XY = 00$	01	10	11	$XY = 00$	01	10	11
S_0	S_2	S_1	S_2	S_0	0	1	1	1
S_1	S_2	S_2	S_0	–	0	1	1	–
S_2	S_1	S_2	S_0	S_1	0	0	0	1

Table 1.66. *Reduced state table for example 3*

(a)

	A	B	C	D	E
B	✓				
C	CF	EF			
D	BE	✓	✓		
E	✗	✓	✗	✗	
F	✗	CE	✓	✗	AB CD

(b)

	A	B	C	D	E
B	✓				
C	CF	EF			
D	BE	✓	✓		
E	✗	✓	✗	✗	
F	✗	✗	✓	✗	AB CD

Table 1.67. *a) Implication table based on the state table; b) implication table after a pass*

When the final implication table has been obtained, the set of maximal compatibility classes can be determined as follows:

1) Begin with the right-most column. If the states are compatible, write down the corresponding compatibility class. Otherwise, enter the compatibility class that only implies the state associated with this column.

2) Continue with the next column on the left. If the state associated with the present column is compatible with all the states in the previously determined set of compatible states, it must be added to this set to form a larger compatibility class. If this state is only compatible with a subset of the previously determined set of compatible states, it must be added to this subset to form a new compatibility class. List out all the compatibility classes that are not included in an already formed compatibility class, including the compatibility class implying only the state associated with this column if it is incompatible with any other state.

3) Repeat steps 1 and 2 until the left-most column is reached. The set of maximal compatibility classes consists of the compatibility classes obtained in the last step.

The same procedure can be used to determine the maximal incompatibility classes by using the concept of incompatibility instead of compatibility.

In the case of example 3, we have:

– maximal compatibility classes:

Column E:	(EF)			
Column D:	(EF)	(D)		
Column C:	(CD)	(CF)	(EF)	
Column B:	(BCD)	(BE)	(CF)	(EF)
Column A:	(ABCD)	(BE)	(CF)	(EF)

The elements of the set of maximal compatibility classes are of the form: $(ABCD)$, (BE), (CF) and (EF).

– maximal incompatibility classes:

Column E:	(E)					
Column D:	(DF)	(DE)				
Column C:	(CE)	(DF)	(DE)			
Column B:	(BF)	(CE)	(DF)	(DE)		
Column A:	(AF)	(AE)	(BF)	(CE)	(DF)	(DE)

The set of maximal compatibility classes is made up of the following pairs of states: (AF), (AE), (BF), (CE), (DF) and (DE).

1.6. State encoding

The hardware cost for the implementation of finite state machines depends on the state encoding.

For a finite state machine with N_E states, at least $\lceil \log_2(N_E) \rceil$ bits or binary variables are required to code each state, where $\lceil x \rceil$ represent the smaller integer, which is equal to or greater than x.

The number of flip-flops, N_B, required for the implementation of a finite state machine with N_E states is such that:

$$2^{N_B - 1} < N_E < 2^{N_B} \qquad [1.38]$$

The number of possible ways to assign 2^{N_B} combinations of binary variables to N_E states or the number of possible ways to encode states is given by[1]:

$$P = \frac{2^{N_B}!}{(2^{N_B} - N_E)!} \qquad [1.39]$$

where $2^{N_B} > N_E$. However, not all these codes are unique as the variables may be permuted in $N_B!$ ways. In addition, as each variable may be complemented, there are 2^{N_B} ways of complementing the set of N_B variables. Thus, the number of unique possible ways of encoding states is reduced to:

$$U = \frac{2^{N_B}!}{(2^{N_B} - N_E)!} \cdot \frac{1}{N_B! 2^{N_B}} = \frac{(2^{N_B} - 1)!}{(2^{N_B} - N_E)! N_B!} \qquad [1.40]$$

Table 1.68 gives the number of encoding possibilities for state machines that can have up to eight states. The higher the number of states, the more difficult it becomes to identify an encoding method that can be implemented with a minimal number of logic gates and/or the smallest possible propagation delay.

N_E	N_B	P	U
1	0	–	–
2	1	2	1
3	2	24	3
4	2	24	3
5	3	6,720	140
6	3	20,160	420
7	3	40,320	840
8	3	40,320	840

Table 1.68. *Number of state encoding possibilities*

1 In general, the binomial coefficient gives the number of possibilities to form different subsets of N_E elements from 2^{N_B} elements. It is given by $\binom{2^{N_B}}{N_E}$, the value of which is given by $2^{N_B}!/[(2^{N_B} - N_E)! N_E!]$. When it comes to ordered subsets, the $N_E!$ possible permutations must be taken into account and the number of possible choices then has a value of $P = N_E! \times \binom{2^{N_B}}{N_E}$.

In general, the state encoding of a machine should be realized such that there is an increase in the number of adjacent terms in the Karnaugh map associated with each function that is to be simplified. The optimal state encoding can be achieved by:

– minimizing the number of bits that change during the transition from one state to another (for example Gray code, Johnson code);

– assigning the highest priority constraint to the adjacent groupings of 1's in the state functions;

– adopting one-hot encoding (or 1-out-of-n encoding).

Table 1.69 gives some examples of codes for a finite state machine with eight states.

	Codes				
PS	Natural binary	Gray	Johnson	1-out-of-8	Almost 1-out-of-8
S_0	000	000	0000	00000001	0000000
S_1	001	001	0001	00000010	0000001
S_2	010	011	0011	00000100	0000010
S_3	011	010	0111	00001000	0000100
S_4	100	110	1111	00010000	0001000
S_5	101	100	1110	00100000	0010000
S_6	110	101	1100	01000000	0100000
S_7	111	111	1000	10000100	1000000

Table 1.69. *Examples of codes for eight states*

With binary encoding, the number of binary variables required to represent n states is of the form $log_2(n)$. This last expression can also be considered as the minimum number of flip-flops required.

Binary encoding presents the disadvantage of increasing the complexity of the combinational logic section that is required for the decoding of each state, as well as the flip-flop switching activity needed for the state representation. It should be noted that the power consumption increases with an increase in the switching activity. The latter may also prove to be a limitation when trying to achieve a high operating speed.

With Gray encoding, two consecutive states only differ by a single bit. In other words, the transition from one state to another only affects one flip-flop. However, the decoding of the states may become complex, especially when the number of states increases. The number of binary variables is equal to the number of flip-flops, that is $log_2(n)$ for a machine with n states.

Johnson encoding requires $n/2$ bits (or flip-flops) in order to represent n states. Only one bit changes between two consecutive states.

One-hot encoding (or 1-out-of-n encoding) is characterized by the fact that for each state, a single bit is set to 1 while all other bits are at 0. Each internal state of the machine is identified by a single bit and represented by a single flip-flop. Thus, n flip-flops are required to implement a state machine with n states. One-hot encoding is suitable for implementations based on CPLDs or FPGAs, which generally have a large number of flip-flops.

The synthesis of the combinational logic section of the machine is simpler because each term in the logic equation for the NS of each flip-flop has exactly one variable.

A machine that uses one-hot encoding can have quite a high operating speed as the speed is not dependent on the number of states but, rather, on the number of transitions required to move from one state to the other.

Almost-one-hot encoding corresponds to one-hot encoding that begins with a null code, all the bits of which are set to 0. The null code is frequently used to represent the initial state, which in practice may be achieved by applying the appropriate signals to the asynchronous reset inputs of flip-flops. Thus, two bits need to be considered to fully decode a given state.

EXAMPLE 1.4.– Using 1-out-of-3 code to represent the states of a 101 sequence detector (Mealy model).

The state table of the 101 sequence detector based on the Mealy model is shown in Table 1.70. Using 1-out-of-3 code to represent the states, we obtain the transition table given in Table 1.71. The implementation of the state machine requires three flip-flops. Because each state is characterized by one of the three flip-flops set to 1, the logic equations of the NSs and the output can be deduced by analyzing the state diagrams shown in Figure 1.42.

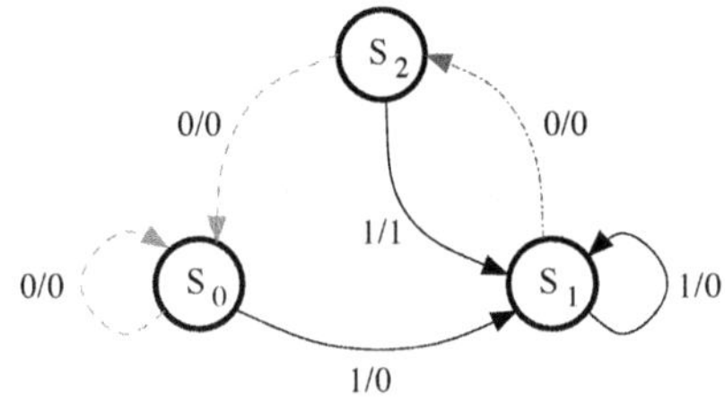

Figure 1.42. *State diagram (Mealy model)*

PS	NS		Output Y	
	$X = 0$	1	$X = 0$	1
S_0	S_0	S_1	0	0
S_1	S_2	S_1	0	0
S_2	S_0	S_1	0	1

Table 1.70. *State table of the sequence detector*

PS	NS		Output	
ABC	$A^+B^+C^+$		Y	
	$X = 0$	1	$X = 0$	1
001	001	010	0	0
010	100	010	0	0
100	001	010	0	1

Table 1.71. *Transition table of the sequence detector*

Flip-flop A, associated with the state S_2, is set to 1 if the machine is in the state S_1, which is represented by flip-flop B, and the input X takes the logic level 0 (see the dash-dot line in Figure 1.42). Thus:

$$A^+ = \overline{X} \cdot B \qquad [1.41]$$

Flip-flop B is set to 1 if the machine is in state S_2 and the input X assumes the logic level 1, or the machine is in the state S_1 and the input X takes the logic level 1; or if the machine is in the state S_0, which is represented by flip-flop C, and the input X is set to 1 (see the full line in Figure 1.42). Hence:

$$B^+ = X \cdot A + X \cdot B + X \cdot C = X(A + B + C) \qquad [1.42]$$

Flip-flop C is set to 1 if the machine is in the state S_2 and the input X takes the logic level 0; or if the machine is in the state S_0 and the input X is set to 0 (see the dash-dash line in Figure 1.42). Thus:

$$C^+ = \overline{X} \cdot A + \overline{X} \cdot A = \overline{X}(A + C) \qquad [1.43]$$

The output Y is set to 1 only if the machine is in the state S_2 and the input X is set to 1. We then have:

$$Y = X \cdot A \qquad [1.44]$$

We can also determine the expressions for A^+, B^+, C^+ and Y by using Karnaugh maps, which are filled out based on the transition table, as shown in Figures 1.43–1.46. Considering these functions to be four-variable functions: only six of the 16 possible combinations, $XABC$, are used; 0000 and 1000 are the forbidden states as they stem from the almost 1-out-of-3 code; and the remaining combinations are don't-care states.

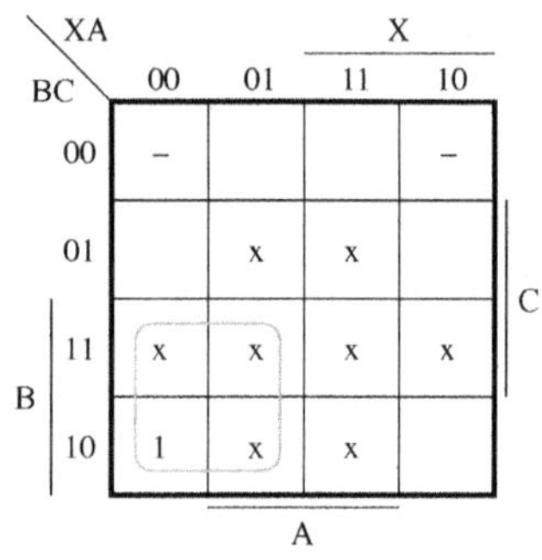

Figure 1.43. *Function* $A^+ = \overline{X} \cdot B$

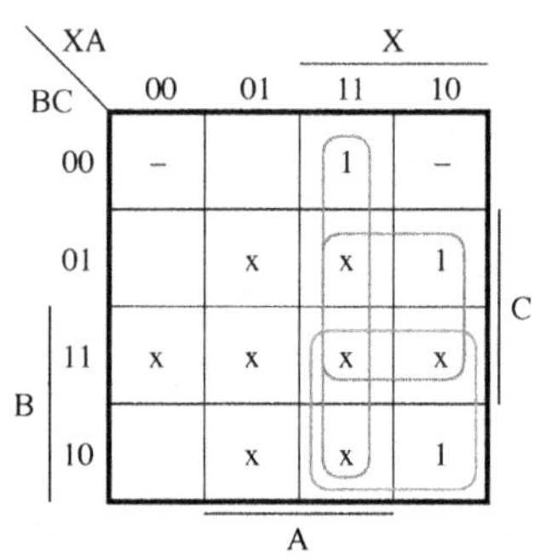

Figure 1.44. *Function* $B^+ = X$

Figure 1.47 depicts the logic circuit of the 101 sequence detector based on the Mealy state machine whose states are represented using the 1-out-of-3 code.

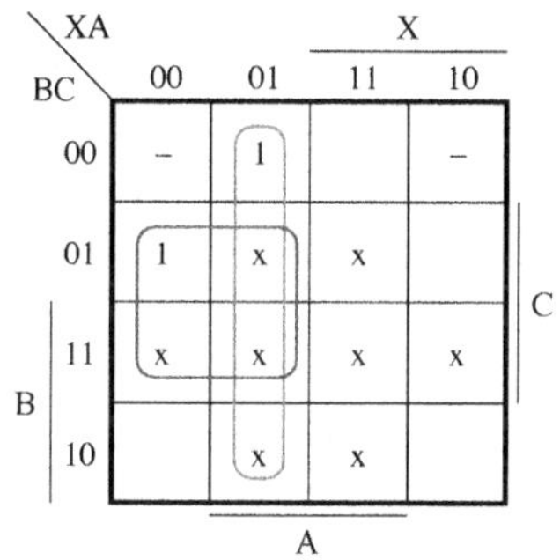

Figure 1.45. *Function* $C^+ = \overline{X} \cdot \overline{B}$

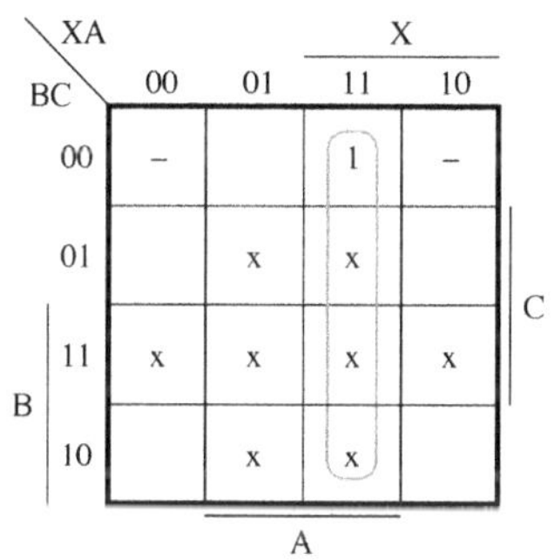

Figure 1.46. *Output* $Y = X \cdot A$

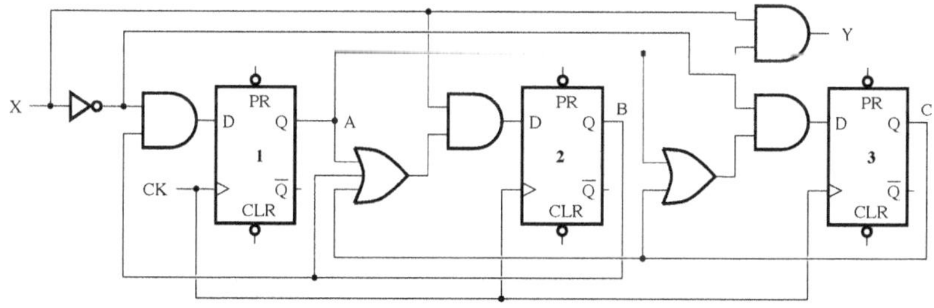

Figure 1.47. *Logic circuit of the 101 sequence detector (Mealy state machine whose states are represented using the 1-out-of-3 code)*

1.7. Transformation of Moore and Mealy state machines

Finite state machines are generally designed on the basis of either the Moore or the Mealy model. However, it is possible to convert one of these models into the other.

The number of states in the Moore model is higher or equal to the number of states of the equivalent Mealy model.

Figure 1.48 shows examples of the transformation of a Moore model to a Mealy model for a section of a finite state machine.

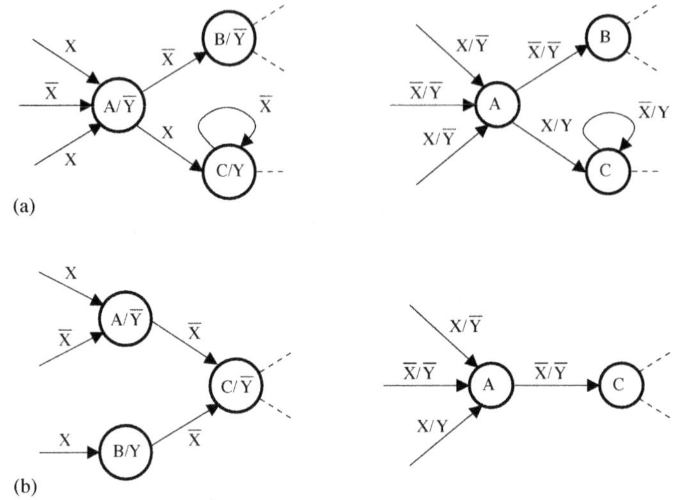

Figure 1.48. *Transformation examples of a Moore model to a Mealy model*

To obtain the Mealy model from the Moore model, the output generated at a given state is attached along with the input condition on the transition arcs that are directed toward that state.

If the transitions from two different states and for the same input condition lead to the same state, it is possible to eliminate a state.

Referring to Figure 1.48(b), two transitions lead to the state C under the input condition $\overline{X}$. Thus, either state A or state B can be eliminated.

Figure 1.49 gives examples of transformation of a Mealy model to a Moore model for a section of a finite state machine.

If all transitions to one state of the Mealy state machine are associated with an identical output, this output should be allocated to the NS of the equivalent Moore state machine.

If all outputs associated with transitions to a state of the Mealy state machine are not identical, an additional state must be used.

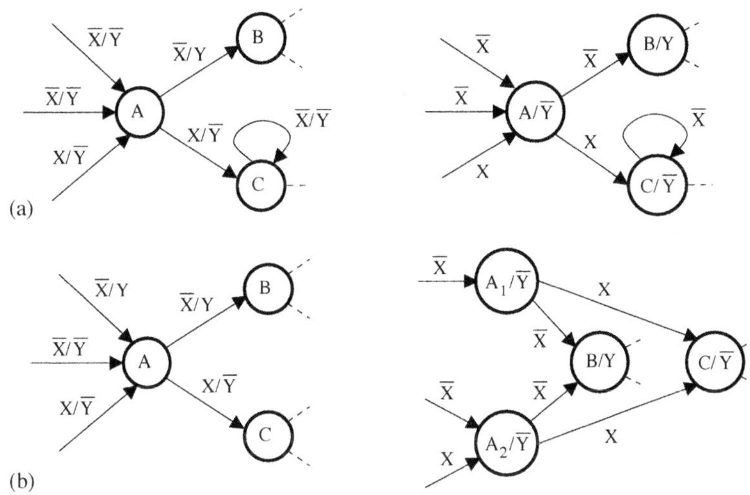

Figure 1.49. *Transformation examples of a Mealy model to a Moore model*

1.8. Splitting finite state machines

After the synthesis of a finite state machine, it may be necessary to split the machine if it cannot be implemented using a single programmable logic component, or must be implemented using several different platforms (CPLD or FPGA, microprocessor or microcontroller). One approach to accomplish this consists in using intermediate states. Splitting a finite state machine can help to achieve a compromise between the number of flip-flops required to code the states and the complexity of the logic functions needed to generate the NSs.

1.8.1. *Rules for splitting*

When choosing to split a finite state machine, we can encounter either the case of a single transition or that of several transitions:

– *Case of a single transition* (see Figure 1.50): The transition between a state S_i, called the source, and a state S_j, called the destination, that yields the output Y when the input condition X is verified can be split by using two intermediate idle states, S_{r1} and S_{r2}, and another output Z_i. After the split, the transition from the state S_i to the state S_{r1} takes place when the condition X is satisfied and results in the generation of the outputs Y and Z_i. We then have an unconditional hold on the state S_{r1}. Starting from the state S_{r2}, there is either a transition to the state S_j if the condition $X \cdot Z_i$

is verified, or the PS is held if the logic expression $\overline{X \cdot Z_i}$ is true. In both cases, the output is considered to have a don't-care state.

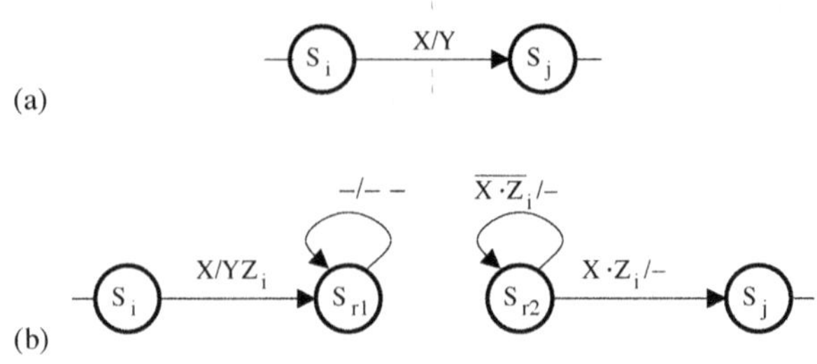

Figure 1.50. *Splitting rule: case of a single transition*

– *Case of several transitions* (see Figure 1.51): The aim is to develop rules applicable when several transitions are associated with the same source or destination. The transitions from the same source (for example state S_i) are grouped under a single transition that enables the state machine to move from this source state to an intermediate idle state and occurs when one of the conditions associated with each of these transitions is verified. The transitions that lead to the same destination state (for example state S_k) are replaced by a single transition that allows the state machine to change from an intermediate idle state to the destination state and is triggered when the logical OR function of conditions associated with each of these transitions is true. The holding condition for an intermediate idle state corresponds to the inverse of the logical OR function of all conditions for the exit transition from this state.

1.8.2. *Example 1*

Consider the finite state machine whose state diagram is shown in Figure 1.52(a). This machine has six states, an input X and an output Y. It is to be split into two four-state machines that can communicate between themselves.

Figure 1.52(b) depicts the result of splitting a machine into two submachines. The submachine with the intermediate idle state, S_{r1}, has three outputs, Y, Z_1 and Z_2, while the machine with the intermediate idle state, S_{r2}, only generates two outputs, Y and Z_3.

Figure 1.53 shows the implementation of the machine in its split form, as shown in Figure 1.52(b). At any given instant only one submachine is active and the other is in a standby state. As the different outputs are combined using an OR logic gate, the output of the inactive submachine must be set to zero so as not to affect the active submachine.

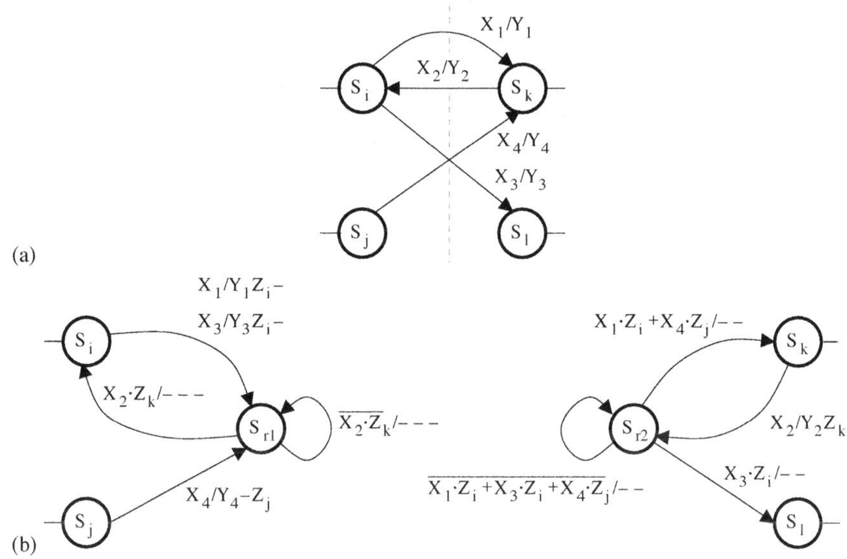

Figure 1.51. *Splitting rule: case of several transitions*

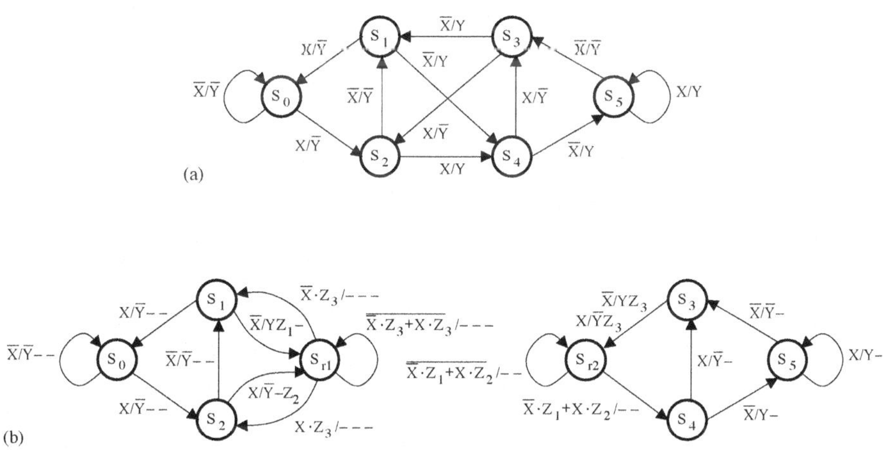

Figure 1.52. *State diagrams: a) initial finite state machine; b) finite state machine obtained after the splitting*

The choice of 1-out-of-4 code to represent the states leads to the use of four flip-flops for the implementation of each submachine. Tables 1.72 and 1.73 present the state encoding with the outputs for the flip-flops being Q_1, Q_2, Q_3 and Q_4 for submachine 1, and Q_A, Q_B, Q_C and Q_D for submachine 2.

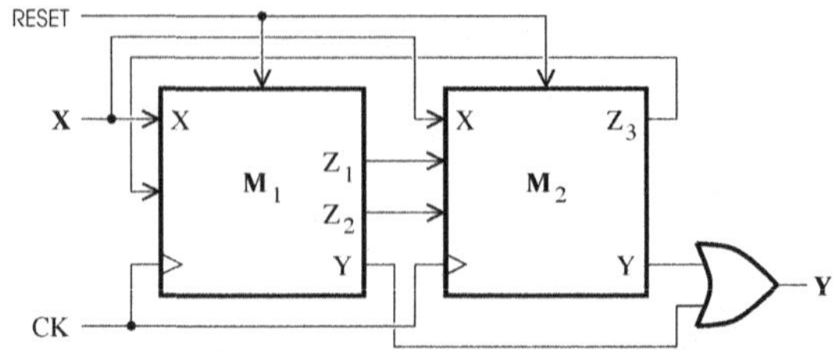

Figure 1.53. *Implementation of the state machine*

	Q_4	Q_3	Q_2	Q_1
S_{r1}	1	0	0	0
S_2	0	1	0	0
S_1	0	0	1	0
S_0	0	0	0	1

Table 1.72. *Encoding the states of submachine 1*

	Q_D	Q_C	Q_B	Q_A
S_{r2}	1	0	0	0
S_5	0	1	0	0
S_4	0	0	1	0
S_3	0	0	0	1

Table 1.73. *Encoding the states of submachine 2*

As the number of flip-flops can be high with a code of the 1-out-of-n type, it may be necessary to add a circuit section to stop the flip-flop switching by synchronizing the clock signal, and consequently, reducing the power consumption of the circuit. Such an implementation is illustrated in Figure 1.54, where additional logic gates and flip-flops allow the deactivation of the clock signal of the submachine that is in the standby state.

The activation function for the clock signal of the submachine 1 and 2 can be defined as follows:

$$A_1 = \overline{Q_4 \cdot \overline{Z_3}} = \overline{Q_4} + Z_3 \quad [1.45]$$

and:

$$A_2 = \overline{Q_D \cdot \overline{Z_1 + Z_2}} = \overline{Q_D} + Z_1 + Z_2 \qquad [1.46]$$

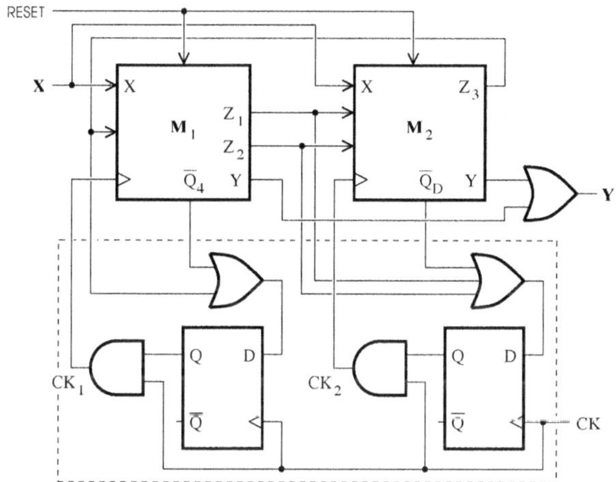

Figure 1.54. *Implementation using a clock signal synchronization system*

It should be noted that one of the terms, Q_4 or Q_D, is set to 1 to indicate that one of the submachines is in the standby state. Additionally, a submachine can only exit the standby state when the other submachine sets one of the signals, Z_1, Z_2 and Z_3, to 1. The clock signal is deactivated for submachine i when A_i $(i = 1, 2)$ is set to 0.

The functions A_i are derived so that there is a correspondence between the last period of the clock signal to be deactivated and the first period of the clock signal to be activated. This results in an overlapping period between the signals, as shown in the timing diagram represented in Figure 1.55.

1.8.3. *Example 2*

A finite state machine is characterized by the state diagram represented in Figure 1.56(a), where X denotes the input signal. Split this machine into three submachines, with the states S_0, S_1 and S_2; S_3, S_4 and S_5; and S_6, S_7 and S_8, respectively.

After the split, we obtain three submachines that can communicate with each other, as shown in Figure 1.56(b), where S_{r1}, S_{r2} and S_{r3} represent the intermediate idle states. Only one submachine is active at a time and the other two are in the idle state. Using 1-out-of-4 code to represent the states of each submachine we obtain the implementation shown in Figure 1.57. As the transition to a given state results in the output Q_i ($i = 1, 2, 3, 4$) of only one of the flip-flops of a submachine being set at 1, the signals Z_4, Z_5, Z_6 and Z_8 can be obtained directly from the appropriate outputs, Q_i.

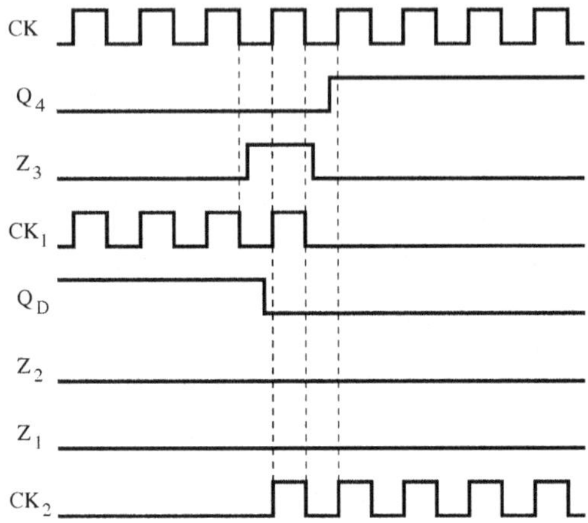

Figure 1.55. *Timing diagram*

1.9. Sequence detector implementation based on a programmable circuit

In general, a serial communication protocol uses a binary sequence to mark the beginning and end of a frame (or each word to be transmitted).

Use a PAL (or *programmable array logic*) with versatile outputs to implement a binary sequence detector that serially receives data at the input X and sets the output Y to 1 only if the sequence 01111110 has been identified.

The operation of a 01111110 sequence detector can be described by the state diagram shown in Figure 1.58 or by the state table given in Table 1.74. The detector has eight states that can be represented using three-bit Gray code. Table 1.75 presents the transition table.

For implementation using D flip-flops, the characteristic equation is of the form, $Q^+ = D$, and the logic equation for each input is obtained by constructing the corresponding Karnaugh map based on the transition table. Karnaugh maps shown in

Figures 1.59–1.61 can be used to determine the logic equations of the variables A^+, B^+ and C^+, while the logic equation of the output Y is obtained based on the Karnaugh map shown in Figure 1.62. Thus:

$$D_A = A^+ = X \cdot A \cdot C + X \cdot B \cdot \overline{C} \qquad [1.47]$$

$$D_B = B^+ = X \cdot \overline{A} \cdot C + X \cdot B \cdot \overline{C} \qquad [1.48]$$

$$D_C = C^+ = \overline{X} + A \cdot B + \overline{A} \cdot \overline{B} \cdot C \qquad [1.49]$$

and:

$$Y = \overline{X} \cdot A \cdot \overline{B} \cdot \overline{C} \qquad [1.50]$$

where $A = Q_A$, $B = Q_B$ and $C = Q_C$. The PAL implementing the sequence detector is illustrated in Figure 1.63. The different outputs are configured by assigning the appropriate logic states to the bits S_1 and S_0.

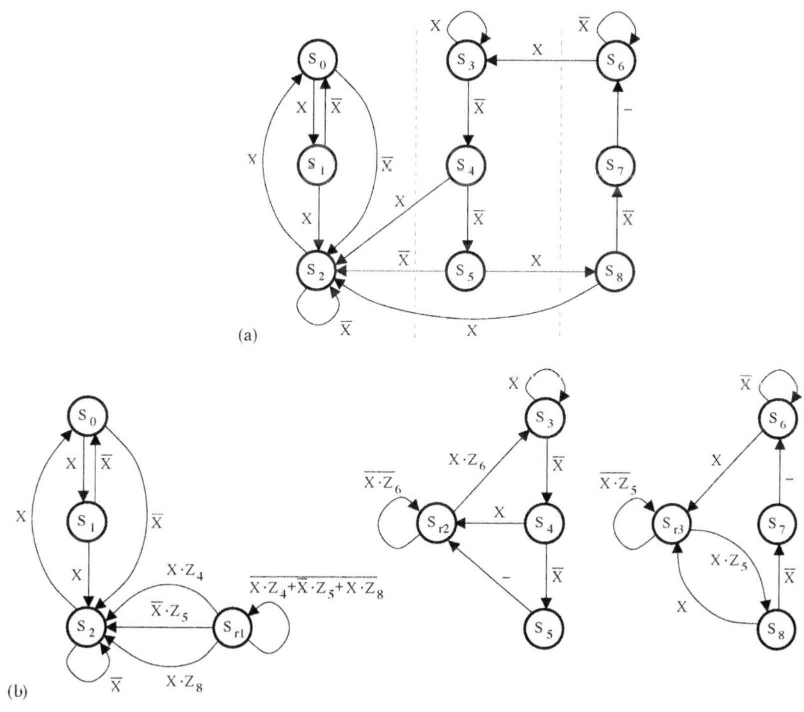

Figure 1.56. *State diagrams: a) initial finite state machine; b) finite state machines obtained after the splitting*

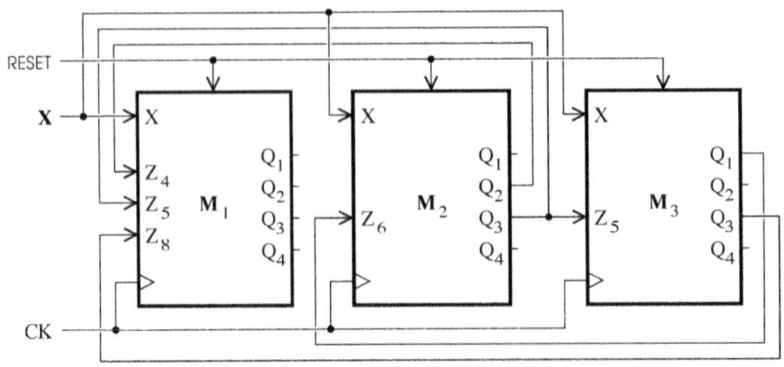

Figure 1.57. *Implementation of the state machine*

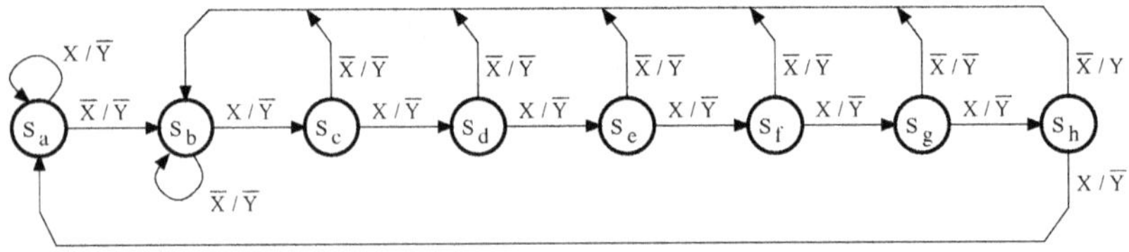

Figure 1.58. *State diagram of the sequence detector*

PS	NS		Output, Y	
	$X=0$	$X=1$	$X=0$	$X=1$
S_a	S_b	S_a	0	0
S_b	S_b	S_c	0	0
S_c	S_b	S_d	0	0
S_d	S_b	S_e	0	0
S_e	S_b	S_f	0	0
S_f	S_b	S_g	0	0
S_g	S_b	S_h	0	1
S_h	S_b	S_a	1	0

Table 1.74. *State table of the sequence detector*

1.10. Practical considerations

To ensure the proper operation of a finite state machine, the outputs should not be affected by undesirable transient signals, which may be caused by propagation delays of logic gates in the output combinational circuit section or race conditions between the state variables.

PS	NS, $A^+B^+C^+$		Output, Y	
ABC	$X = 0$	$X = 1$	$X = 0$	$X = 1$
000	001	000	0	0
001	001	011	0	0
011	001	010	0	0
010	001	110	0	0
110	001	111	0	0
111	001	101	0	0
101	001	100	0	0
100	001	000	1	0

Table 1.75. *Transition table of the sequence detector*

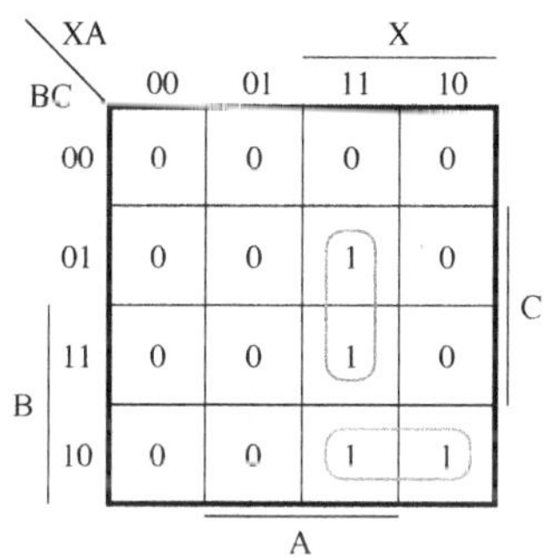

Figure 1.59. *Function A^+*

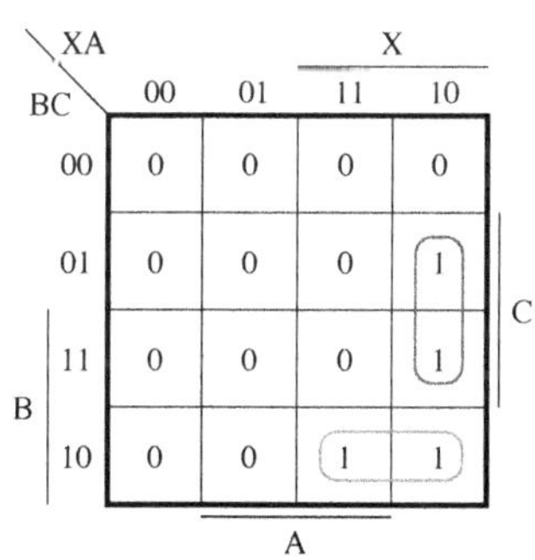

Figure 1.60. *Function B^+*

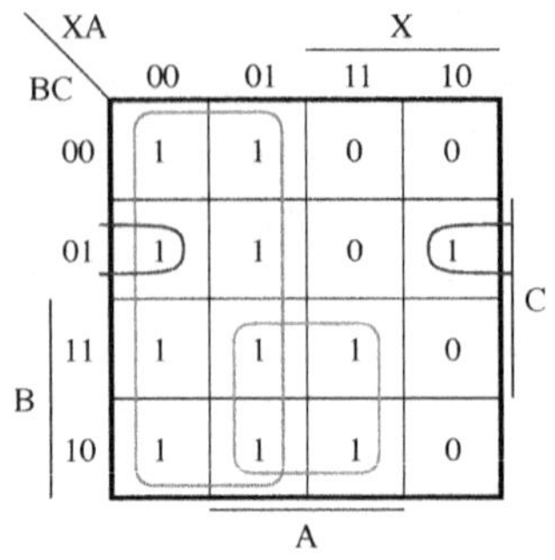

Figure 1.61. *Function* C^+

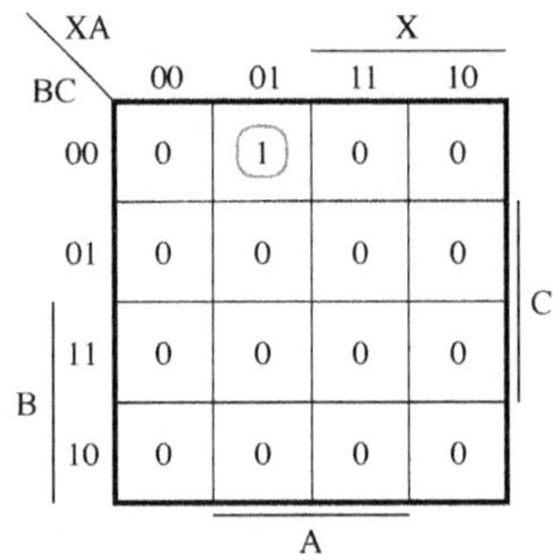

Figure 1.62. *Output* Y

1.10.1. *Propagation delays and race conditions*

A critical race condition is caused by a transition between two states leading to a change in at least two state variables. In fact, when the change in n $(n \geq 2)$ state variables required for the transition from one state to another cannot be simultaneous, there are $n$possible paths. This implies a race between the state variables and the race may be non-critical or critical.

In the case of a non-critical race, each path associated with a transition leads to the same destination state. However, for a critical race, the destination state depends on the path associated with the transition or the order in which the variables change. This can result in the generation of undesirable transient signals at the output of a machine.

For a transition that can involve only two paths, the race may not result in the generation of a transient signal if, for each output, the type of operation associated with the original state is different from that of the destination state.

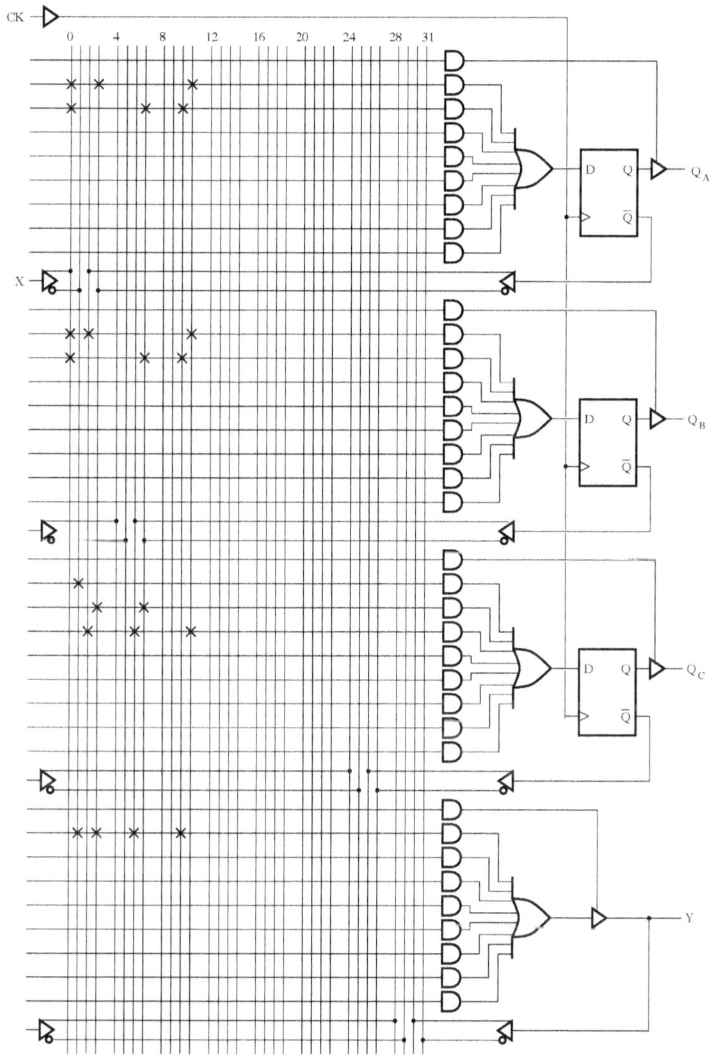

Figure 1.63. *Sequence detector implemented by programming a PAL*

In general, one of the following approaches can be used to eliminate transient signals caused by a critical race condition:

– judiciously assign a value to a don't-care state that is found on a race path;

– use another code (for example Gray code) to represent states;

– filter the output signal;

– insert an intermediate state (for example adding a state to a machine that initially had only three states);

– increase the number of state variables, as is the case for a machine based on a code of the 1-out-of-n type.

Unlike other approaches, the first two may not result in an increase in the number of components and a reduction in the operating speed.

Transient signals caused by a critical race condition usually appear at times just after the active edge of the clock signal. They can, therefore, be filtered by connecting a flip-flop triggered by the opposite edge of the clock signal at the output of the finite state machine. Thus, if the flip-flops of the state machine are triggered by the rising edge of the clock signal, the flip-flop of the filter must be activated by the falling edge of the clock signal, as shown in Figure 1.64(a). The timing diagram showing the suppression of transient signals is depicted in Figure 1.64(b). It should be noted that the filter introduces an additional delay of half a period.

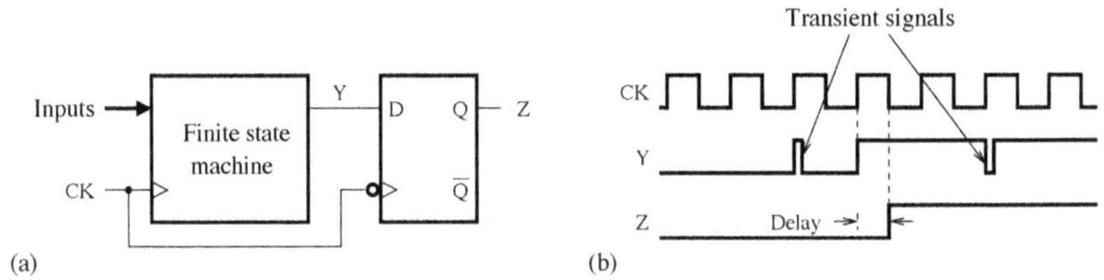

Figure 1.64. *a) Suppression of transient signals by filtering; b) timing diagram*

1.10.2. *Timing specifications*

To ensure the proper working of a finite state machine, certain timing constraints must be satisfied. In the case of a synchronous machine, the clock signal is used to synchronize the flip-flops. As the sequential logic section and the combinational logic section exhibit a propagation delay, it takes a period of time before the output of the state machine acquires its final value after each change of the input signal.

The timing characteristics of a logic circuit are illustrated in Figure 1.65.

The set-up time, t_{su}, is the minimum amount of time the data signal must be held constant before the active edge of the clock signal.

The hold time, t_h, is the minimum amount of time the data signal must stay constant after the active edge of the clock signal.

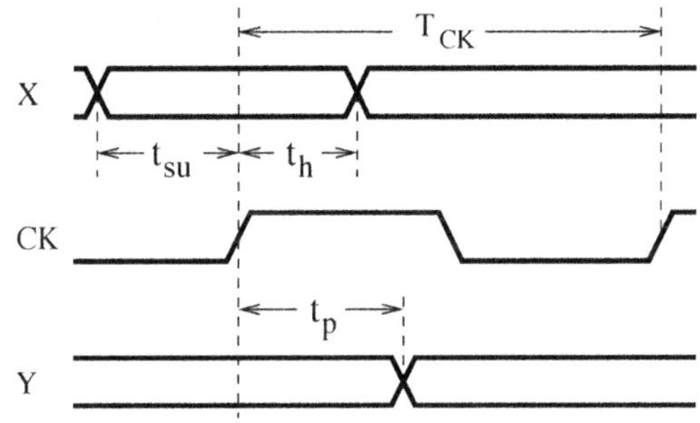

Figure 1.65. *Illustration of the timing characteristics of a logic circuit*

The propagation time, t_p, corresponds to the time that elapses between the instant when the signal is applied at the input and the instant when the output acquires its final value. This depends on the operating conditions of the logic circuit.

In general, component manufacturers specify a time window around the sampling instant and wherein the input signal must stay constant to prevent the circuit from entering a metastable state. After the sampling instant, the sequence of the data signal must not change for a certain time that is lower than the propagation delay from the input to the output of the circuit. Thus, for the proper operation of a synchronous circuit, the following relationships must be verified:

$$T_{CK} > t_{p,\max} + t_{su} \quad [1.51]$$

$$t_h < t_{p,\min} \quad [1.52]$$

where T_{CK} is the clock signal period, $t_{p,\min}$ and $t_{p,\max}$ represent the minimal and maximal values of the propagation delay, respectively. The first relationship can be used to determine the maximum operating frequency, that is $f_{CK} = 1/T_{CK}$. The second relationship is independent of the clock signal period and is always verified for circuits whose hold time is equal to zero.

The maximum frequency of the clock signal is obtained by assuming that there is no timing margin. However, a good rule of thumb consists of adding a timing margin of about 10% of the minimal period of the clock signal to the set-up time to take into account various fluctuation phenomena (jitter, timing skew) that may affect the clock signal. Furthermore, a timing margin greater than zero for the hold-time means there is no violation of the constraint on the hold time.

The timing analysis of the logic circuit of a finite state machine consists of identifying the slowest path that determines the maximum frequency of the clock signal and the fastest path that sets the timing margin for the hold time.

Consider the 101 sequence detector shown in Figure 1.66, where X represents the input and Y is the output. Determine the maximum frequency of the clock signal, assuming that a timing margin of 10% of the minimal period of the clock signal has been added to the set-up time. What is the value of the timing margin for the hold time?

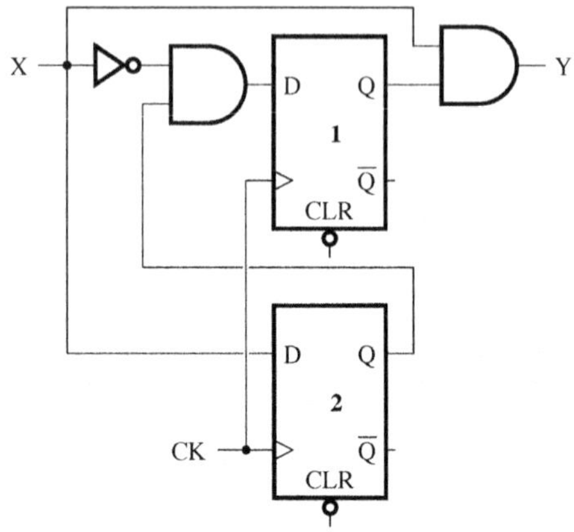

Figure 1.66. *Logic circuit of the 101 sequence detector*

From datasheets, the following parameters can be obtained:

– D flip-flop of the type 74LS74A:

$$t_{su} = 20 \text{ ns}, t_h = 5 \text{ ns}, t_{p,\max} = 40 \text{ ns}, t_{p,\min} = 13 \text{ ns};$$

– AND gate of the type 74LS08:

$$t_{p,\max} = 20 \text{ ns}, t_{p,\min} = 8 \text{ ns};$$

– NOT gate of the type 74LS04:

$$t_{p,\max} = 5 \text{ ns}, t_{p,\min} = 1 \text{ ns}.$$

The path with the highest propagation delay goes through the NOT gate, the AND gate (that is $t_{p\text{comb},\max} = (5 + 20)$ ns for the combinational logic section of the circuit), and the flip-flop 1 (that is a delay of $t_{pbd,\max} + t_{su}$). Taking into consideration the timing margin, $t_{su,\text{marg}}$, that has been added to the hold time, the minimal period of the clock signal is given by:

$$T_{CK,\min} = t_{\text{pbd},\max} + t_{p\text{comb},\max} + t_{su} + t_{su,\text{marg}} \qquad [1.53]$$

Using $t_{su,\text{marg}} = T_{CK,\min}/10$, we can obtain:

$$T_{CK,\min} = (10/9)(t_{pbd,\max} + t_{p\text{comb},\max} + t_{su}) \quad [1.54]$$

$$= (10/9)(40 + 25 + 20) = 94.44 \text{ ns} \quad [1.55]$$

and

$$f_{CK,\max} = 1/T_{CK,\min} = 1/(94.44 \times 10^{-9}) = 10.58 \text{ MHz} \quad [1.56]$$

The path with the smallest propagation delay goes through flip-flop 2 whose input is not connected to any logic gate. That is to say, $t_{p\text{comb},\min} = 0$. We can, therefore, write:

$$t_h + t_{h,\text{marg}} = t_{pbd,\min} + t_{p\text{comb},\min} \quad [1.57]$$

or:

$$t_{h,\text{marg}} = t_{\text{pbd},\min} + t_{p\text{comb},\min} - t_h \quad [1.58]$$

$$= 13 + 0 - 5 = 8 \text{ ns} \quad [1.59]$$

Because $t_{h,\text{marg}} > 0$, there is no violation of the constraint related to hold time.

For the logic circuit shown in Figure 1.67, where the Sync flip-flop is used to ensure the synchronization of the input signal X, determine the maximum frequency of the clock signal and the temporal margin for the hold time. The clock signal has a duty cycle of 50% and from datasheets, we can obtain:

– D flip-flop of the type 74LS74A:

$t_{su} = 20$ ns, $t_h = 5$ ns, $t_{p,\max} = 40$ ns, $t_{p,\min} = 13$ ns;

– AND gate of the type 74LS08:

$t_{p,\max} = 20$ ns, $t_{p,\min} = 8$ ns;

– OR gate of the type 74LS32:

$t_{p,\max} = 22$ ns, $t_{p,\min} = 14$ ns;

– NOT gate of the type 74LS04:

$t_{p,\max} = 5$ ns, $t_{p,\min} = 1$ ns.

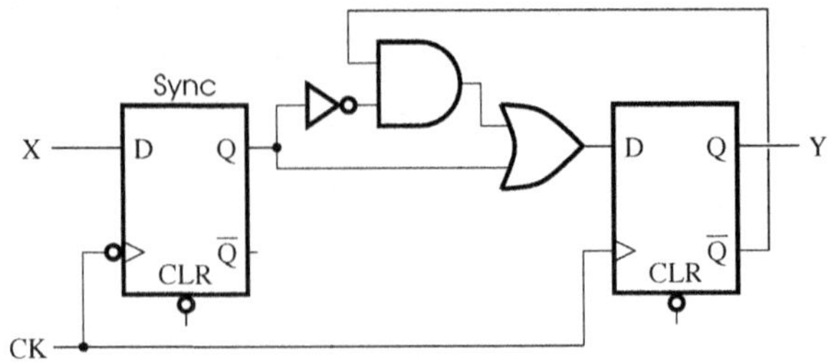

Figure 1.67. *Finite state machine with a synchronization stage*

As the flip-flops are triggered by the opposite edges of the clock signal, we can consider that the input signal, X, is delayed by $T_{CK,\min}/2$ by the Sync flip-flop. The slowest path across the combinational logic section passes through the inverter, the AND gate, and the OR gate, that is $t_{pcomb,\max} = (5 + 20 + 22)$ ns. The minimal period of the clock signal can be expressed as follows:

$$T_{CK,\min} = T_{CK,\min}/2 + t_{pcomb,\max} + t_{pbd,\max} + t_{su} \quad [1.60]$$

Hence:

$$T_{CK,\min} = 2(t_{pcomb,\max} + t_{pbd,\max} + t_{su}) \quad [1.61]$$

$$= 2(47 + 40 + 20) = 214 \text{ ns} \quad [1.62]$$

The maximum frequency of the clock signal is then given by:

$$f_{CK,\max} = 1/T_{CK,\min} = 1/(214 \times 10^{-9}) = 4.67 \text{ MHz} \quad [1.63]$$

The fastest path leading to the input of the second flip-flop passes through the AND gate and the OR gate, that is $t_{pcom,\min} = (8 + 14)$ ns. Thus, the hold time must satisfy the following relationships:

$$t_h + t_{h,marg} = t_{pbd,\min} + t_{pcomb,\min}. \quad [1.64]$$

or, equivalently:

$$t_{h,marg} = t_{pbd,\min} + t_{pcomb,\min} - t_h \quad [1.65]$$

$$= 13 + 22 - 5 = 30 \text{ ns}. \quad [1.66]$$

For a proper operation, the timing margin $t_{h,marg}$ must be greater than zero.

1.11. Exercises

EXERCISE 1.1.– Can we implement the finite state machine described by the state diagram in Figure 1.68 using only one flip-flop and logic gates?

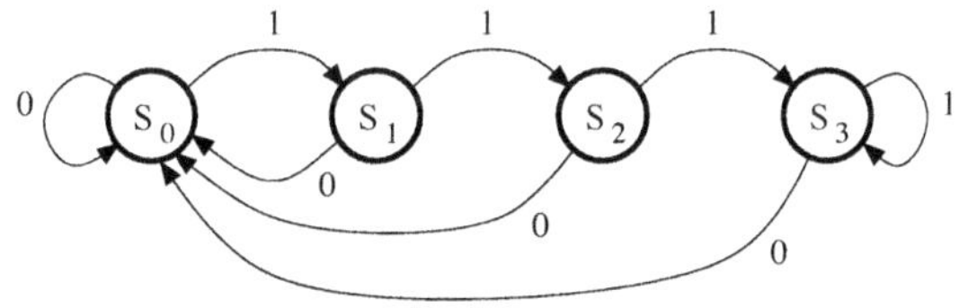

Figure 1.68. *State diagram*

EXERCISE 1.2.– Propose the state diagrams corresponding to the transition table shown in Table 1.76 and to the state table shown in Table 1.77.

PS	NS, A^+B^+		Output Y	
$A\ B$	$X=0$	1	$X=0$	1
0 0	00	10	0	1
0 1	00	00	0	0
1 0	11	01	1	1
1 1	10	10	1	0

Table 1.76. *Transition table*

PS	NS		Output Y
	$X=0$	1	
S_1	S_1	S_2	1
S_2	S_4	S_3	1
S_3	S_4	S_3	0
S_4	S_1	S_2	0

Table 1.77. *State table*

EXERCISE 1.3.– RT (fictional) flip-flop.

The RT (fictional) flip-flop, which has two inputs (R and T) and two outputs (Q and $\overline{Q}$), is characterized by the state table shown in Table 1.78:

a) We wish to implement this RT flip-flop using a D flip-flop, triggered by the rising edge of the clock signal, and logic gates (AND and OR).

– determine the logic equation for the input D;

– deduce the characteristic equation, Q^+, of the RT flip-flop;

– construct the state diagram;

– draw up the truth table;

– represent the logic circuit.

PS Q	NS = Output Q^+ $RT = 00$	01	10	11
0	0	1	0	1
1	0	0	0	1

Table 1.78. *State table*

b) Consider the case where the implementation of the RT flip-flop requires the use of a JK flip-flop triggered by the rising edge of the clock signal and logic gates:

– determine the logic equation for both inputs, J and K;

– represent the logic circuit.

EXERCISE 1.4.– The logic circuit shown in Figure 1.69 represents a three-input flip-flop:

– determine the characteristic equation;

– complete the truth table shown in Table 1.79;

– construct the state diagram.

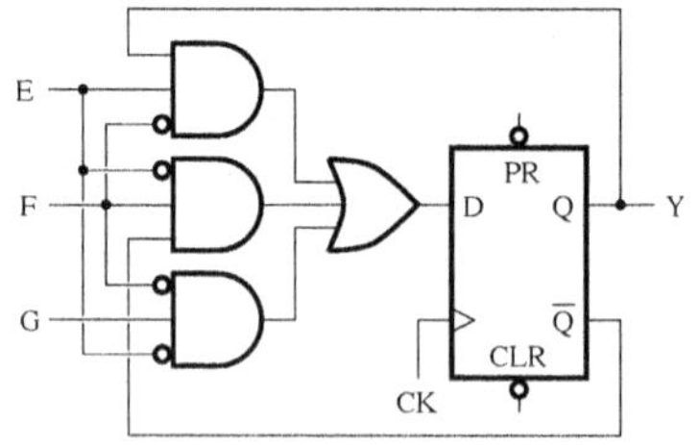

Figure 1.69. *Logic circuit*

E	F	Y
0	0	
0	1	
1	0	
1	1	

Table 1.79. *Truth table*

EXERCISE 1.5.– Consider the finite state machine whose logic circuit is represented in Figure 1.70:

– determine the characteristic equation;

– complete the truth table shown in Table 1.80;

– construct the state diagram;

– what is the role of this finite state machine?

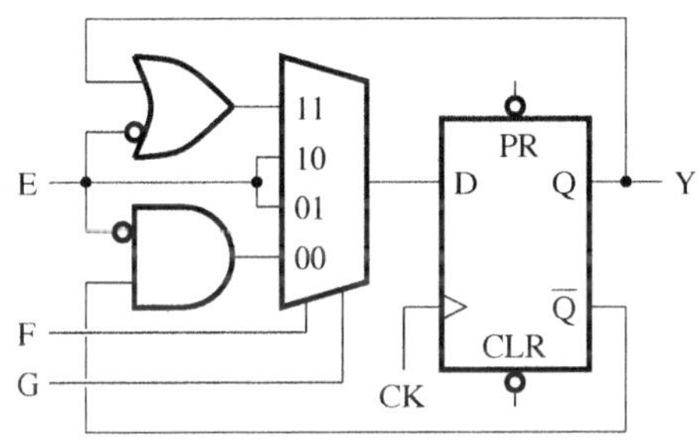

Figure 1.70. *Logic circuit*

F	G	Y
0	0	
0	1	
1	0	
1	1	

Table 1.80. *Truth table*

EXERCISE 1.6.– We wish to implement the 01 sequence detector as a finite state machine based on Mealy and Moore models:

– represent the state diagram;

– draw up the state table;

– implement the logic circuit using D flip-flops and logic gates;

– modify the state diagram in order to enable the recognition of the two binary sequences, 01 and 10.

EXERCISE 1.7.– Implement a finite state machine (Mealy and Moore) whose output, Y, corresponds to the two's complement of the binary number applied sequentially to the input, X, beginning with the least significant bit, as shown in Figure 1.71. Each conversion operation begins with the reset of the state machine, and the Start signal then takes logic state 1. At the end of the conversion, the TC signal is set to 1, causing the output register to enter the hold state.

The two's complement can be obtained by scanning a binary number starting from the least significant bit and complementing only the bits that come after the first bit assuming the logic state 1.

EXERCISE 1.8.– Serial comparator.

A serial comparator can be implemented as shown in Figure 1.72, where two registers contain the numbers, $A = a_0a_1 \cdots a_{n-1}$ and $B = b_0b_1 \cdots b_{n-1}$, to be compared starting from the most significant bit. Initially, the finite state machine and the counter are reset. The Start signal takes the logic state 1 to indicate the beginning of the count cycle and a sequence of comparisons that will end when the output, TC, of the counter is set to 1.

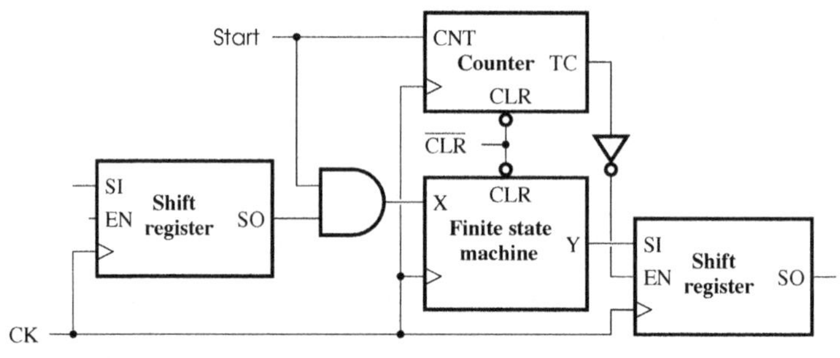

Figure 1.71. *Finite state machine for the sequential generation of two's complement*

Using D flip-flops with an enable signal, EN, and logic gates, implement the finite state machine that is assumed to be characterized by the state table shown in Table 1.81, where a_i and b_i are any two bits of A and B, respectively.

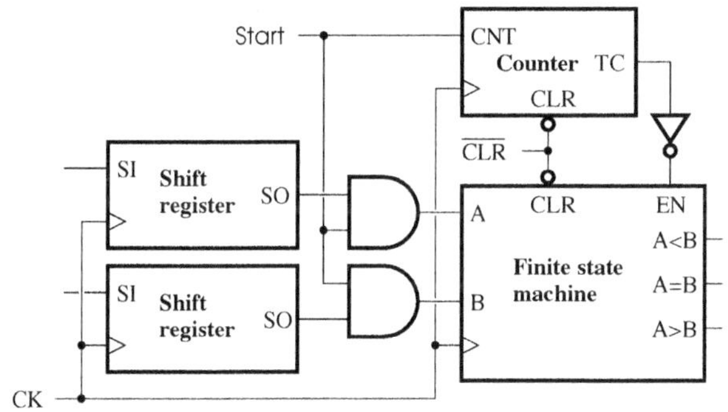

Figure 1.72. *Serial comparator*

	PS	NS				Output		
						$O_{A<B}$	$O_{A=B}$	$O_{A>B}$
		$a_i b_i = 00$	01	10	11			
$A = B$	S_0	S_0	S_2	S_1	S_0	0	1	0
$A > B$	S_1	S_1	S_1	S_1	S_1	0	0	1
$A < B$	S_2	S_2	S_2	S_2	S_2	1	0	0

Table 1.81. *State table (Moore model)*

EXERCISE 1.9.– A shift register can be considered as a Moore state machine whose states are defined by flip-flop outputs.

Construct the state diagram for each of the shift registers shown in Figure 1.73.

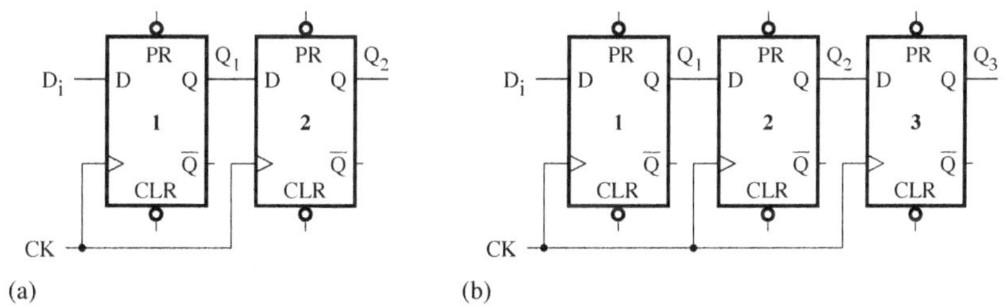

Figure 1.73. *Two-bit a) and three-bit b) shift registers*

EXERCISE 1.10.– Each of the finite state machines, whose state diagram is represented in Figure 1.74, can only operate correctly if the sum rule and the mutual-exclusion requirement are fulfilled.

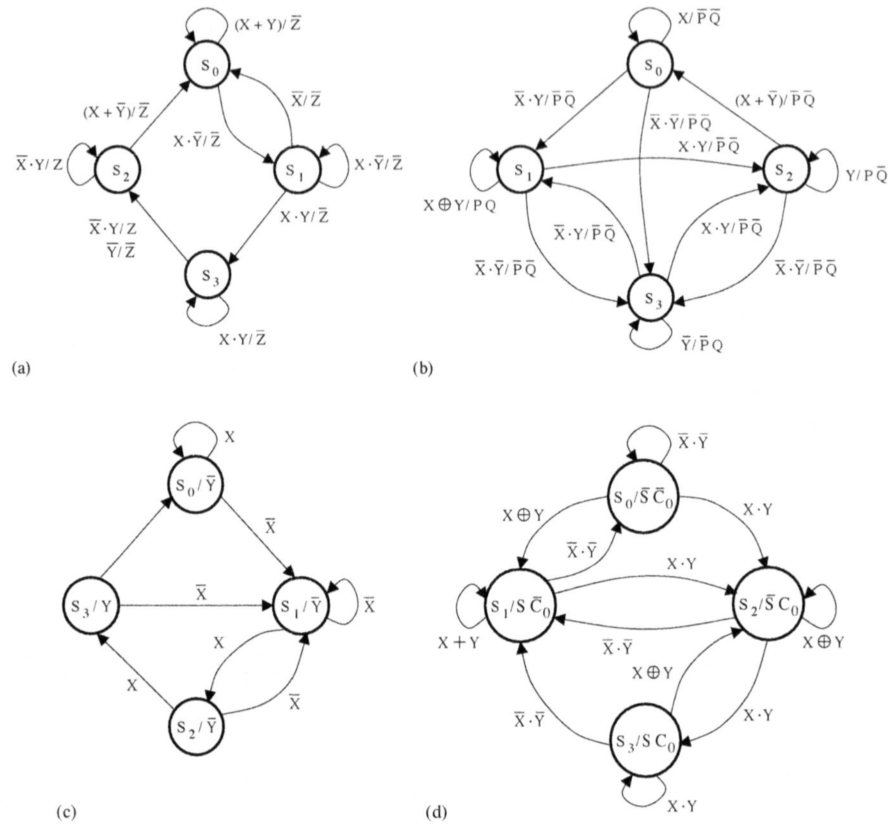

Figure 1.74. *Finite state machines*

The output of the machine shown in Figure 1.74(a) is set to 1 if and only if each of the input sequences, 10, 11 and 01, is detected at least once.

The operation of the state machine represented in Figure 1.74(b) is described by the state table shown in Table 1.82, where X and Y are the inputs, and P and Q denote the outputs.

The machine shown in Figure 1.74(c) operates as a 011 sequence detector.

The machine shown in Figure 1.74(d) is a 1-bit serial adder, and S and C_0 represent the sum and the carry-out.

PS	NS				Outputs PQ			
	$XY = 00$	01	10	11	$XY = 00$	01	10	11
A	D	B	A	A	00	00	00	00
B	D	B	B	C	00	11	11	00
C	D	C	A	C	00	10	00	10
D	D	B	D	C	01	00	01	00

Table 1.82. *State table for machine 2*

Analyze each state diagram to determine the incorrect term and perform the necessary modifications to ensure the proper operation of the state machine.

EXERCISE 1.11.– Consider the finite state machines whose state diagrams are represented in Figure 1.75, where X and Y denote the inputs and the output is designated by Z.

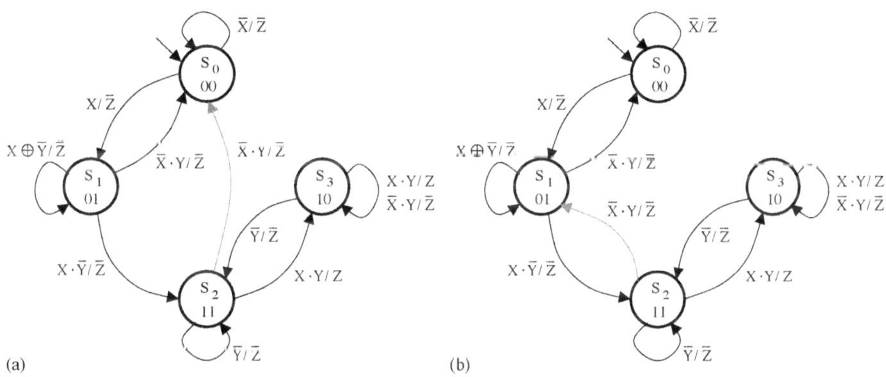

Figure 1.75. *State diagram: a) machine 1; b) machine 2*

Analyzing the state machine 1, show that there is a critical race condition that can affect the transition from the state S_2 to the state S_0 and that is caused by the input condition $\overline{X} \cdot Y$.

Verify that there is no critical race condition that affects the operation of the state machine 2, which is based on the same algorithm as the state machine 1.

EXERCISE 1.12.– Analyze the finite state machine (state table, state diagram) shown in Figure 1.76 and complete the timing diagram given in Figure 1.77.

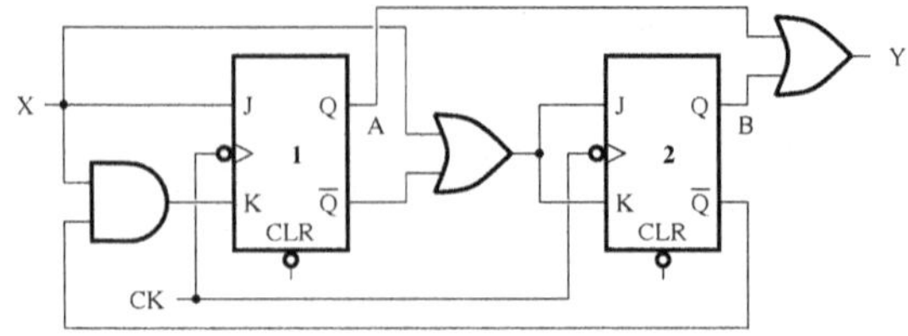

Figure 1.76. *Finite state machine (Moore model)*

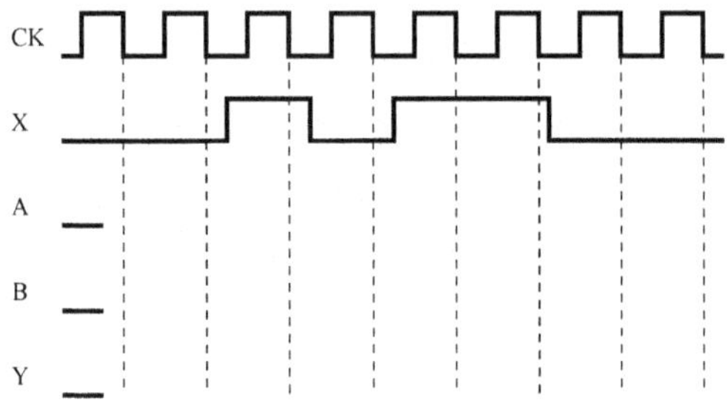

Figure 1.77. *Timing diagram*

EXERCISE 1.13.– Using D flip-flops, implement a synchronous counter, whose output depends on the logic state of a control signal C:

– when $C = 0$, the output sequence is 00, 01, 11;

– when $C = 1$, the output sequence is 00, 11, 01.

EXERCISE 1.14.– Using JK flip-flops, implement a logic circuit whose output is set to 1 when the sequence 010 is detected at the input. We assume that the operation of this circuit is similar to that of a Moore state machine and that there is no overlapping of the bits applied to the input. For the input *010*100, the sequence 010 will, therefore, be detected only once.

EXERCISE 1.15.– Analyze (state diagram, state table) the finite state machine represented in Figure 1.78 and complete the timing diagrams shown in Figures 1.79 and 1.80.

EXERCISE 1.16.– Using two JK flip-flops, implement a counter that operates as follows:

– if the input $X = 0$, the counting is carried out in increasing order, according to the sequence: 0, 1, 2, 3, 3;

– if the input $X = 1$, the counting is carried out in decreasing order, according to the sequence: 3, 2, 1, 0, 0.

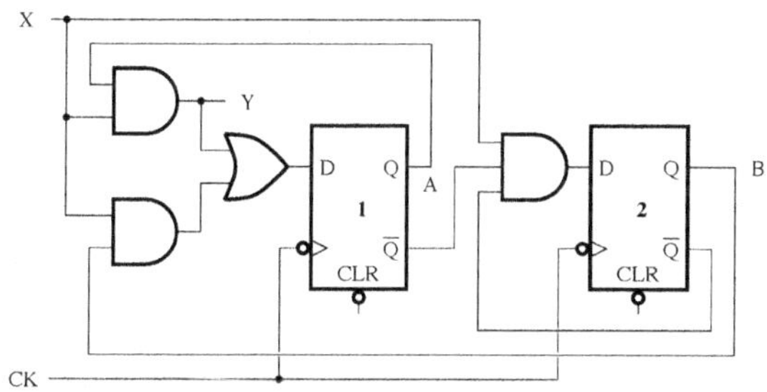

Figure 1.78. *Finite state machine (Mealy model)*

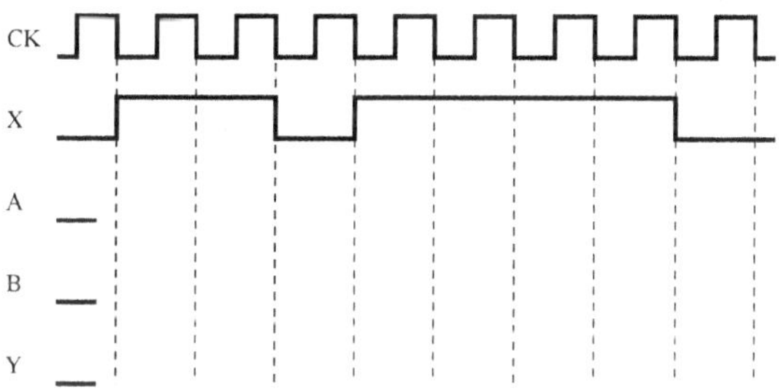

Figure 1.79. *Timing diagram 1*

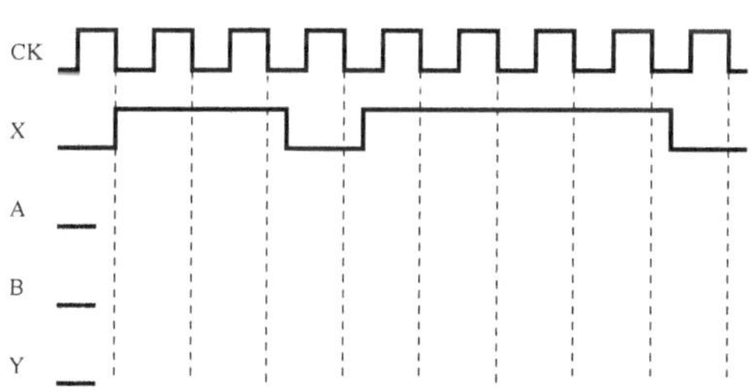

Figure 1.80. *Timing diagram 2*

EXERCISE 1.17.– Implement a Mealy state machine that allows for the detection of the 010 sequence in the following two cases (see Table 1.83):

a) the overlapping of input bits is allowed;

b) the overlapping of input bits is not allowed.

	Input X	1	1	0	1	0	1	0	1	0	0	1	0	0	1	0	0	1	1	0	
(a)	Output Y	0	0	0	0	1	0	1	0	1	0	0	1	0	0	1	0	0	0	0	0
(b)	Output Y	0	0	0	0	1	0	0	0	1	0	0	1	0	0	1	0	0	0	0	0

Table 1.83. *Table illustrating the operation of the detector*

EXERCISE 1.18.– Implement a counter that can periodically generate the sequence 2 6 1 7 5 using:

a) D flip-flops;

b) JK flip-flops.

EXERCISE 1.19.– Use the implication table approach to minimize the number of states of the finite state machine whose state tables are shown in Tables 1.84 and 1.85, where X is the input.

PS	NS		Output Y
	$X=0$	1	
A	A	B	1
B	C	A	0
C	A	D	0
D	C	C	1
E	G	F	1
F	C	E	0
G	E	B	1

Table 1.84. *State table for the finite state machine 1*

PS	NS		Output Y	
	$X=0$	1	$X=0$	1
S_0	S_1	S_2	1	1
S_1	S_3	S_5	1	1
S_2	S_5	S_4	0	0
S_3	S_1	S_6	1	1
S_4	S_5	S_2	0	0
S_5	S_4	S_3	0	0
S_6	S_5	S_6	0	0

Table 1.85. *State table for the finite state machine 2*

EXERCISE 1.20.–

a) use the partitioning method to minimize the number of states of the finite state machines whose state tables are shown in Tables 1.87 and 1.88;

b) Table 1.86 gives the state table of a finite state machine. Minimize the number of states of this machine using the partitioning method;

c) consider the finite state machine whose state table is shown in Table 1.89. Use the partitioning method to minimize the number of states of this machine.

PS	NS				Output Z
	$XY = 00$	01	10	11	
A	A	F	C	B	0
B	A	B	D	H	1
C	C	B	C	D	0
D	C	F	D	D	1
E	G	A	E	D	1
F	F	F	G	B	0
G	G	B	G	E	0
H	F	B	E	H	1

Table 1.86. *State table for the finite state machine 1*

EXERCISE 1.21.– Consider the finite state machine whose state tables are shown in Tables 1.90–1.93.

To simplify the state machine 1:

– construct the merger graph;

– determine the set of maximum compatibility classes;

– construct the compatibility graph;

– draw up the reduced state table.

To simplify each of the state machines 2, 3 and 4:

– construct the implication table;

– determine the set of maximum compatibility classes;

– construct the compatibility graph;

– draw up the reduced state table.

PS	NS		Output Y
	$X = 0$	1	
S_0	S_1	S_2	1
S_1	S_3	S_5	1
S_2	S_5	S_4	0
S_3	S_1	S_6	1
S_4	S_5	S_2	0
S_5	S_4	S_3	0
S_6	S_5	S_6	0

Table 1.87. *State table for the finite state machine 1*

PS	NS		Output Y	
	$X = 0$	1	$X = 0$	1
S_0	S_4	S_3	0	1
S_1	S_5	S_3	0	0
S_2	S_4	S_1	0	1
S_3	S_5	S_1	0	0
S_4	S_2	S_5	0	1
S_5	S_1	S_2	0	0

Table 1.88. *State table for the finite state machine 2*

PS	NS				Output Z			
	$XY = 00$	01	10	11	$XY = 00$	01	10	11
A	A	G	E	H	0	0	1	1
B	F	B	B	D	0	1	0	0
C	F	C	G	H	0	1	0	1
D	H	C	E	D	1	0	1	0
E	A	F	E	D	1	0	1	0
F	F	C	B	A	0	1	0	1
G	F	G	G	D	0	1	0	0
H	H	B	E	H	0	0	1	1

Table 1.89. *State table for the finite state machine 2*

PS	NS		Output Y	
	$X=0$	1	$X=0$	1
A	B	C	–	0
B	D	–	0	–
C	–	E	1	–
D	B	G	0	0
E	F	C	1	1
F	E	D	0	1
G	F	–	1	0

Table 1.90. *State table (machine 1)*

PS	NS		Output Y
	$X=0$	1	
A	D	–	0
B	C	E	1
C	B	G	0
D	A	B	–
E	–	E	0
F	G	B	–
G	F	–	0

Table 1.91. *State table (machine 2)*

EXERCISE 1.22.– Transform each of the finite state machines based on Moore model (see Figure 1.81) into the equivalent Mealy model.

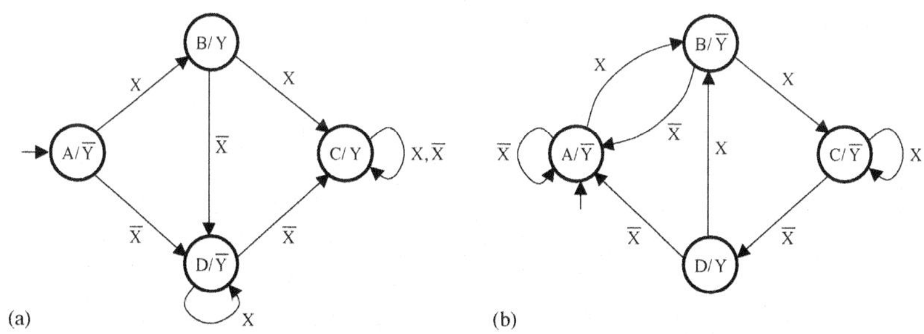

Figure 1.81. *Moore model: a) machine 1; b) machine 2*

PS	NS				Output Z
	$XY = 00$	01	10	11	
A	A	B	–	D	–
B	A	B	C	–	0
C	–	B	C	H	0
D	–	–	G	D	–
E	E	F	–	D	1
F	E	F	G	G	1
G	D	F	G	H	1
H	A	–	–	H	0

Table 1.92. *State table (machine 3)*

PS	NS				Output Z			
	$XY = 00$	01	10	11	$XY = 00$	01	10	11
A	G	A	–	H	–	1	–	1
B	G	–	B	D	0	–	0	0
C	C	F	E	–	0	–	–	–
D	–	A	E	D	–	–	–	0
E	C	–	E	D	–	–	1	–
F	G	F	–	D	–	1	–	–
G	G	A	B	–	0	–	0	–
H	–	–	E	H	–	–	1	1

Table 1.93. *State table (machine 4)*

EXERCISE 1.23.– For Figures 1.82 and 1.83, transform each of the Mealy model based finite state machines into the equivalent Moore state machine.

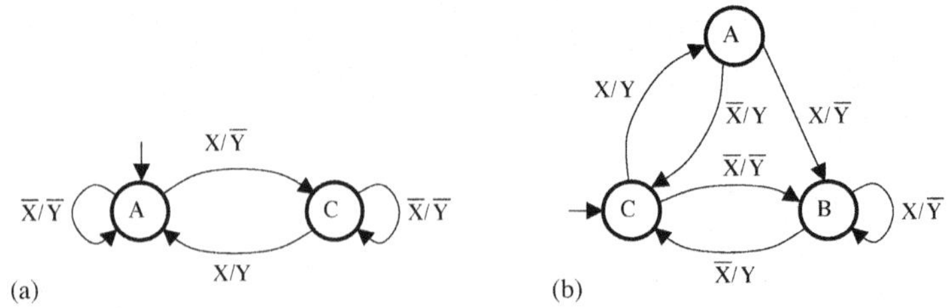

Figure 1.82. *Mealy model: a) machine 1; b) machine 2*

EXERCISE 1.24.– Split each of the finite state machines (modulo 6 counter and 010 and 1001 sequence detector) into two machines with an identical number of states and that can communicate with each other, as shown in Figures 1.84 and 1.85. It is assumed that the states are represented using a 1-out-of-n code.

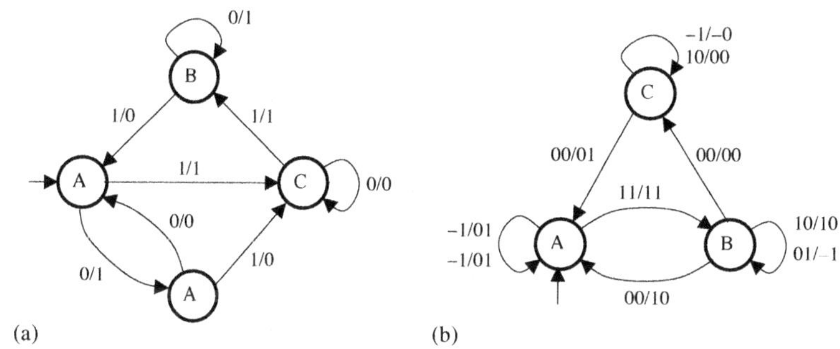

Figure 1.83. *Mealy model: a) machine 1; b) machine 2*

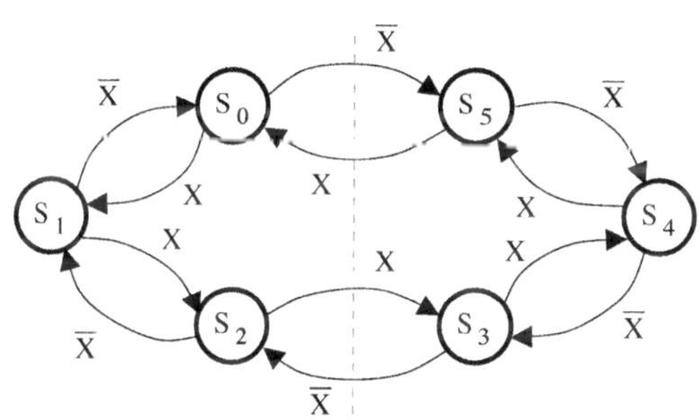

Figure 1.84. *Modulo 6 counter*

EXERCISE 1.25.– Consider the finite state machines whose logic circuit, shown in Figure 1.86, is based on D flip-flops.

Complete the signals A and B of the timing diagram given in Figure 1.87.

EXERCISE 1.26.– Consider the logic circuit shown in Figure 1.88, representing a finite state machine implemented using JK flip flops and logic gates

Complete the signals, A and B, and the output signal, Y, of the timing diagram given in Figure 1.89.

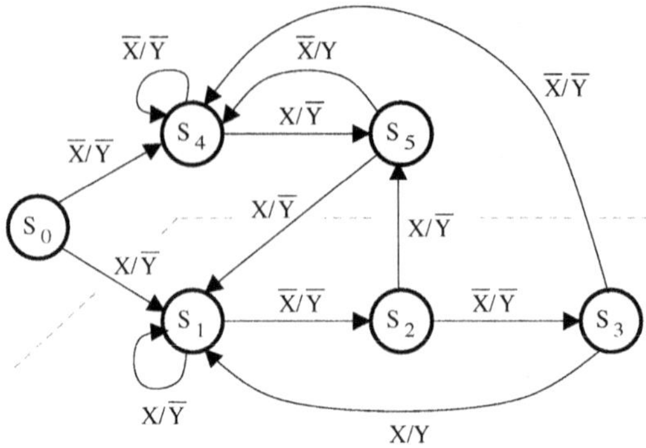

Figure 1.85. *State diagram of the 010 and 1001 sequence detector*

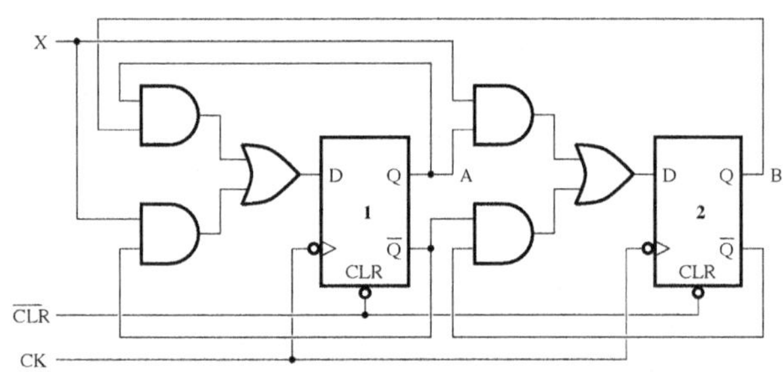

Figure 1.86. *Logic circuit*

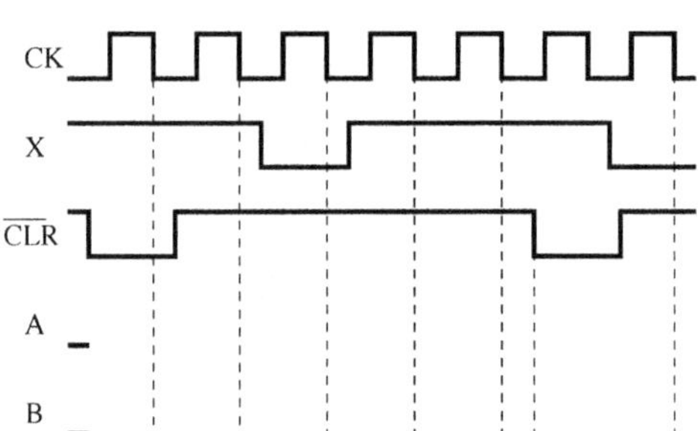

Figure 1.87. *Timing diagram*

EXERCISE 1.27.– Using T flip-flops and logic gates, implement a Mealy state machine that has two inputs, X and Y, and one output, Z, and which operates as follows:

– $Z = X^{-} \cdot Y$ (AND operation) until the input Y assumes the logic state 1; thereafter, $Z = X^{-} + Y$ (OR operation), with X^{-} being the previous state of X;

– the next transition of the input Y to the logic state 1 causes the output to revert to the AND operation and this switching of the output between the AND and OR operations, continues whenever the input Y takes to logic state 1.

EXERCISE 1.28.– Median filter.

Using D flip-flops and logic gates, implement a median filter that can be described as a Moore state machine whose role is to replace at the output each input bit 0 located between two 1 bits by a bit of value 1.

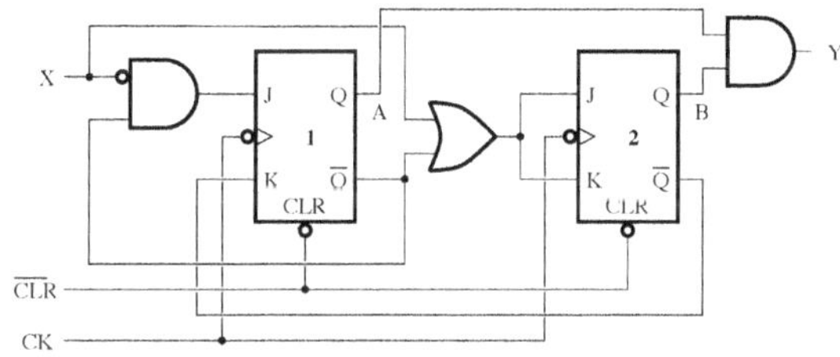

Figure 1.88. *Logic circuit*

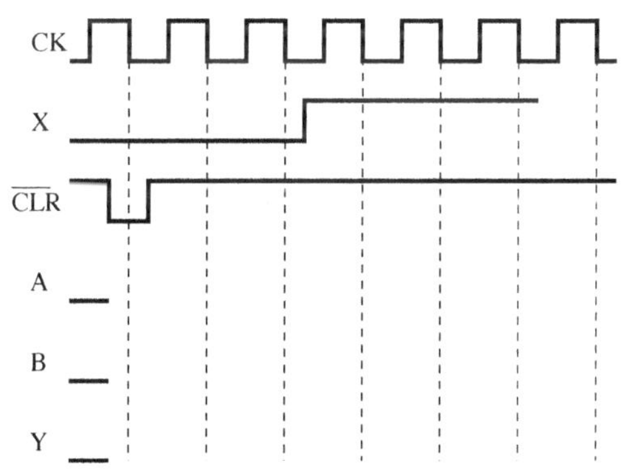

Figure 1.89. *Timing diagram*

EXERCISE 1.29.– Bus arbiter.

Consider the bus arbiter that allocates the bus to one of the two elements, A or B, prioritizing element A when two simultaneous requests occur. In the block diagram shown in Figure 1.90, the bus arbiter provides the control signals for two three-state buffers connected to a common data bus.

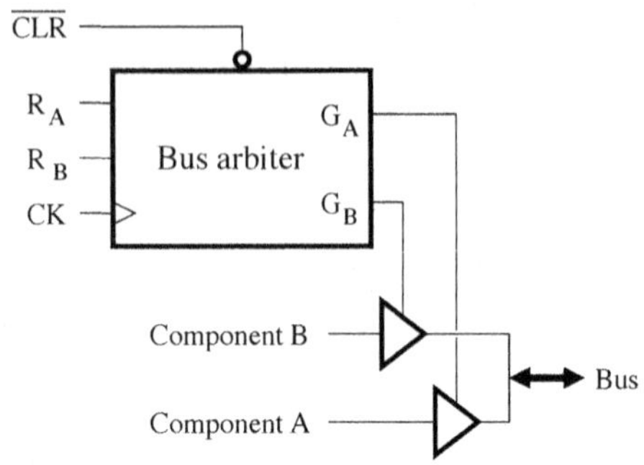

Figure 1.90. *Bus arbiter*

The bus arbiter must be implemented as a synchronous Moore state machine, which operates as follows: when the two inputs R_A and R_B are set to 0, the machine returns to the initial state or is held in the initial state. When the inputs R_A and R_B take the binary combination 10 or 11, the machine goes to the state where the bus is allocated to element A and is held in this state as long as the input R_A remains at 1. On the other hand, when the combination 01 is assigned to the inputs R_A and R_B, the machine moves from the initial state to the state where the bus is allocated to the component B and remains in this state as long as the input R_B remains at 1. The transition between the two states where the bus is allocated to one of the two components is caused by assigning either the combination 01 or the combination 10 to the inputs R_A and R_B.

– construct the state diagram for the state machine;

– using Gray code to represent the states, determine the logic equations for the implementation using D flip-flops;

– represent the logic circuit of the state machine.

EXERCISE 1.30.– Robot ant.

Consider a robot ant whose legs are controlled by servo motors, which are controlled by a finite state machine. Two antennae attached to the front part of the robot act as sensors to indicate possible contacts with the different sections of the walls of a labyrinth. The labyrinth exit is recognized by a indicator light detector that sets the signal $\overline{EN}$ to 1, thereby disabling the outputs of the finite state machine.

Using D flip-flops, implement a finite state machine that allows the robot ant to find a way out of the labyrinth shown in Figure 1.91. It is assumed that the robot ant moves forward by trying to keep the wall to its right after each control pulse, and that a Moore state machine with the following characteristics is used:

– inputs: signals from the left and right antennae, L and R, set to 1 whenever there is a contact with a wall section;

– outputs: control signals that are used to initiate a forward movement, F, and a slight rotation to the left or right, TL or TR.

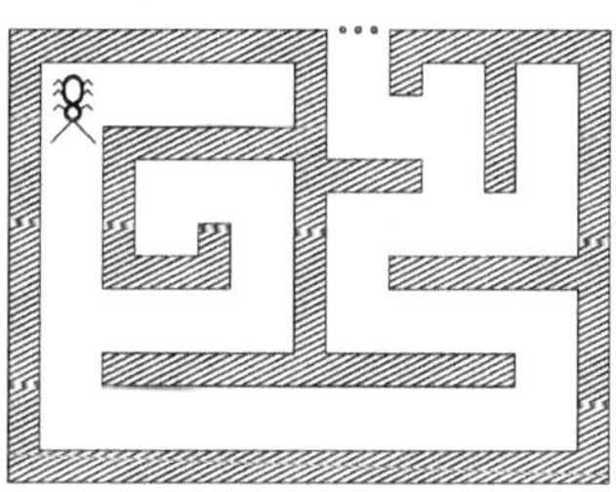

Figure 1.91. *Labyrinth*

1.12. Solutions

SOLUTION 1.1.– The state diagrams shown in Figure 1.92 are equivalent and a machine with two states can be implemented using a single flip-flop and logic gates.

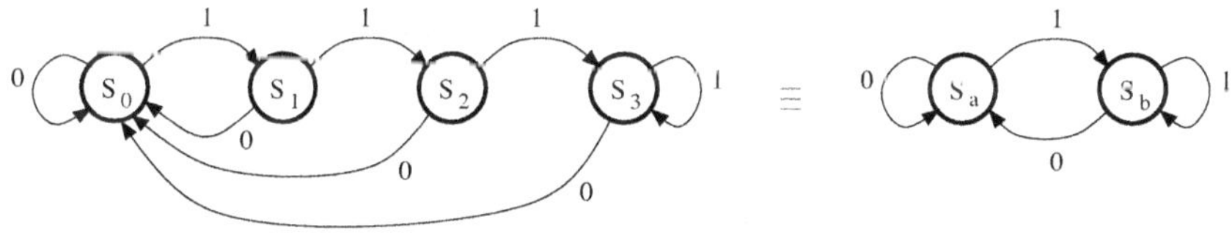

Figure 1.92. *State diagram*

SOLUTION 1.2.– The state diagram corresponding to each state table is represented in Figure 1.93.

Tables 1.94 and 1.95 give bit sequences (input, states and output) that illustrate the operation of each finite state machine. The output sequence is obtained in response to a given input binary sequence. In some cases, it is possible to predict the behavior of the state machine even when the input is not yet known.

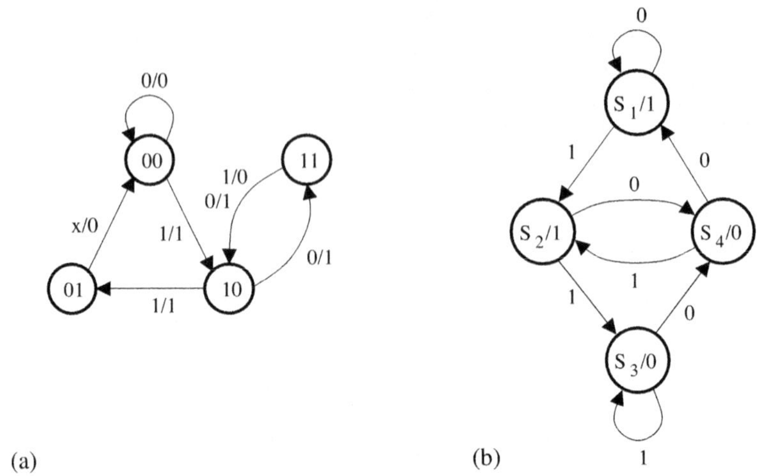

Figure 1.93. *State diagram*

Input X	0 1 0 0 1 1 1 0
AB	00 → 00 → 10 → 11 → 10 → 01 → 00 → 10 → 11 → 10 → - 1 → - 0
Output Y	0 1 1 1 1 0 1 1 - 1

Table 1.94. *Illustration of the operation of the state machine a*

Input X	0 1 0 1 1 1 0 0 0 0
PS	$S_1 \to S_1 \to S_2 \to S_4 \to S_2 \to S_3 \to S_3 \to S_4 \to S_1 \to S_1 \to S_1$
Output Y	1 1 1 0 1 0 0 0 1 1 1 1

Table 1.95. *Illustration of the operation of the state machine b*

SOLUTION 1.3.– (RT (Fictional) Flip-Flop).

a) The state table can be used to construct the transition table shown in Table 1.96. For an implementation based on a D flip-flop, we have $Q^+ = D$, and Figure 1.95(a) depicts the Karnaugh map obtained from the transition table. The logic equation for the input D can be written as follows:

$$Q^+ = D = \overline{R} \cdot T\overline{Q} + R \cdot T\overline{Q} + R \cdot T \cdot Q = T(R + \overline{Q}) \qquad [1.67]$$

Q	R	T	Q^+
0	0	0	0
0	0	1	1
0	1	0	1
0	1	1	1
1	0	0	0
1	0	1	0
1	1	0	1
1	1	1	1

Table 1.96. *Transition table of the flip-flop*

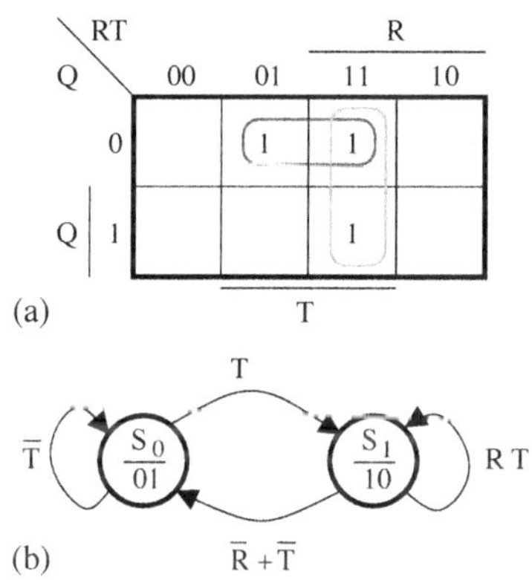

Figure 1.94. *a) Karnaugh map; b) state diagram*

The RT flip-flop has two states: S_0 where $Q = 0$ and $\overline{Q} = 1$, and S_1 where $Q = 1$ and $\overline{Q} = 0$. Analyzing the state table, we can deduce the conditions:

– to remain in the state S_0:

$$\overline{R} \cdot \overline{T} + R \cdot \overline{T} = (\overline{R} + R)\overline{T} = \overline{T}$$

– for the transition from S_0 to S_1:

$$R \cdot T + R \cdot T = (\overline{R} + R)T = T$$

– to remain in the state S_1:

$$R \cdot T$$

– for the transition from S_1 to S_0:

$$\overline{R} \cdot T + \overline{R} \cdot T + R \cdot \overline{T} = \overline{R}(\overline{T} + T) + R \cdot \overline{T} = \overline{R} + R \cdot \overline{T} = \overline{R} + \overline{T}$$

The state diagram is represented in Figure 1.94(b).

R	T	Q^+
0	0	0
0	1	$\overline{Q}$
1	0	0
1	1	1

Table 1.97. *Truth table*

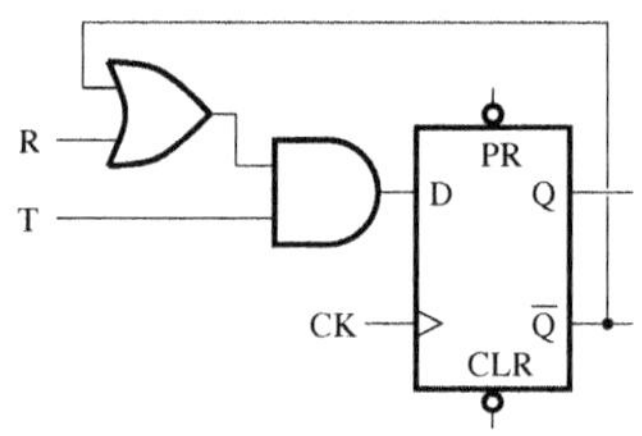

Figure 1.95. *Logic circuit*

Table 1.97 depicts the truth table. The logic circuit of the RT flip-flop is given in Figure 1.95.

b) For the implementation based on a JK flip-flop, the transition table is obtained from the state table using the JK flip-flop excitation table as shown in Table 1.98.

From the Karnaugh maps shown in Figure 1.96, we can obtain the logic equations for the inputs J and K. Thus:

$$J = T \quad \text{and} \quad K = \overline{R} + \overline{T} = \overline{R \cdot T} \qquad [1.68]$$

Figure 1.97 depicts the logic circuit of the RT flip-flop.

SOLUTION 1.4.– Analyzing the logic circuit of the finite state machine, we can obtain:

$$Q^+ = D \qquad [1.69]$$

$$= E \cdot \overline{F} \cdot Q + \overline{E} \cdot F \cdot \overline{Q} + \overline{E} \cdot \overline{F} \cdot G \qquad [1.70]$$

Q	R	T	Q^+	J	K
0	0	0	0	0	x
0	0	1	1	1	x
0	1	0	1	0	x
0	1	1	1	1	x
1	0	0	0	x	1
1	0	1	0	x	1
1	1	0	1	x	1
1	1	1	1	x	0

Table 1.98. *Excitation table of the flip-flop*

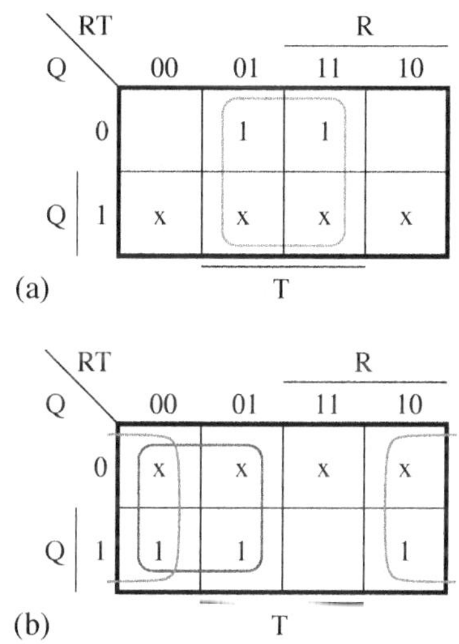

Figure 1.96. *Karnaugh maps: a) J; b) K*

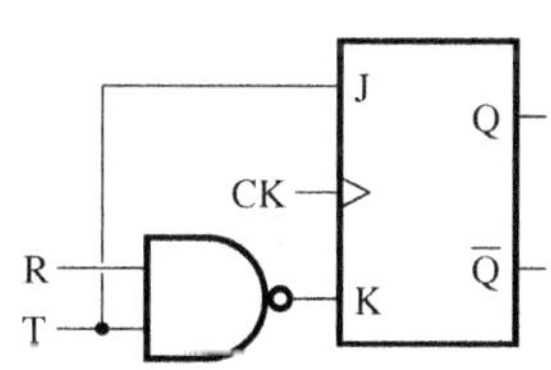

Figure 1.97. *Logic circuit*

E	F	Y
0	0	G
0	1	$\overline{Q}$
1	0	Q
1	1	0

Table 1.99. *Truth table*

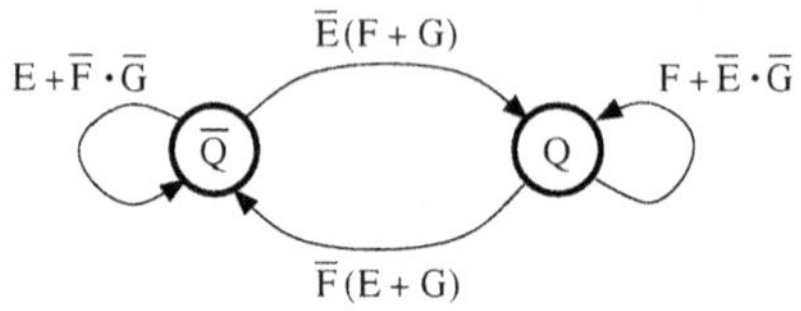

Figure 1.98. *State diagram*

The characteristic equations can be used to construct the truth table shown in Table 1.99.

The state diagram is represented in Figure 1.98.

SOLUTION 1.5.– The characteristic equation of the finite state machine is obtained as follows:

$$Q^+ = D \quad [1.71]$$

$$= (Q + \overline{E})F \cdot G + E \cdot F \cdot \overline{G} + E \cdot \overline{F} \cdot G + Q \cdot \overline{E} \cdot \overline{F} \cdot \overline{G} \quad [1.72]$$

F	G	Y
0	0	$\overline{Q} \cdot \overline{E}$
0	1	E
1	0	E
1	1	$Q + \overline{E}$

Table 1.100. *Truth table*

The truth table obtained from the characteristic equation is shown in Table 1.100.

Figure 1.98 presents the state diagram.

The function table is shown in Table 1.101.

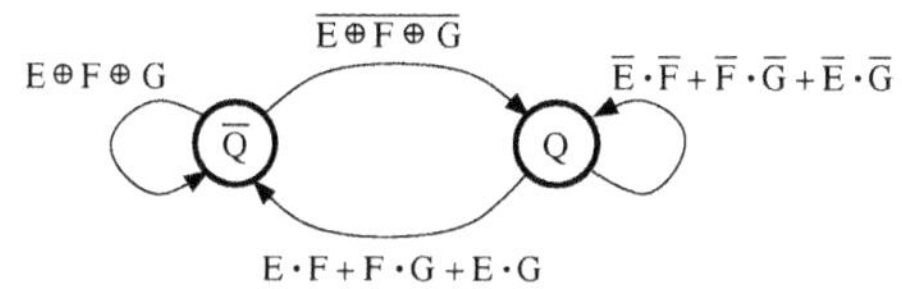

Figure 1.99. *State diagram*

Inputs	Output Y	Observation
3 inputs at 0	$\overline{Q}$	Toggle
2 of 3 inputs at 0	0	Reset
2 of 3 inputs at 1	1	Set
3 inputs at 1	Q	No change

Table 1.101. *Function table*

SOLUTION 1.6.– (Sequence Detector Design).

In the case of the Mealy state machine, the state diagram for the 01 sequence detector is given in Figure 1.100.

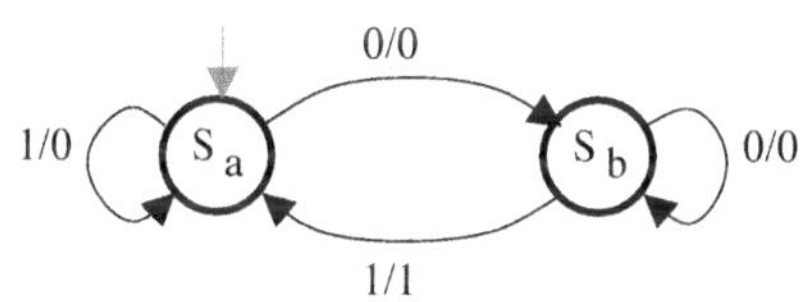

Figure 1.100. *State diagram (Mealy model)*

Table 1.102 depicts the state table.

As the machine has only two states, we can assign the binary code 0 to the state S_a, and 1 to the state S_b. The Karnaugh maps shown in Figures 1.101(a) and (b) are obtained by using the D flip-flop excitation table. We can obtain the following logic equations:

$$Q^+ = D = \overline{X} \quad [1.73]$$

and

$$Y = X \cdot Q \quad [1.74]$$

PS	NS		Output Y	
	$X=0$	1	$X=0$	1
S_a	S_b	S_a	0	0
S_b	S_b	S_a	0	1

Table 1.102. *State table (Mealy model)*

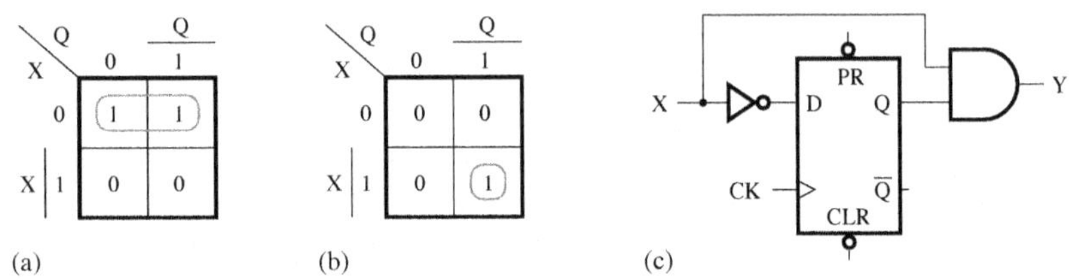

Figure 1.101. *Karnaugh maps for a) Q^+ and b) Y; c) logic circuit*

Figure 1.101(a) presents the logic circuit of the sequence detector.

Figure 1.102 shows the state diagram of the 01 sequence detector based on the Moore model.

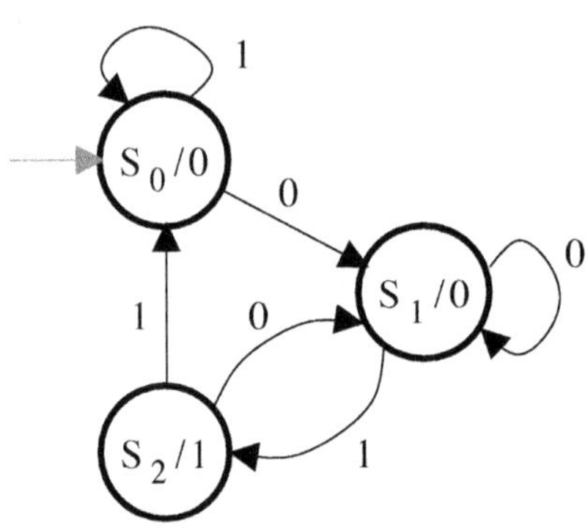

Figure 1.102. *State table (Moore model)*

The state table can be constructed as shown in Table 1.103.

PS	NS		Output Y
	$X = 0$	1	
S_0	S_1	S_0	0
S_1	S_1	S_2	0
S_2	S_1	S_0	1

Table 1.103. *State table (Moore model)*

The codes 00, 01 and 11 are assigned to the states S_0, S_1 and S_2, respectively. Using the excitation table of a D flip-flop, we can obtain the Karnaugh maps represented in Figure 1.103. By grouping together adjacent cells containing 1s, we obtain the following equations:

$$A^+ = X \cdot \overline{A} \cdot B \qquad [1.75]$$

$$B^+ = \overline{X} + \overline{A} \cdot B \qquad [1.76]$$

and

$$Y = A \cdot B \qquad [1.77]$$

Figure 1.104 presents the logic circuit of the 01 sequence detector.

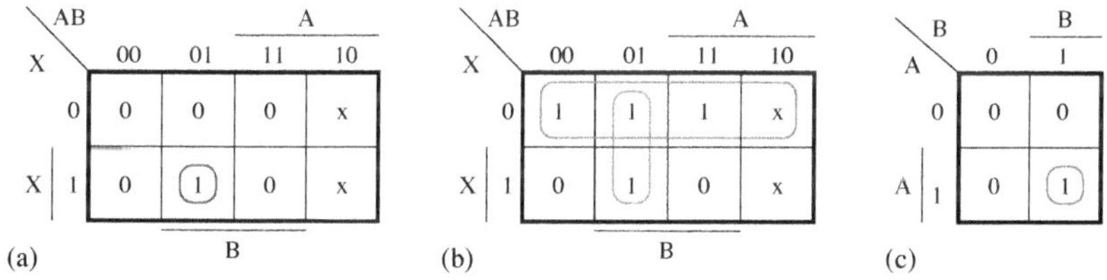

Figure 1.103. *Karnaugh maps: a) A^+; b) B^+; c) Y*

In order to recognize the two binary sequences, 01 and 10, the state machine must operate according to each of the state diagrams shown in Figure 1.105. To satisfy the same requirements, the Moore state machine generally requires a higher number of states as compared to the number of states required by the Mealy state machine.

SOLUTION 1.7.– A finite state machine that generates the two's complement of a number can be based on Mealy or Moore model.

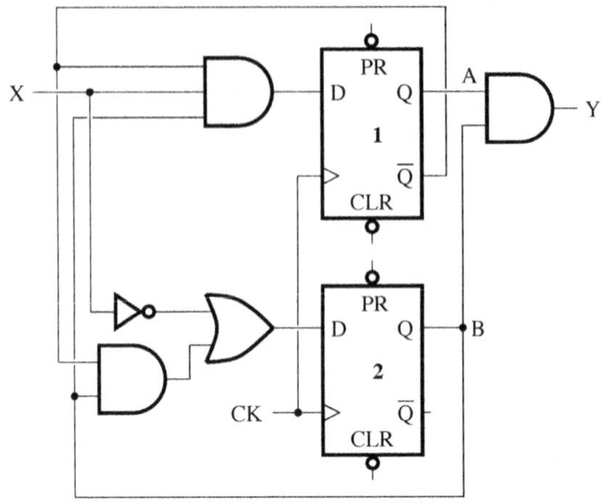

Figure 1.104. *Logic circuit*

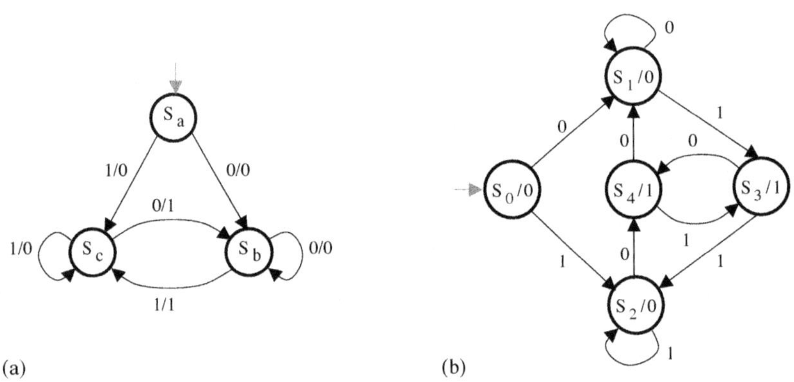

Figure 1.105. *State diagram: a) Mealy model b) Moore model*

The state diagram of the state machine based on Mealy model is represented in Figure 1.106.

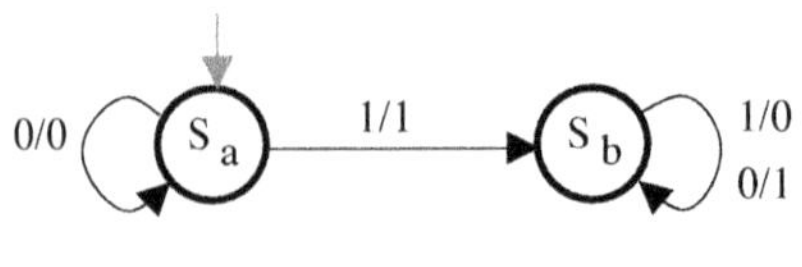

Figure 1.106. *State diagram (Mealy model)*

PS	NS		Output Y	
	$X=0$	1	$X=0$	1
S_a	S_a	S_b	0	1
S_b	S_b	S_b	1	0

Table 1.104. *State table (Mealy model)*

The state table can be constructed as shown in Table 1.104.

The machine has two states that can be represented by 0 and 1. The excitation table of the D flip-flop can be used to construct the Karnaugh maps shown in Figures 1.107(a) and (b). Thus:

$$Q^+ = X + Q \qquad [1.78]$$

$$Y = X \oplus Q \qquad [1.79]$$

The logic circuit of the finite state machine is depicted in Figure 1.107(c).

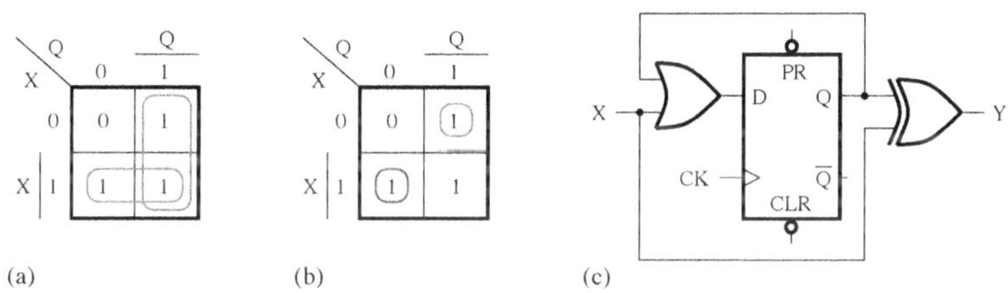

Figure 1.107. *Karnaugh maps for a) Q^+ and b) Y; c) logic circuit*

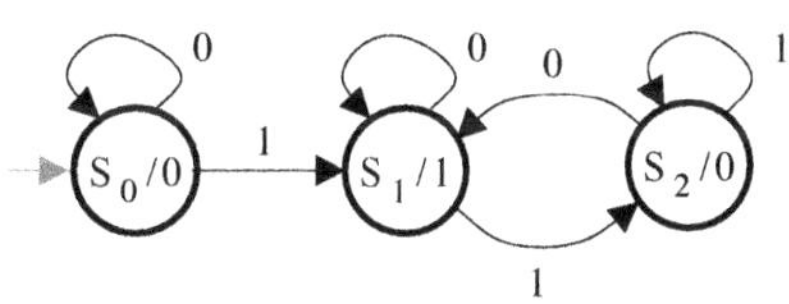

Figure 1.108. *State diagram (Moore model)*

In the case of the Moore model, the finite state machine that generates the two's complement of a number can be described by the state diagram in Figure 1.108.

The state table is represented in Table 1.105.

The machine has three states, S_0, S_1 and S_2, that can be represented by 00, 01 and 11, respectively. The logic equations for the implementation using D flip-flops are obtained from each of the Karnaugh maps shown in Figure 1.109 and can be written as follows:

$$A^+ = X \cdot B \tag{1.80}$$

$$B^+ = X + B \tag{1.81}$$

and:

$$Y = \overline{A} \cdot B \tag{1.82}$$

PS	NS		Output Y
	$X = 0$	1	
S_0	S_0	S_1	0
S_1	S_1	S_2	1
S_2	S_1	S_2	0

Table 1.105. *State table (Moore model)*

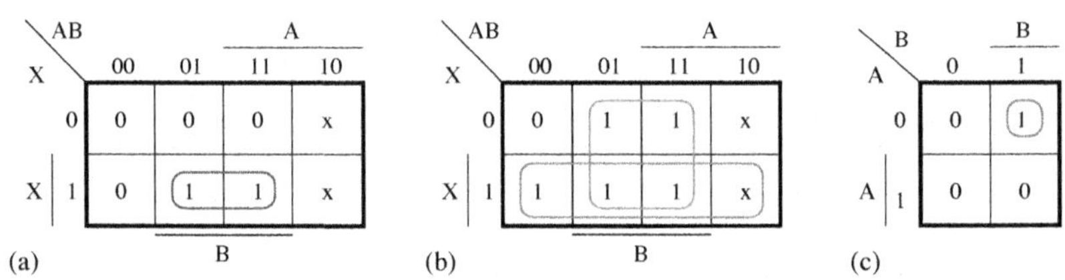

Figure 1.109. *Karnaugh maps: a) A^+; b) B^+; c) Y*

Figure 1.110 depicts the logic circuit for the state machine.

SOLUTION 1.8.– (Serial Comparator).

The transition table shown in Table 1.106 is obtained by assigning the binary codes 00, 01 and 10 to the states, S_0, S_1, and S_2, respectively, of the state table.

Using the excitation table for the D flip-flop, it is possible to construct, based on the transition table, the Karnaugh maps in Figure 1.111, where the code 11 is associated with don't-care states. The logic equations that characterize the NSs are given by:

$$X^+ = X + \overline{a_i} \cdot b_i \cdot \overline{Y} \tag{1.83}$$

$$Y^+ = Y + a_i \cdot \overline{b_i} \cdot \overline{X} \tag{1.84}$$

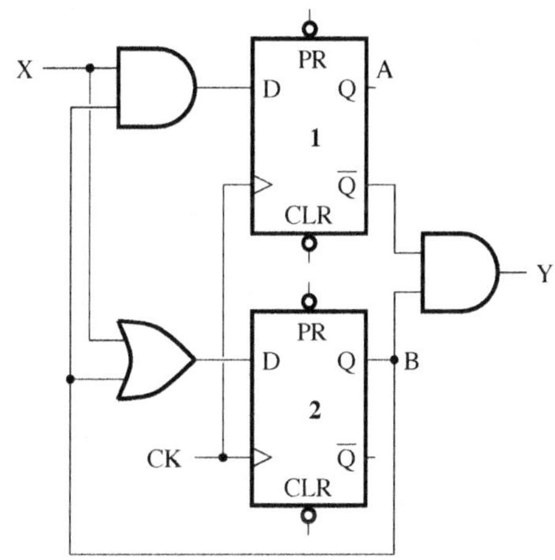

Figure 1.110. *Logic circuit*

PS	NS				Outputs		
	X^+Y^+				$O_{A<B}$	$O_{A=B}$	$O_{A>B}$
XY	u_ib_i – 00	01	10	11			
00	00	10	01	00	0	1	0
01	01	01	01	01	0	0	1
10	10	10	10	10	1	0	0
11	x	x	x	x	0	0	0

Table 1.106. *Transition table (Moore model)*

The output logic equations are derived from the Karnaugh maps shown in Figure 1.112 as follows:

$$O_{A>B} = \overline{X} \cdot Y \qquad [1.85]$$

$$O_{A=B} = \overline{X} \cdot \overline{Y} \qquad [1.86]$$

$$O_{A<B} = X \cdot \overline{Y} \qquad [1.87]$$

The logic circuit of the finite state machine is represented in Figure 1.113, where EN denotes the enable signal.

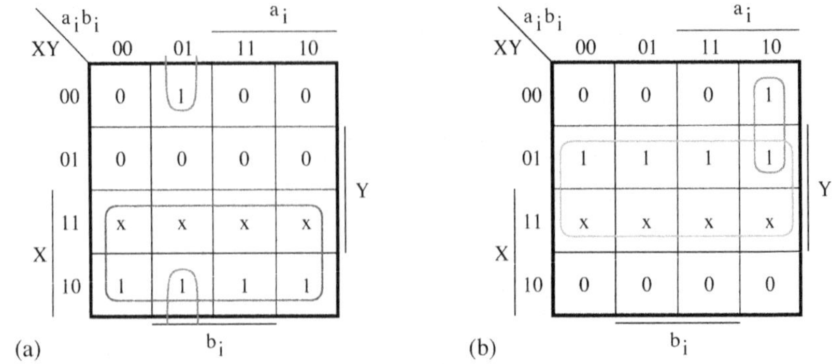

Figure 1.111. *Karnaugh maps: a) X^+; b) Y^+*

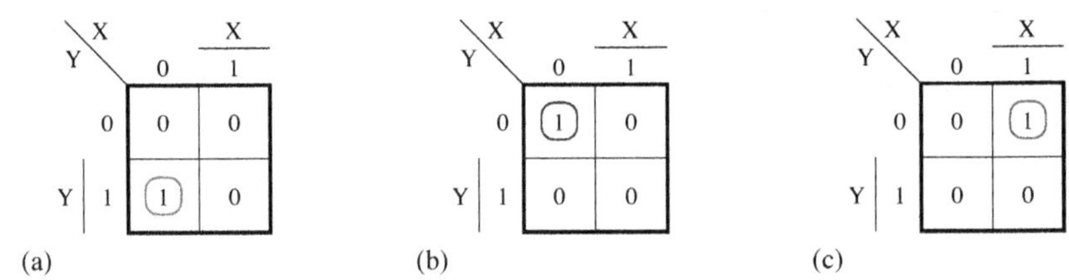

Figure 1.112. *Karnaugh maps: a) $O_{A>B}$; b) $O_{A=B}$; c) $O_{A<B}$*

SOLUTION 1.9.– (State Diagram of a Shift Register).

The state diagram of the two-bit shift register is shown in Figure 1.114(a), where the states correspond to the different Q_1Q_2 combinations and each transition is determined by the value applied to the input D_i.

Figure 1.114(b) depicts the state diagram of the three-bit shift register. The state machine has eight states that are defined by the different combinations of $Q_1Q_2Q_3$ and it moves from one state to another based on the value applied to the input D_i.

We can see that the complexity of the state diagram is not directly related to the complexity of the logic circuit.

SOLUTION 1.10.– (State Diagram: Sum Law and Mutual-Exclusion Requirement).

Each of the state diagrams shown in Figure 1.115 has been modified so as to satisfy the sum law and mutual exclusion requirement.

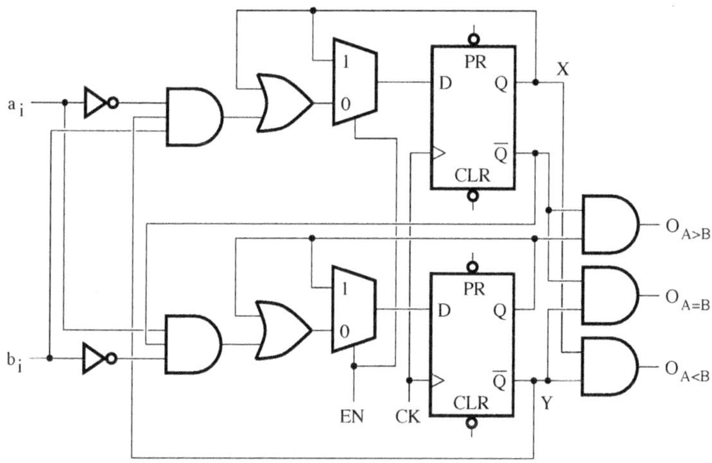

Figure 1.113. *Logic circuit of the serial comparator*

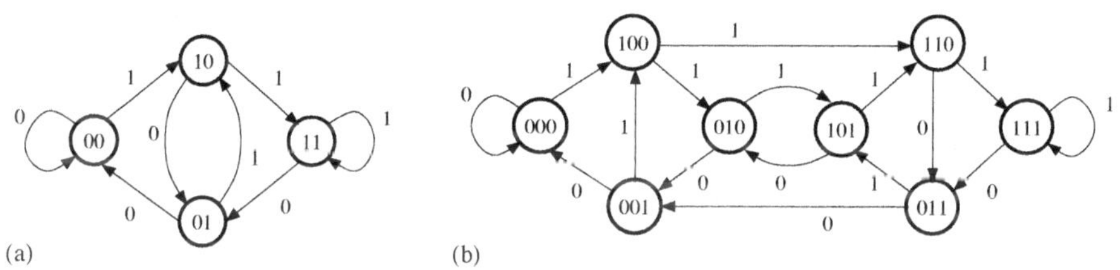

Figure 1.114. *State diagram: a) two-bit register; b) three-bit register*

SOLUTION 1.11.– (Critical Race Condition).

For the state machine 1, the transition from the state S_2 (11) to the state S_0 (00) requires the modification of two state variables. If the state of the first variable changes before that of the second, the transition takes place via the state S_3 (10), and the machine wrongly remains in this state as the holding condition, Y, is verified. If, on the other hand, the second variable changes state before the first, the transition takes place via the state S_1 (01) and the state machine moves correctly to the state S_0. As it is difficult in practice to predict the path the machine will take, this is a critical race condition.

In the case of the state machine 2, each transition requires the modification of only one variable and there is, therefore, no critical race condition.

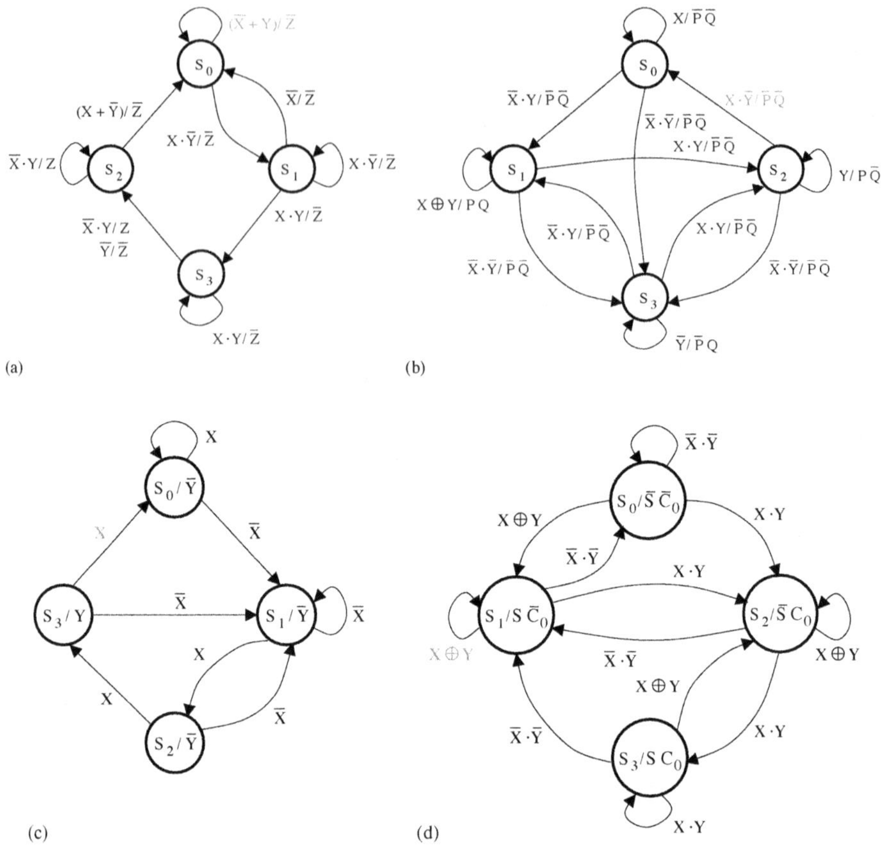

Figure 1.115. *Finite State Machine*

SOLUTION 1.12.– (Analysis of a Finite State Machine (Moore Model)).

The logic equations for the inputs J and K obtained from the logic circuit of the state machine are given by:

$$J_1 = X \quad [1.88]$$

$$K_1 = X \cdot \overline{B} \quad [1.89]$$

$$J_2 = K_2 = X + \overline{A} \quad [1.90]$$

For the output, we have:

$$Y = A + B \quad [1.91]$$

It is possible to represent four states with two flip-flops. These states are encoded with two variables, A and B. The truth table of the JK flip-flop can be used to obtain the NSs (A^+ and B^+) for the state machine.

	PS	NS, A^+B^+		Output
	$A\,B$	$X = 0$	1	Y
S_1	0 0	0 1	1 1	0
S_2	0 1	0 0	1 0	1
S_3	1 0	1 0	0 1	1
S_4	1 1	1 1	1 0	1

Table 1.107. *State table*

We can also draw up the state table as shown in Table 1.107 by using the characteristic equation for the JK flip-flop, that is:

$$Q^+ = J \cdot \overline{Q} + \overline{K} \cdot Q$$

Thus

$$\begin{aligned} A^+ &= J_1 \cdot \overline{A} + \overline{K_1} \cdot A \\ &= X \cdot \overline{A} + \overline{X \cdot \overline{B}} \cdot A \qquad [1.92] \\ &= X \cdot \overline{A} + \overline{X} \cdot A + A \cdot B \end{aligned}$$

$$\begin{aligned} B^+ &= J_2 \cdot \overline{B} + \overline{K_2} \cdot B \\ &= (X + \overline{A}) \cdot \overline{B} + \overline{(X + \overline{A})} \cdot B \\ &= \overline{X} \cdot A \cdot B + X \cdot \overline{B} + \overline{A} \cdot \overline{B} \qquad [1.93] \end{aligned}$$

To complete the description of the finite state machine, Figure 1.116 presents the state diagram and a timing diagram is illustrated in Figure 1.117.

SOLUTION 1.13.– (Synchronous Counter).

The counter to be implemented can be described as a Moore state machine:

– the outputs of the flip-flops correspond to the outputs of the counter;

– there are four states (S_1, S_2, S_3 and S_4).

Figure 1.118 shows the state diagram of the counter.

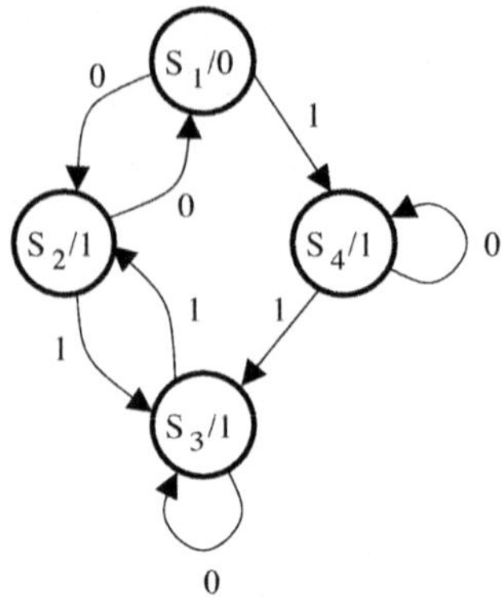

Figure 1.116. *State diagram*

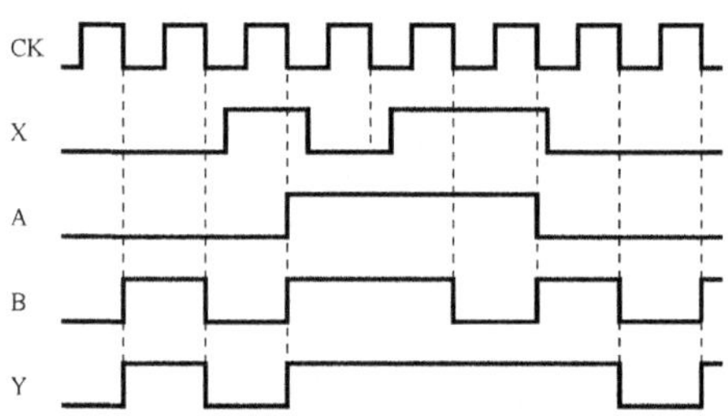

Figure 1.117. *Timing diagram*

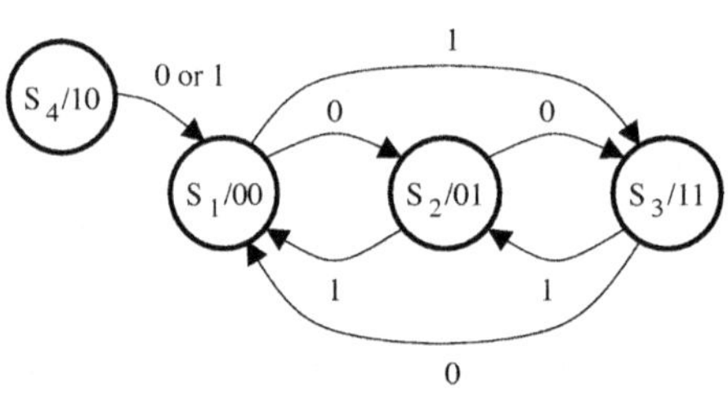

Figure 1.118. *State table of the counter*

The state table is represented in Table 1.108.

By assigning a binary code to each state, we can obtain the transition table shown in Table 1.109.

PS	NS	
	$C = 0$	1
S_1	S_2	S_3
S_2	S_3	S_1
S_3	S_1	S_2
S_4	S_1	S_1

Table 1.108. *State table*

PS	NS, A^+B^+	
AB	$C = 0$	1
00	01	11
01	11	00
11	00	01
10	00	00

Table 1.109. *Transition table*

The logic equations for the NSs, obtained based on the Karnaugh maps in Figures 1.119 and 1.120, are given by:

$$A^+ = D_1 = \overline{A} \cdot B \cdot \overline{C} + \overline{A} \cdot \overline{B} \cdot C \qquad [1.94]$$

$$B^+ = D_2 = A \cdot B \cdot C + \overline{A} \cdot \overline{B} \cdot C + \overline{A} \cdot \overline{C} \qquad [1.95]$$

In the expression for B^+, the term $\overline{A} \cdot \overline{B} \cdot C$, which also appears in A^+, has not been reduced to allow the sharing of a logic gate.

The logic circuit for the counter is represented in Figure 1.121.

SOLUTION 1.14.– (Design of a 010 Sequence Detector (Moore Model)).

The 010 sequence detector can be described by the state diagram shown in Figure 1.122, which is based on a Moore model.

Table 1.110 presents the state table, while Table 1.111 shows the transition table obtained by assigning a binary code to each state.

From Table 1.112, Karnaugh maps can be obtained as shown in Figures 1.123–1.126.

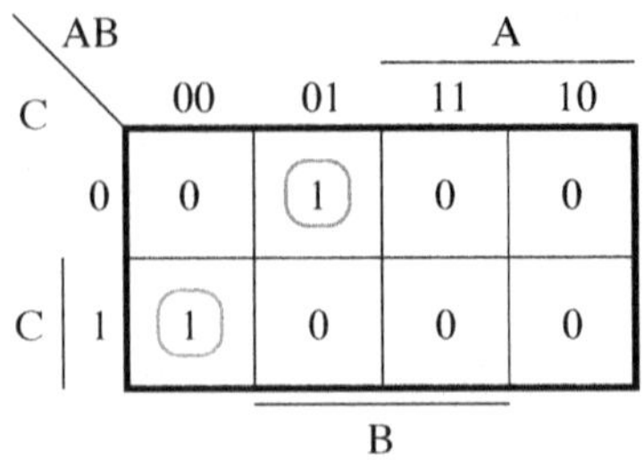

Figure 1.119. *Function A^+*

$$A^+ = D_1 = \overline{A} \cdot B \cdot \overline{C} + \overline{A} \cdot \overline{B} \cdot C$$

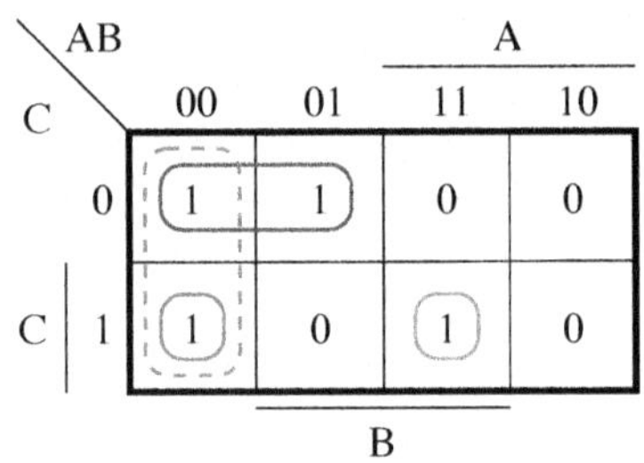

Figure 1.120. *Function B^+*

$$B^+ = D_2 = A \cdot B \cdot C + \overline{A} \cdot \overline{B} \cdot C + \overline{A} \cdot \overline{C}$$

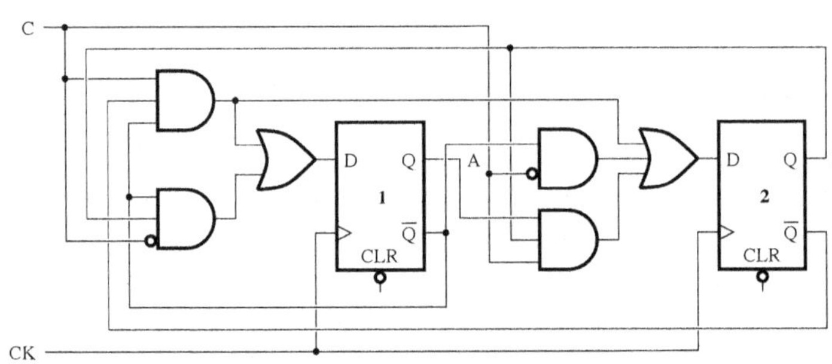

Figure 1.121. *Logic circuit of the counter (outputs: A and B)*

The logic equations for the flip-flop inputs can be written as follows:

$$J_1 = A + B \cdot X \qquad [1.96]$$

$$K_1 = B + X \qquad [1.97]$$

$$J_2 = \overline{X} \qquad [1.98]$$

$$K_2 = X \qquad [1.99]$$

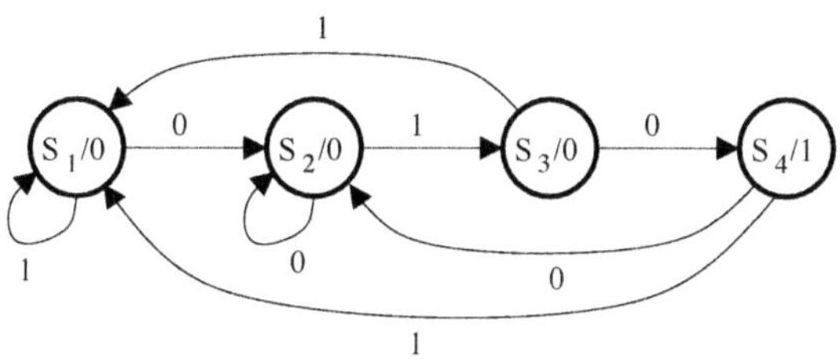

Figure 1.122. *State diagram of the 010 sequence detector*

PS	NS		Output
	$X = 0$	1	Y
S_1	S_2	S_1	0
S_2	S_2	S_3	0
S_3	S_4	S_1	0
S_4	S_2	S_1	1

Table 1.110. *State table*

PS	NS, A^+B^+		Output
AB	$X = 0$	1	Y
00	01	00	0
01	01	10	0
10	11	00	0
11	01	00	1

Table 1.111. *Transition table*

For the output of the detector, we obtain:

$$Y = A \cdot B \qquad [1.100]$$

Figure 1.127 depicts the logic circuit of the 010 sequence detector.

X	$A = Q_1$	$B = Q_2$	$A^+ = Q_1^+$	$B^+ = Q_2^+$	J_1	K_1	J_2	K_2
0	0	0	0	1	0	x	1	x
0	0	1	0	1	0	x	x	0
0	1	0	1	1	x	0	1	x
0	1	1	0	1	x	1	x	0
1	0	0	0	0	0	x	0	x
1	0	1	1	0	1	x	x	1
1	1	0	0	0	x	1	0	x
1	1	1	0	0	x	1	x	1

Table 1.112. *Table that can be used to derive the logic expressions for the flip-flop inputs*

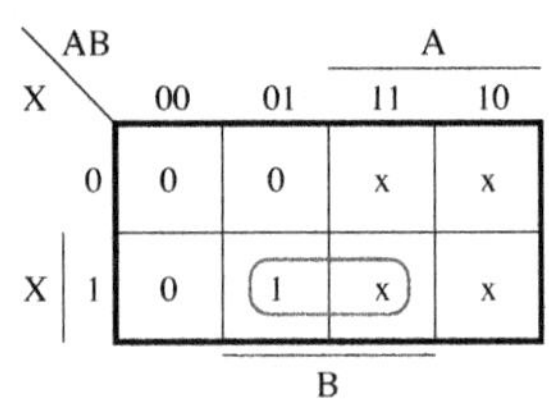

Figure 1.123. *Input* J_1

$J_1 = B \cdot X$

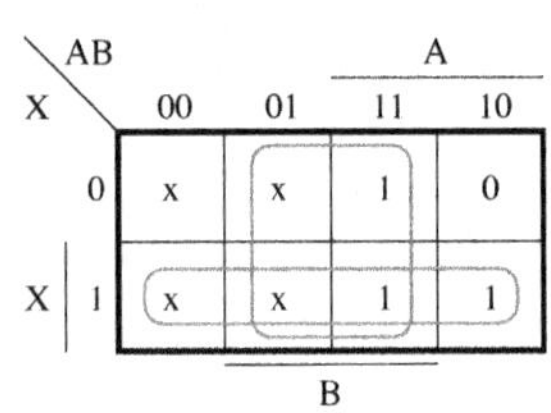

Figure 1.124. *Input* K_1

$K_1 = B + X$

SOLUTION 1.15.– (Analysis of a Finite State Machine (Mealy Model)).

Analyzing the logic circuit of the finite state machine, we can obtain the logic equation for the *D* input of each flip-flop as follows:

$$D_1 = A^+ = X \cdot A + X \cdot B \quad [1.101]$$

$$D_2 = B^+ = X \cdot \overline{A} \cdot \overline{B} \quad [1.102]$$

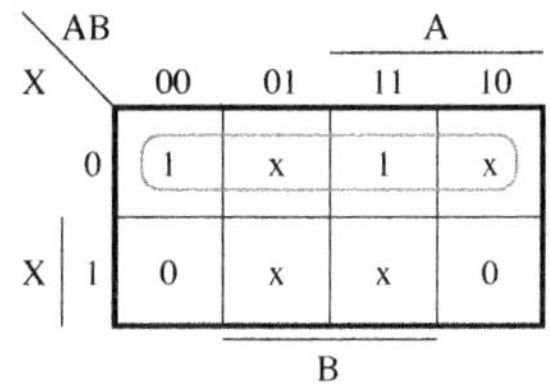

Figure 1.125. *Input* J_2
$J_2 = \overline{X}$

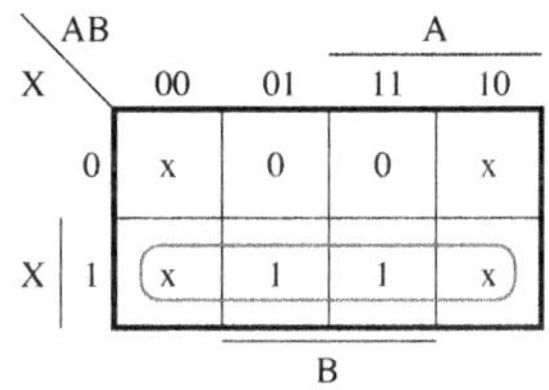

Figure 1.126. *Input* K_2
$K_2 = X$. *For a color version of this figure, see www.iste.co.uk/ndjountche/electronics3.zip*

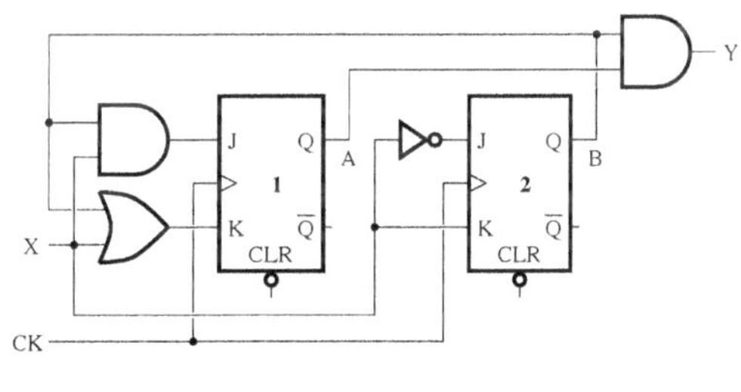

Figure 1.127. *Logic circuit of the* 010 *sequence detector*

We have, for the output:

$$Y = X \cdot A \qquad [1.103]$$

Table 1.12 depicts the state table constructed on the basis of the logic equations for the flip-flop inputs and output.

The state diagram is represented in Figure 1.128:

	PS	NS, A^+B^+		Output Y	
	AB	$X=0$	1	$X=0$	1
S_1	00	00	01	0	0
S_2	01	00	10	0	0
S_3	10	00	10	0	1
S_4	11	00	10	0	1

Table 1.113. *State table*

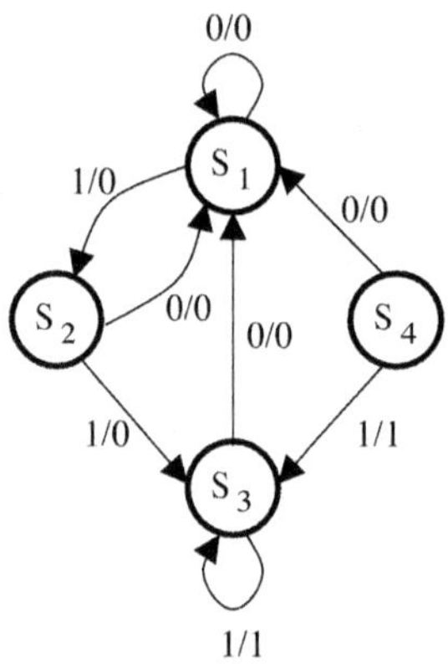

Figure 1.128. *State diagram*

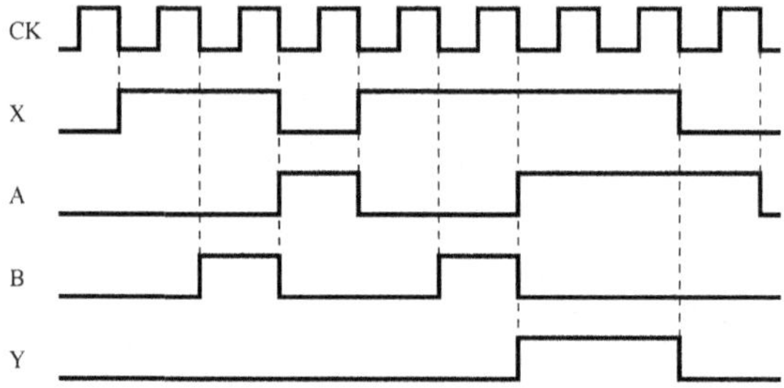

Figure 1.129. *Timing diagram 1. For a color version of this figure, see www.iste.co.uk/ndjountche/electronics3.zip*

– the output, Y, is set to 1 when the input signal, X, has been at the logic state 1 for three consecutive clock cycles;

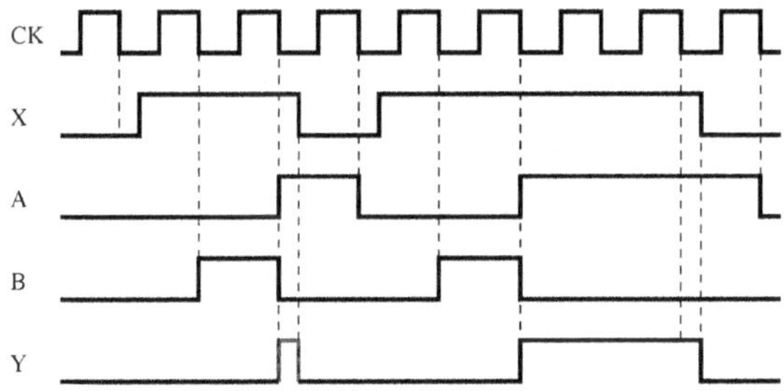

Figure 1.130. *Timing diagram 2 (synchronization problem). For a color version of this figure, see www.iste.co.uk/ndjountche/electronics3.zip*

– the machine never reaches the state S_4 and can only take this state initially;

– with an input X whose logic state is 0, we can realize the condition for the state machine to remain in the state S_1. The initialization to S_1 is, thus, not necessary.

Figure 1.129 depicts a timing diagram of the finite state machine.

Because this is a Mealy state machine, the timing diagram can be affected by the propagation delays of logic gates, as shown in Figure 1.130.

SOLUTION 1.16.– (Bidirectional Counter).

The bidirectional counter can be described by the transition table shown in Table 1.114.

	PS		NS		Inputs			
X	A	B	A^+	B^+	J_1	K_1	J_2	K_2
0	0	0	0	1	0	x	1	x
0	0	1	1	0	1	x	x	1
0	1	0	1	1	x	0	1	x
0	1	1	1	1	x	0	x	0
1	0	0	0	0	0	x	0	x
1	0	1	0	0	0	x	x	1
1	1	0	0	1	x	1	1	x
1	1	1	1	0	x	0	x	1

Table 1.114. *Transition table*

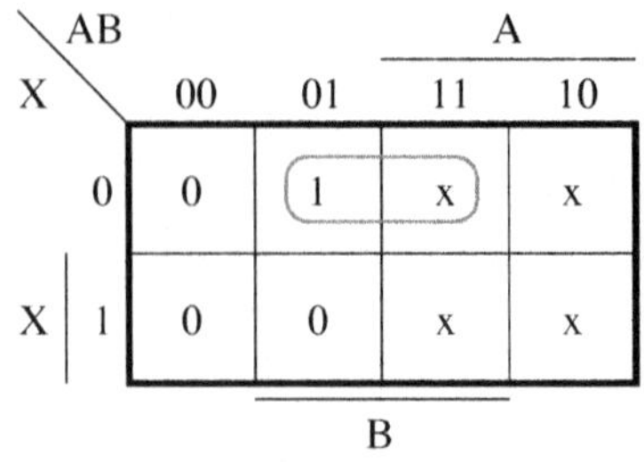

Figure 1.131. *Input* J_1
$J_1 = \overline{X} \cdot B$

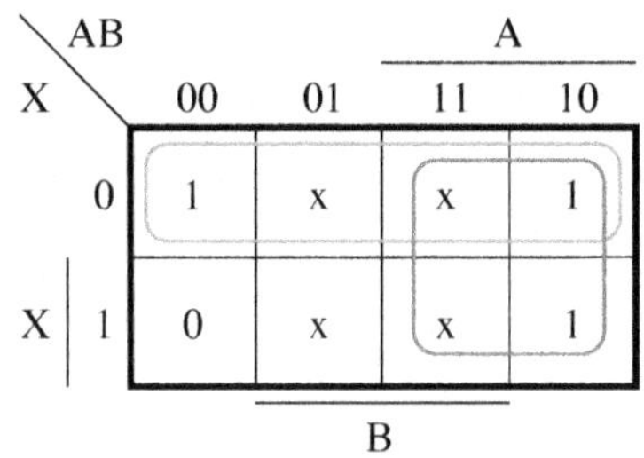

Figure 1.132. *Input* J_2
$J_2 = \overline{X} + A$

Figures 1.131–1.134 present Karnaugh maps obtained based on the transition table. The logic equations for the inputs of the JK flip-flops are, therefore, given by:

$$J_1 = \overline{X} \cdot B \quad [1.104]$$

$$K_1 = X \cdot \overline{B} \quad [1.105]$$

$$J_2 = \overline{X} + A \quad [1.106]$$

and:

$$K_2 = X + \overline{A} \quad [1.107]$$

SOLUTION 1.17.– (Design of a 010 Sequence Detector (Mealy Model)).

The Mealy state machine that can be used for the detection of the binary sequence 010 is described either by the state table shown in Table 1.115 (detection with overlap, or case a), or by the state table shown in Table 1.116 (detection without overlap, or case b).

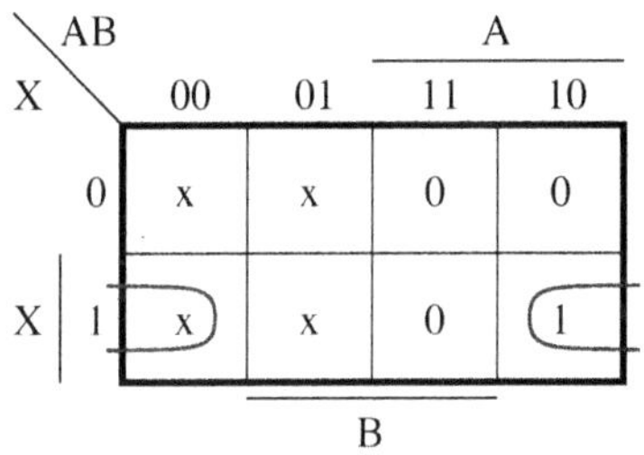

Figure 1.133. *Input K_1*
$K_1 = X \cdot \overline{B}$

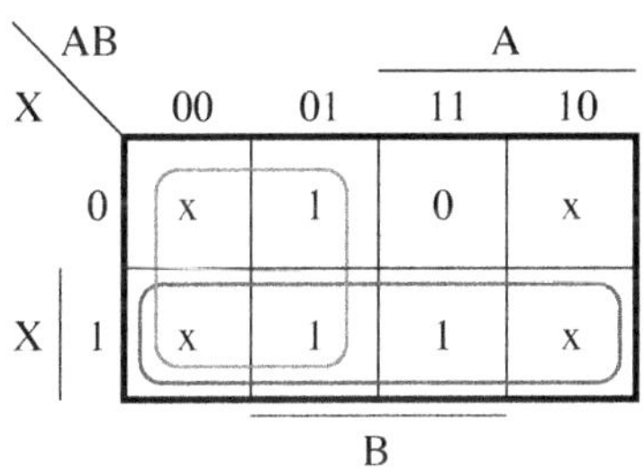

Figure 1.134. *Input K_2*
$K_2 = X + \overline{A}$

PS	NS		Output Y	
	$X = 0$	1	$X = 0$	1
S_1	S_2	S_1	0	0
S_2	S_2	S_3	0	0
S_3	S_2	S_1	1	0

Table 1.115. *State table (case a)*

PS	NS		Output Y	
	$X = 0$	1	$X = 0$	1
S_1	S_2	S_1	0	0
S_2	S_2	S_3	0	0
S_3	S_1	S_1	1	0

Table 1.116. *State table (case b)*

Two bits must be used to encode three states. The state S_4, which is unused, must be inserted in the state diagram in such a way as not to affect the operation of the detector.

The state diagrams for cases a and b are shown in Figures 1.135 and 1.136, respectively.

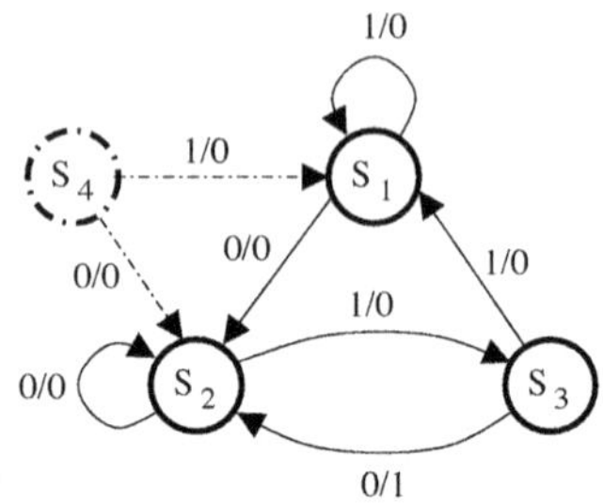

Figure 1.135. *State diagram (case a)*

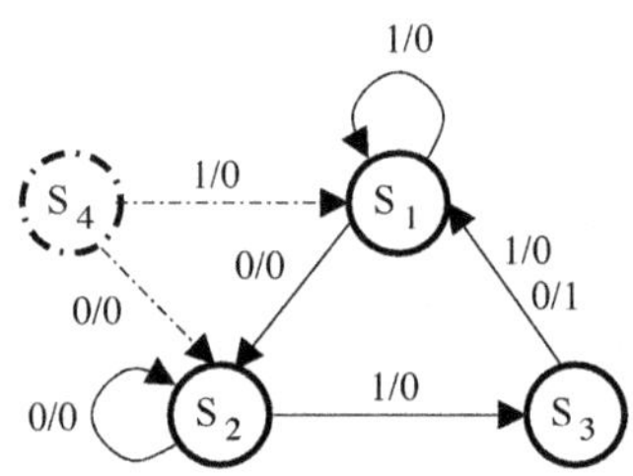

Figure 1.136. *State diagram (case b)*

X	$A = Q_1$	$B = Q_2$	$A^+ = Q_1^+$	$B^+ = Q_2^+$	J_1	K_1	J_2	K_2
0	0	0	0	1	0	x	1	x
0	0	1	0	1	0	x	x	0
0	1	0	0	1	x	1	1	x
0	1	1	0	1	x	1	x	0
1	0	0	0	0	0	x	0	x
1	0	1	1	0	1	x	x	1
1	1	0	0	0	x	1	0	x
1	1	1	0	0	x	1	x	1

Table 1.117. *Table that can be used to derive the logic expressions for the J and K inputs (case a)*

X	$A = Q_1$	$B = Q_2$	$A^+ = Q_1^+$	$B^+ = Q_2^+$	J_1	K_1	J_2	K_2
0	0	0	0	1	0	x	1	x
0	0	1	0	1	0	x	x	0
0	1	0	0	0	x	1	0	x
0	1	1	0	1	x	1	x	0
1	0	0	0	0	0	x	0	x
1	0	1	1	0	1	x	x	1
1	1	0	0	0	x	1	0	x
1	1	1	0	0	x	1	x	1

Table 1.118. *Table that can be used to derive the logic expressions for the J and K inputs (case b)*

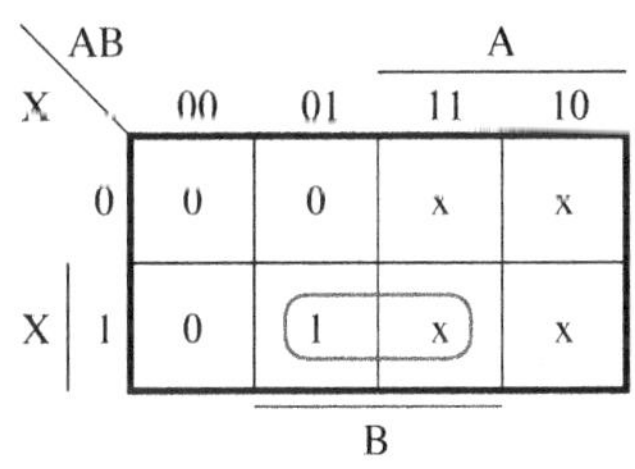

Figure 1.137. *Input J_1*
$J_1 = X \cdot B$

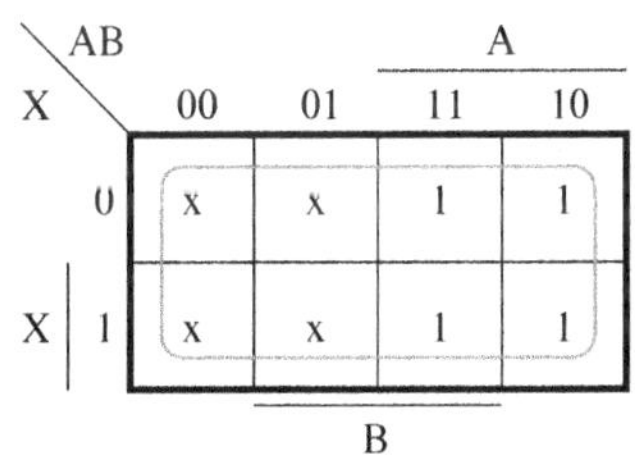

Figure 1.138. *Input K_1*
$K_1 = 1$

The tables that can be used to derive the logic expressions for the flip-flop inputs, J and K, are represented in Tables 1.117 and 1.118. These tables are used to construct Karnaugh maps shown in Figures 1.137–1.141. We thus have:

$$J_1 = X \cdot B \quad [1.108]$$

$$K_1 = 1 \quad [1.109]$$

$$J_2 = \begin{cases} \overline{X} & \text{for case a} \\ \overline{X} \cdot \overline{A} & \text{for case b} \end{cases} \quad [1.110]$$

and:

$$K_2 = X \qquad [1.111]$$

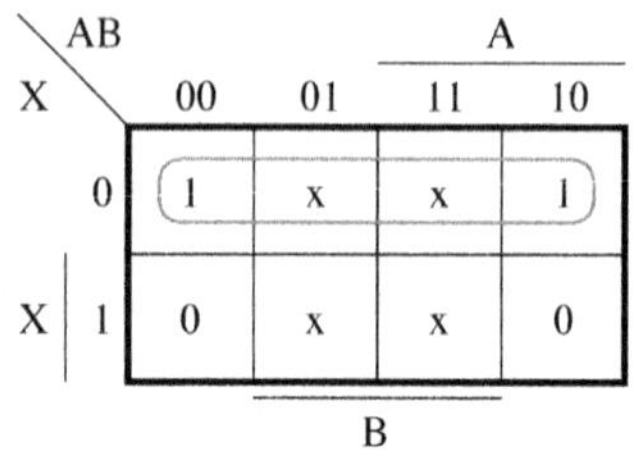

Figure 1.139. *Input J_2 (a)*
$J_2 = \overline{X}$

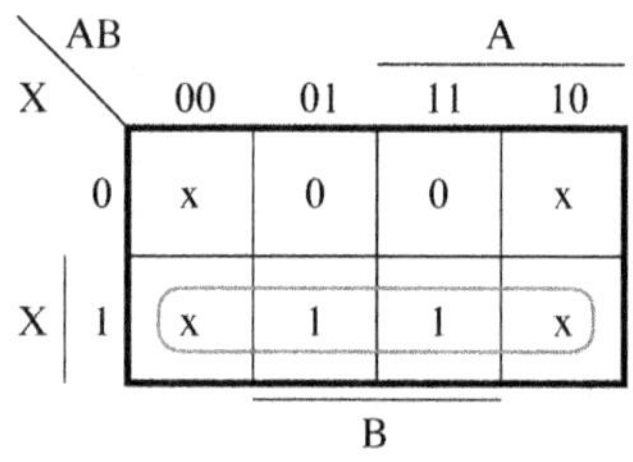

Figure 1.140. *Input K_2*
$K_2 = X$

The logic equation for the output is obtained from the Karnaugh map shown in Figure 1.142 and can be written as follows:

$$Y = \overline{X} \cdot A \cdot \overline{B} \qquad [1.112]$$

SOLUTION 1.18.– (Counter Generating the Sequence 2 6 1 7 5).

Three flip-flops are needed, and the maximum number of states that can then be represented is eight. The three unused states are associated with don't-care states, x, of NSs and inputs.

Karnaugh maps that can be used to derive the logic equations of the flip-flop inputs are constructed from the transition table presented in Table 1.119.

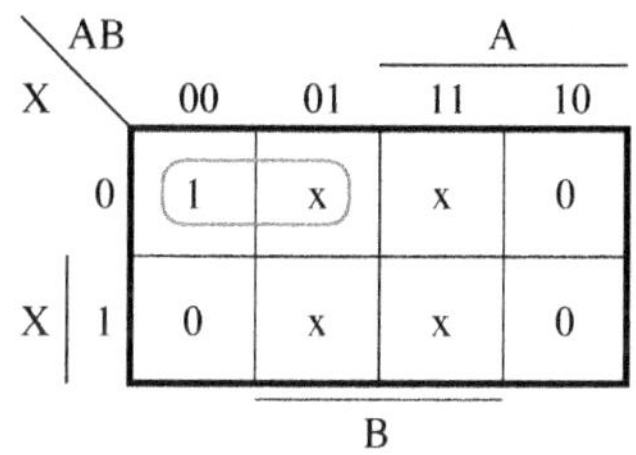

Figure 1.141. *Input J_2 (b)*
$J_2 = \overline{X} \cdot \overline{A}$

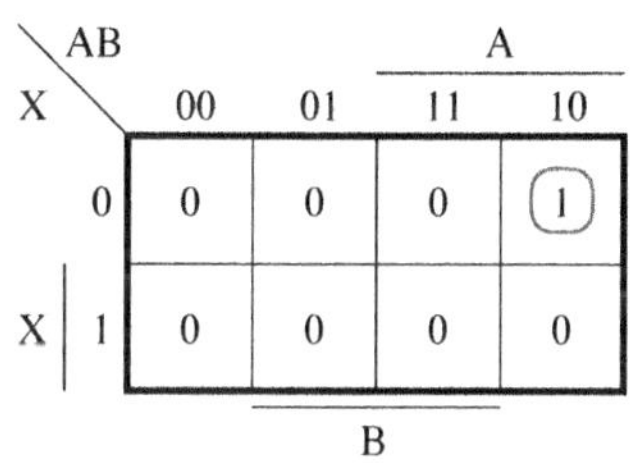

Figure 1.142. *Output Y*
$Y = \overline{X} \cdot A \cdot \overline{B}$

PS			NS			Input					
A	B	C	A^+	B^+	C^+	J_1	K_1	J_2	K_2	J_3	K_3
0	0	0	x	x	x	x	x	x	x	x	x
0	0	1	1	1	1	1	x	1	x	x	0
0	1	0	1	1	0	1	x	x	0	0	x
0	1	1	x	x	x	x	x	x	x	x	x
1	0	0	x	x	x	x	x	x	x	x	x
1	0	1	0	1	0	x	1	1	x	x	1
1	1	0	0	0	1	x	1	x	1	1	x
1	1	1	1	0	1	x	0	x	1	x	0

Table 1.119. *Table that can be used to derive the logic expressions of the flip-flop inputs*

For the state machine implementation using D flip-flops, Karnaugh maps are represented in Figures 1.143–1.145. The input logic equations are given by:

$$D_1 = A^+ = \overline{A} + B \cdot C \qquad [1.113]$$

$$D_2 = B^+ = \overline{A} + \overline{B} \qquad [1.114]$$

$$D_3 = C^+ = A \cdot B + \overline{A} \cdot \overline{B} \quad \text{or} \quad D_3 = C^+ = A \cdot B + \overline{A} \cdot C \qquad [1.115]$$

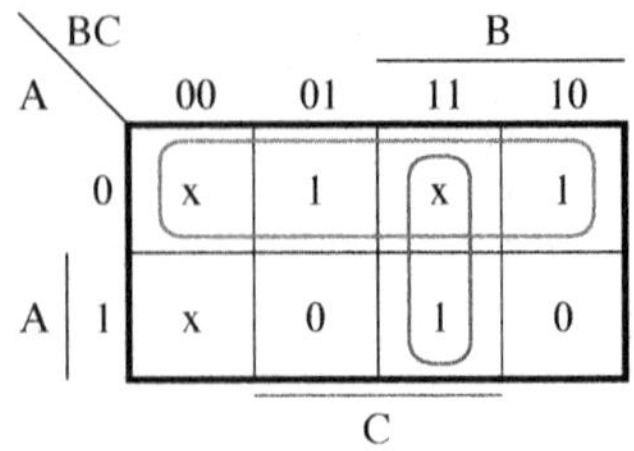

Figure 1.143. *Inputs* D_1
$D_1 = \overline{A} + B \cdot C$

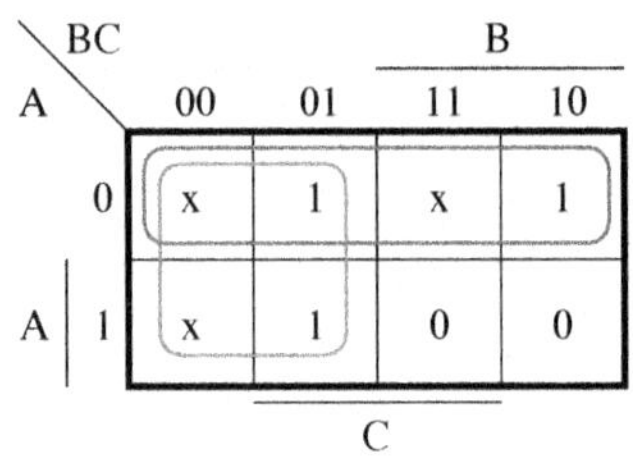

Figure 1.144. *Input* D_2
$D_2 = \overline{A} + \overline{B}$

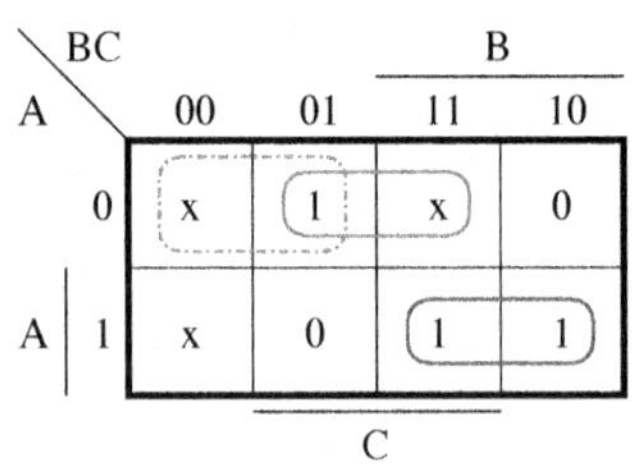

Figure 1.145. *Input* D_3
$D_3 = A \cdot B + \overline{A} \cdot \overline{B}$

Substituting the sequences not used for counting, we obtain:

– PS: 0 (000), $D_1 = 1$, $D_2 = 1$, $D_3 = 1$ or $D_3 = 0 \Rightarrow$ NS: 111 (7) or 110 (6);

– PS: 3 (011), $D_1 = 1$, $D_2 = 1$, $D_3 = 0$ or $D_3 = 1 \Rightarrow$ NS: 110 (6) or 111 (7);

– PS: 4 (100), $D_1 = 0$, $D_2 = 1$, $D_3 = 0 \Rightarrow$ NS: 010 (2).

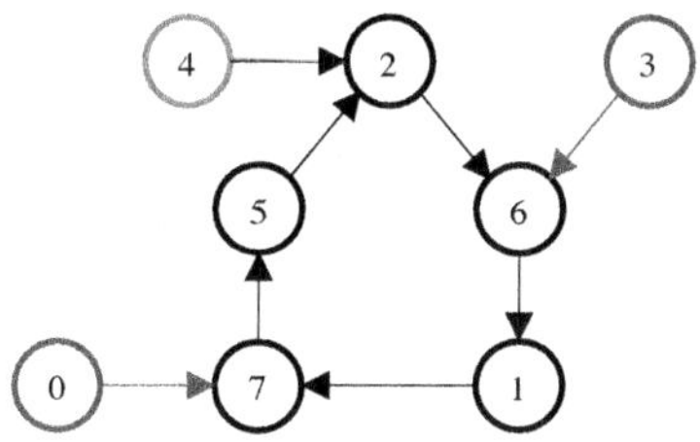

Figure 1.146. *State diagram*
$D_3 = A \cdot B + \overline{A} \cdot \overline{B}$

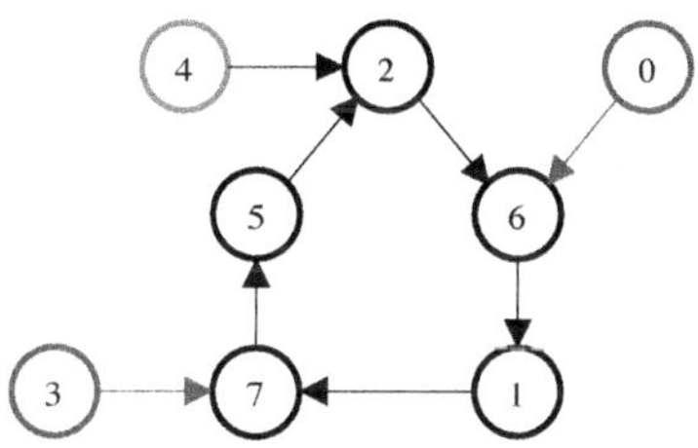

Figure 1.147. *State diagram*
$D_3 = A \cdot B + \overline{A} \cdot C$

The two possible state diagrams are shown in Figures 1.146 and 1.147.

The Karnaugh maps, in the case of the JK flip-flops, are depicted in Figures 1.148–1.153. The logic equations for the inputs J and K can be written as follows:

$$J_1 = 1 \quad \text{and} \quad K_1 = B + \overline{C} \qquad [1.116]$$

$$J_2 = 1 \quad \text{and} \quad K_2 = A \qquad [1.117]$$

$$J_3 = A \quad \text{and} \quad K_3 = A \cdot \overline{B} \qquad [1.118]$$

Substituting the sequences not used for counting, we obtain:

– PS: 0 (000), $J_1 = K_1 = 1$, $J_2 = 1$, $K_2 = 0$, $J_3 = K_3 = 0 \Rightarrow$ NS: 110 (6);

– PS: 3 (011), $J_1 = 1$, $K_1 = 0$, $J_2 = 1$, $K_2 = 0$, $J_3 = K_3 = 0 \Rightarrow$ NS: 111 (7);

– PS: 4 (100), $J_1 = K_1 = 1$, $J_2 = K_2 = 1$, $J_3 = K_3 = 1 \Rightarrow$ NS: 011 (3).

Figure 1.154 shows the state diagram of the counter.

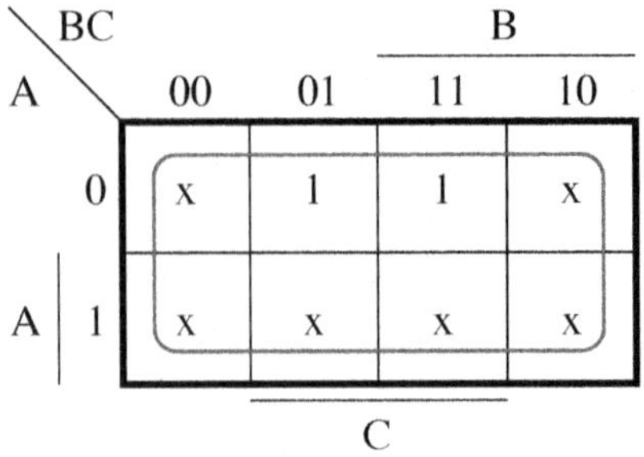

Figure 1.148. *Input J_1*

$J_1 = 1$

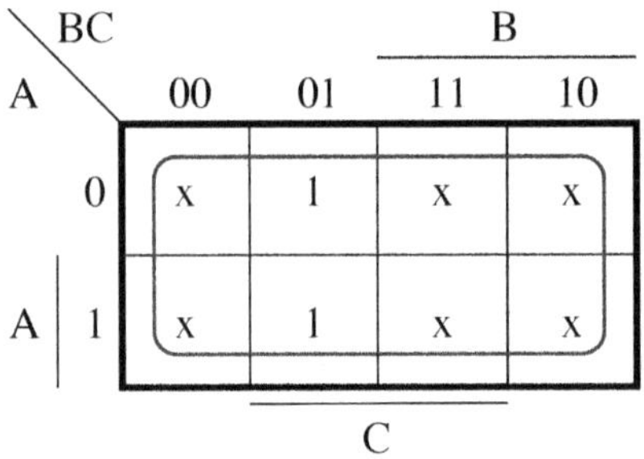

Figure 1.149. *Input J_2*

$J_2 = 1$

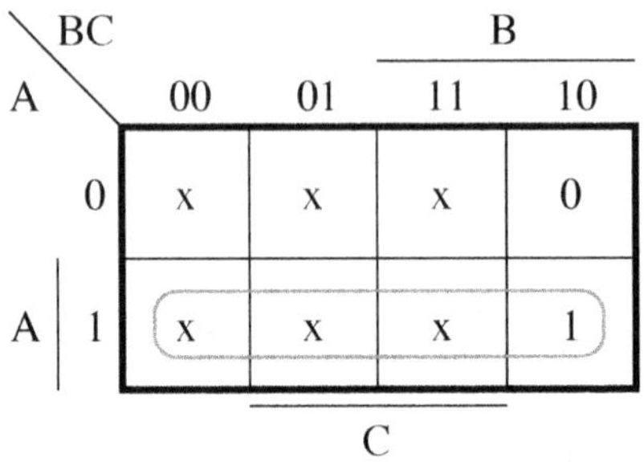

Figure 1.150. *Input J_3*

$J_3 = A$

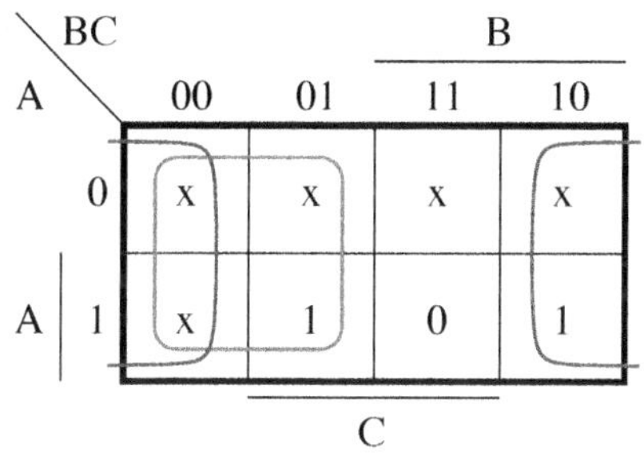

Figure 1.151. *Input* K_1
$K_1 = \overline{B} + \overline{C}$

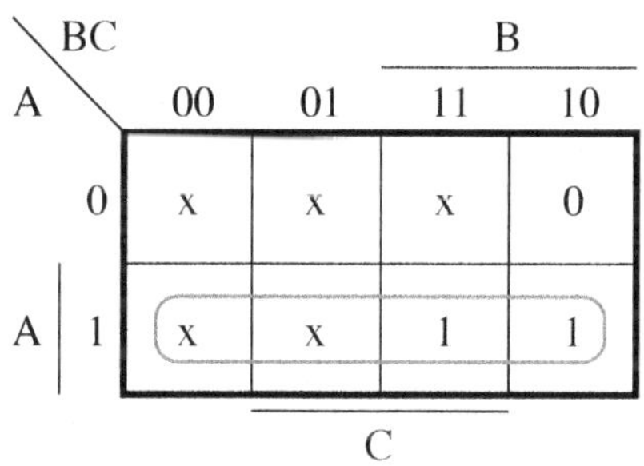

Figure 1.152. *Input* K_2
$K_2 = A$

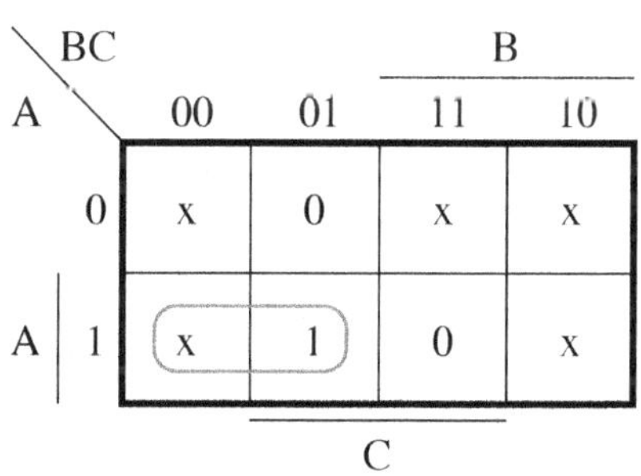

Figure 1.153. *Input* K_3
$K_3 = A \cdot \overline{B}$

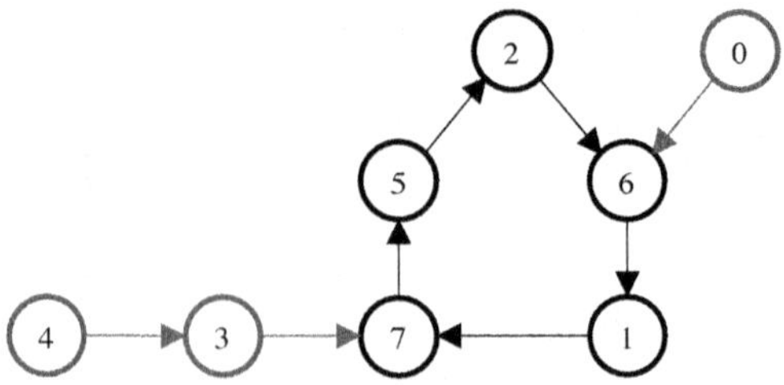

Figure 1.154. *State diagram (JK flip-flops)*

SOLUTION 1.19.– Minimizing the number of states using the implication method.

The state table shown in Table 1.120 can be used to construct the implication table represented in Table 1.122. After a single marking pass, we obtain the implication table shown in Table 1.123. As there are no more possibilities for marking, we can conclude that the states A, E and G are equivalent ($A \equiv E \equiv G$), as are the states B and F ($B \equiv F$). Table 1.124 presents the reduced state table of the state machine 1.

PS	NS		Output Y
	$X = 0$	1	
A	A	B	1
B	C	A	0
C	A	D	0
D	C	C	1
E	G	F	1
F	C	E	0
G	E	B	1

Table 1.120. *State table of the state machine 1*

The implication table corresponding to the state table shown in Table 1.121 is represented in Table 1.125, where the pair of states $S_5 - S_3$, which is found in a cell implicating these same states, is redundant and can be eliminated. As some cells marked with a cross, because they are associated with states leading to different outputs, can also lead to other markings, we can construct the implication table shown in Table 1.126. Taking into account the different marking possibilities that become apparent, the resulting implication table can be set up as shown in

Table 1.127. We can, thus, establish the following equivalent relationships: $S_1 \equiv S_2 \equiv S_4$ and $S_3 \equiv S_5$. Assuming that:

$$A = S_0 \quad [1.119]$$

$$B = S_1 = S_2 = S_4 \quad [1.120]$$

$$C = S_3 = S_5 \quad [1.121]$$

$$D = S_6 \quad [1.122]$$

we obtain, for the state machine 2, the reduced state table shown in Table 1.128.

PS	NS		Output Y	
	$X = 0$	1	$X = 0$	1
S_0	S_1	S_2	1	1
S_1	S_3	S_5	1	1
S_2	S_5	S_4	0	0
S_3	S_1	S_6	1	1
S_4	S_5	S_2	0	0
S_5	S_4	S_3	0	0
S_6	S_5	S_6	0	0

Table 1.121. *State table of the state machine 2*

	A	B	C	D	E	F
B	X					
C	X	C – A A – D				
D	A – C B – C	X	X			
E	A – G B – F	X	X	C – G C – F		
F	X	A – E	A – C D – E	X	X	
G	A – E	X	X	C – E C – B	G – E F – B	X

Table 1.122. *Implication table according to the state table*

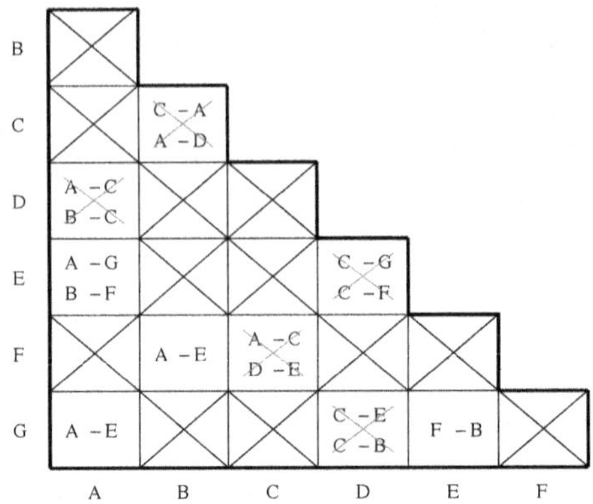

Table 1.123. *Implication table after the first marking pass*

PS	NS $X = 0$	 1	Output Y
A	A	B	1
B	C	A	0
C	A	D	0
D	C	C	1

Table 1.124. *Reduced state table for the state machine 1*

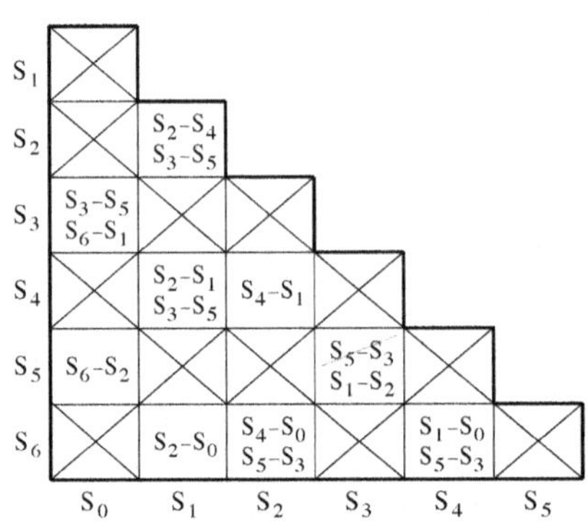

Table 1.125. *Implication table according to the state table*

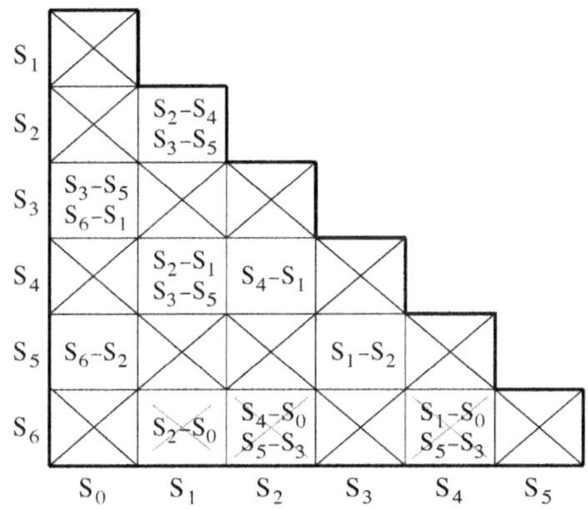

Table 1.126. *State table after one marking pass*

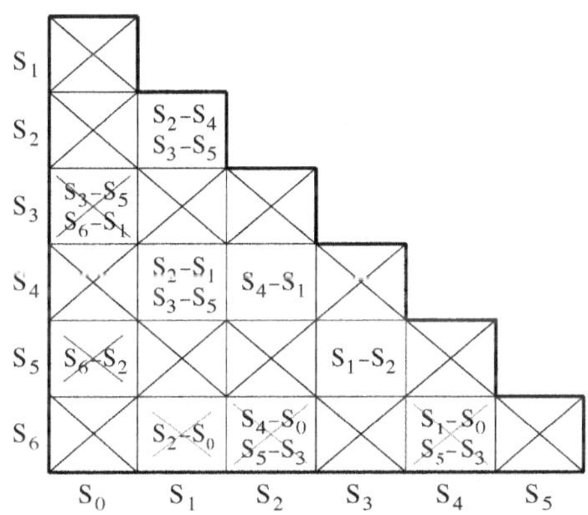

Table 1.127. *Implication table after two marking pass*

PS	NS		Output Y	
	$X = 0$	1	$X = 0$	1
A	C	D	1	1
B	B	C	0	1
C	C	B	1	1
D	A	C	0	1

Table 1.128. *Reduced state table for the state machine 2*

PS	NS		Output Y
	$X=0$	1	
S_0	S_1	S_2	1
S_1	S_3	S_5	1
S_2	S_5	S_4	0
S_3	S_1	S_6	1
S_4	S_5	S_2	0
S_5	S_4	S_3	0
S_6	S_5	S_6	0

Table 1.129. *State table of the state machine 1*

SOLUTION 1.20.– (Minimizing the Number of States Using the Partitioning Method):

a) Table 1.131 summarizes the different steps necessary for the determination of equivalent states from the state table shown in Table 1.129. Thus, states S_0 and S_3 are equivalent, just like the states S_2, S_4 and S_6. Assuming that:

$$A = S_0 = S_3 \quad [1.123]$$

$$B = S_1 \quad [1.124]$$

$$C = S_2 = S_4 = S_6 \quad [1.125]$$

$$D = S_5 \quad [1.126]$$

we can obtain the reduced state table in Table 1.132.

PS	NS		Output Y	
	$X=0$	1	$X=0$	1
S_0	S_4	S_3	0	1
S_1	S_5	S_3	0	0
S_2	S_4	S_1	0	1
S_3	S_5	S_1	0	0
S_4	S_2	S_5	0	1
S_5	S_1	S_2	0	0

Table 1.130. *State table of the state machine 2*

		Blocks to be formed
P_0	$(S_0S_1S_2S_3S_4S_5S_6)$	
Output Y	1 1 0 1 0 0 0	$S_0S_1S_3$ and $S_2S_4S_5S_6$
P_1	$(S_0S_1S_3)(S_2S_4S_5S_6)$	
NS		
$X=0$	$S_1S_3S_1$ $S_5S_5S_4S_5$	
$X=1$	$S_2S_5S_6$ $S_4S_2S_3S_6$	$S_2S_4S_6$ and S_5
P_2	$(S_0S_1S_3)(S_2S_4S_6)(S_5)$	
NS		
$X=0$	$S_1S_3S_1$ $S_5S_5S_5$ S_4	
$X=1$	$S_2S_5S_6$ $S_4S_2S_6$ S_3	S_0S_3 and S_1
P_3	$(S_0S_3)(S_1)(S_2S_4S_6)(S_5)$	
NS		
$X=0$	S_1S_1 S_3 $S_5S_5S_5$ S_4	
$X=1$	S_2S_6 S_5 $S_4S_2S_6$ S_3	
$P_4=P_3$	$(S_0S_3)(S_1)(S_2S_4S_6)(S_5)$	

Table 1.131. *Determination of equivalent states using the partitioning method (state machine 1)*

PS	NSI		Output Y
	$X=0$	1	
A	B	C	1
B	A	D	1
C	D	C	0
D	C	A	0

Table 1.132. *Reduced state table (state machine 1)*

The state machine 2 is described by the state table shown in Table 1.130. Based on Table 1.133, the states S_0 and S_2 are equivalent, as are the states S_1 and S_3. Assuming that:

$$A = S_0 = S_2 \quad [1.127]$$

$$B = S_1 = S_3 \quad [1.128]$$

$$C = S_4 \quad [1.129]$$

$$D = S_5 \quad [1.130]$$

we can obtain the reduced state table shown in Table 1.134.

	CI	Blocks to be formed
P_0	$(S_0S_1S_2S_3S_4S_5)$	
Output Y		
$X=0$	0 0 0 0 0 0	
$X=1$	1 0 1 0 1 0	$S_0S_2S_4$ and $S_1S_3S_5$
P_1	$(S_0S_2S_4)(S_1S_3S_5)$	
NS		
$X=0$	$S_4S_4S_2$ $S_5S_5S_1$	
$X=1$	$S_3S_1S_5$ $S_3S_1S_2$	S_1S_3 and S_5
P_2	$(S_0S_2S_4)(S_1S_3)(S_5)$	
NS		
$X=0$	$S_4S_4S_2$ S_5S_5 S_1	
$X=1$	$S_3S_1S_5$ S_3S_1 S_2	S_0S_2 and S_4
P_3	$(S_0S_2)(S_4)(S_1S_3)(S_5)$	
NS		
$X=0$	S_4S_4 S_2 S_5S_5 S_1	
$X=1$	S_3S_1 S_5 S_3S_1 S_2	
$P_4=P_3$	$(S_0S_2)(S_4)(S_1S_3)(S_5)$	

Table 1.133. *Determination of equivalent states using the partitioning method (state machine 2)*

PS	NS		Output Y	
	$X=0$	1	$X=0$	1
A	C	B	0	1
B	D	B	0	0
C	A	D	0	1
D	B	A	0	0

Table 1.134. *Reduced state table (state machine 2)*

b) The state table for the state machine 1 is represented in Table 1.135. Forming partitions as shown in Table 1.136, we can deduce that the equivalent states are as follows: A and F, C and G, B and H and D and E. The reduced state table is represented in Table 1.137.

PS	NS $XY = 00$	01	10	11	Output Z
A	A	F	C	B	0
B	A	B	D	H	1
C	G	B	C	D	0
D	C	F	D	D	1
E	G	A	E	D	1
F	F	F	G	B	0
G	G	B	G	E	0
H	F	B	E	H	1

Table 1.135. *State table of the state machine 1*

		Blocks to be formed
P_0	$(ABCDEFGH)$	
Output Z	0 1 0 1 1 0 0 1	$ACFG$ and $BDEH$
P_1	$(ACFG)(BDEH)$	
NS		
$XY = 00$	$AGFG\ ACGF$	
$XY = 01$	$FBFB\ BFAB$	AF and CG
$XY = 10$	$CCGG\ DDEE$	
$XY = 11$	$BDBE\ HDDH$	
P_2	$(AF)(CG)(BDEH)$	
NS		
$XY = 00$	$AF\ CC\ ACGF$	BH and DE
$XY = 01$	$FF\ BB\ BFAB$	BH and DE
$XY = 10$	$CG\ CG\ DDEE$	
$XY = 11$	$BB\ DE\ HDDH$	
P_3	$(AF)(CG)(BH)(DE)$	
NS		
$XY = 00$	$AF\ GG\ AF\ CG$	
$XY = 01$	$FF\ BB\ BB\ FA$	
$XY = 10$	$CG\ CG\ DE\ DE$	
$XY = 11$	$BB\ DE\ HH\ DD$	
$P_4 = P_3$	$(AF)(CG)(BH)(DE)$	

Table 1.136. *Determination of equivalent states using the partitioning method (state machine 1)*

PS	NS				Output
					Z
	$XY = 00$	01	10	11	
A	A	A	C	B	0
B	A	B	D	B	1
C	C	B	C	D	0
D	C	A	D	D	1

Table 1.137. *Reduced state table (state machine 1)*

c) For the state machine 2, whose state table is shown in Table 1.138, the construction in Table 1.139 allows to determine the following equivalent states: A and H, B and G, C and F and D and E. Table 1.140 presents the reduced state table.

PS	NS				Output			
					Z			
	$XY = 00$	01	10	11	$XY = 00$	01	10	11
A	A	G	E	H	0	0	1	1
B	F	B	B	D	0	1	0	0
C	F	C	G	H	0	1	0	1
D	H	C	E	D	1	0	1	0
E	A	F	E	D	1	0	1	0
F	F	C	B	A	0	1	0	1
G	F	G	G	D	0	1	0	0
H	H	B	E	H	0	0	1	1

Table 1.138. *State table of the state machine 2*

SOLUTION 1.21.– Simplification of the finite state machines whose state tables are shown in Tables 1.141–1.144.

Figure 1.155(a) shows the merger graph for the state machine 1 that can be used to construct the simplified merger graph for compatible states, as shown in Figure 1.155(b), and the simplified merger graph for incompatible states, as shown in Figure 1.155(c). The pairs of compatible states are as follows: (AB), (AC), (AD), (BD), (CE) and (CG), while the set of maximal compatibility classes is formed of (ABD), (AC), (CE), F and (CG). Figure 1.155(d) presents the compatibility graph. Replacing (ABD) by S_0, (CE) by S_1, (F) by S_2 and (CG) by S_3 we obtain the reduced state table shown in Table 1.146.

		Blocks to be formed
P_0	$(ABCDEFGH)$	
Output Y		
$XY = 00$	0 0 0 1 1 0 0 0	$ABCFGH$ and DE
$XY = 01$	0 1 1 0 0 1 1 0	$ADEH$ and $BCFG$
$XY = 10$	1 0 0 1 1 0 0 1	$ADEH$ and $BCFG$
$XY = 11$	1 0 1 0 0 1 0 1	$ACFH$ and $BDEG$
P_1	$(AH)(BG)(CF)(DE)$	
NS		
$XY = 00$	AH FF FF HA	
$XY = 01$	GB BG CC CF	
$XY = 10$	EE BG GB EE	
$XY = 11$	HH DD HD DD	
$P_2 = P_1$	$(AH)(BG)(CF)(DE)$	

Table 1.139. *Determination of equivalent states using the partitioning method (state machine 2)*

PS	NS				Output Z			
	$XY = 00$	01	10	11	$XY = 00$	01	10	11
A	A	B	D	A	0	0	1	1
B	C	B	B	D	0	1	0	0
C	C	C	B	A	0	1	0	1
D	A	C	D	D	1	0	1	0

Table 1.140. *Reduced state table (state machine 2)*

In order to determine the pairs of compatible states for the state machine 2, we can construct the implication tables represented in Tables 1.145(a)–(c). The pairs of compatible states are as follows: (AD), (AE), (AF), (AG), (CE), (DF), (DG), (EG) and (FG). Figures 1.156(a) and (b) show the simplified merger graphs for the compatible and incompatible states, respectively. The set of maximal compatibility classes consists of $(ADFG)$, (AEG), (B), (CE) and (EG). Figure 1.156(c) presents the compatibility graph. Assuming that $S_0 = ADFG$, $S_1 = B$, $S_2 = CE$ and $S_3 = EG$, we can obtain the reduced state table shown in Table 1.147.

PS	NS		Output Y	
	$X=0$	1	$X=0$	1
A	B	C	–	0
B	D	–	0	–
C	–	E	1	–
D	B	G	0	0
E	F	C	1	1
F	E	D	0	1
G	F	–	1	0

Table 1.141. *State table (state machine 1)*

PS	NS		Output Y
	$X=0$	1	
A	D	–	0
B	C	E	1
C	B	G	0
D	A	B	–
E	–	E	0
F	G	B	–
G	F	–	0

Table 1.142. *State table (state machine 2)*

PS	NS				Output Z
	$XY=00$	01	10	11	
A	A	B	–	D	–
B	A	B	C	–	0
C	–	B	C	H	0
D	–	–	G	D	–
E	E	F	–	D	1
F	E	F	G	G	1
G	D	F	G	H	1
H	A	–	–	H	0

Table 1.143. *State table (state machine 3)*

PS	NS				Output Z			
	$XY = 00$	01	10	11	$XY = 00$	01	10	11
A	G	A	–	H	–	1	–	1
B	G	–	B	D	0	–	0	0
C	C	F	E	–	0	–	–	–
D	–	A	E	D	–	–	–	0
E	C	–	E	D	–	–	1	–
F	G	F	–	D	–	1	–	–
G	G	A	B	–	0	–	0	–
H	–	–	E	H	–	–	1	1

Table 1.144. *State table (state machine 4)*

NOTE 1.7.– Referring to the implication table shown in Table 1.145(c), we can also determine the maximal compatibility or incompatibility classes as follows:

– maximal compatibility classes:

Column F:	(FG)			
Column E:	(EG)	(FG)		
Column D:	(DFG)	(EG)		
Column C:	(CE)	(DFG)	(EG)	
Column B:	(B)	(CE)	(DFG)	(EG)
Column A:	(ADFG)	(AEG)	(B)	(CE)

– maximal incompatibility classes:

Column F:	(F)					
Column E:	(EF)					
Column D:	(DE)	(EF)				
Column C:	(CG)	(CF)	(CD)	(DE)	(EF)	
Column B:	(BCG)	(BCF)	(BCD)	(BDE)	(BEF)	
Column A:	(ABC)	(BCG)	(BCF)	(BCD)	(BDE)	(BEF)

Tables 1.148(a) and (b) give the implication tables for the state machine 3. The simplified merger graphs for the compatible and incompatible states are shown in Figures 1.157(a) and (b), respectively. For the compatible pairs, we have: (AG), (AH), (BF), (BG), (CE), (CH) and (DE). The set of maximal compatibility

classes consists of the following elements: $(ABCH)$, (ADH), (DEF) and (DEG). Figure 1.157(c) presents the compatibility graph. The reduced state table, as illustrated in Table 1.150, where $S_0 = (ABCH)$, $S_1 = (DEF)$, $S_2 = (DG)$ and $S_3 = (DH)$, is obtained by merging the rows of the initial state table.

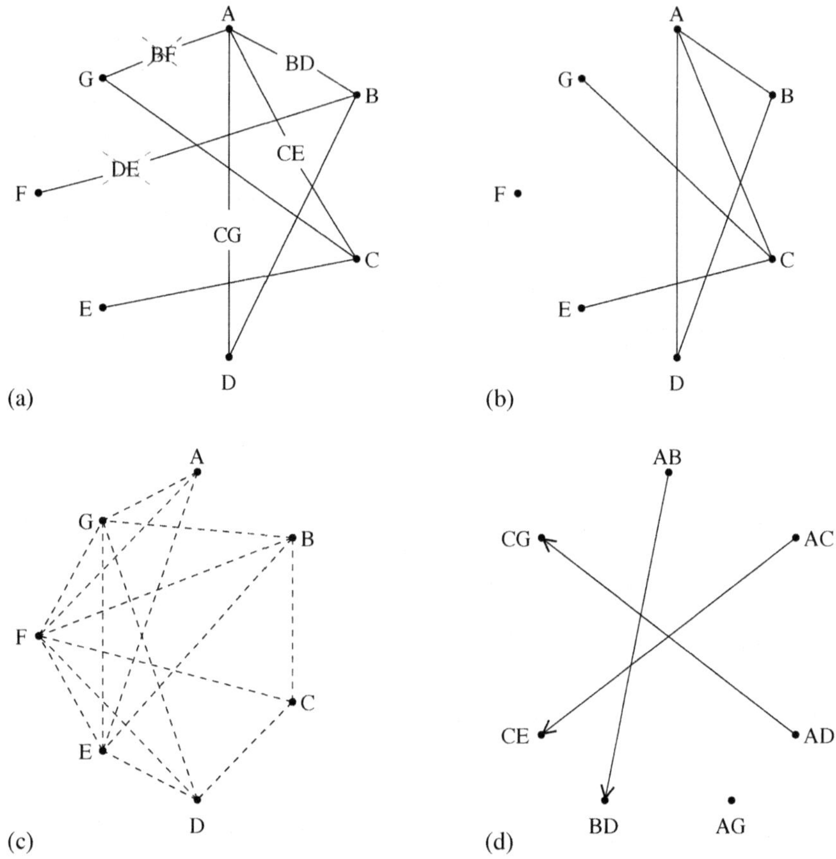

Figure 1.155. *a) Merger graph; b) simplified merger graph for compatible states; c) simplified merger graph for incompatible states; d) compatibility graph*

	A	B	C	D	E	F
B	×					
C	BD	×				
D	√	AC BE	AB BG			
E	√	×	EG	BE		
F	DG	CG BE	BG	AG	BE	
G	DF	×	BF	AF	√	√

(a)

	A	B	C	D	E	F
B	×					
C	BD	×				
D	√	×	×			
E	√	×	EG	×		
F	DG	×	×	AG	×	
G	DF	×	BF	AF	√	√

(b)

	A	B	C	D	E	F
B	×					
C	×	×				
D	√	×	×			
E	√	×	EG	×		
F	DG	×	×	AG	×	
G	DF	×	×	AF	√	√

(c)

Table 1.145. *Implication table: a) based on the state table; b) after one marking pass; c) after two marking pass*

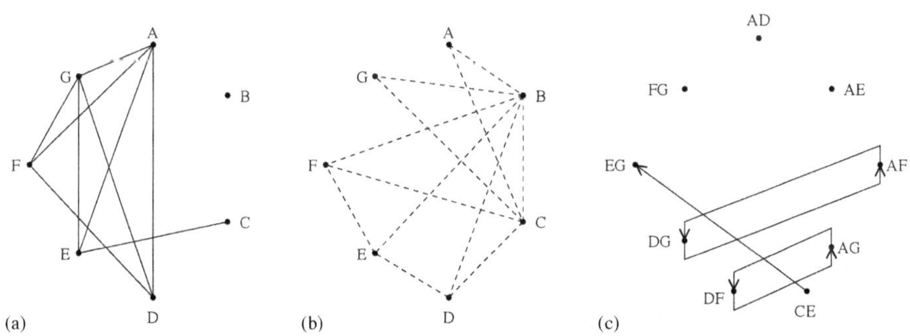

Figure 1.156. *a) Simplified merger graph for compatible states; b) simplified merger graph for incompatible states; c) compatibility graph*

PS	NS		Output Y	
	$X=0$	1	$X=0$	1
S_0	S_0	S_3	0	0
S_1	S_2	S_1	1	1
S_2	S_1	S_0	0	1
S_3	S_2	S_1	1	0

Table 1.146. *Reduced state table (state machine 1)*

PS	NS		Output Y
	$X=0$	1	
S_0	S_0	S_1	0
S_1	S_2	S_2/S_3	1
S_2	S_1	S_3	0
S_3	S_0	S_3	0

Table 1.147. *Reduced state table (state machine 2)*

(a)

	A	B	C	D	E	F	G
B	✓						
C	DH	✓					
D	✓	CG	CG DH				
E	AE BF	×	×	✓			
F	AE BF DG	×	×	DG	DG		
G	AD BF DH	×	×	DH	DE DH	DE GH	
H	DH	✓	✓	✓	×	×	×

(b)

	A	B	C	D	E	F	G
B	✓						
C	DH	✓					
D	✓	×	×				
E	×	×	×	✓			
F	×	×	×	DG	DG		
G	×	×	×	DH	DE DH	×	
H	DH	✓	✓	✓	×	×	×

Table 1.148. *Implication table a) based on the state table and b) after one marking pass*

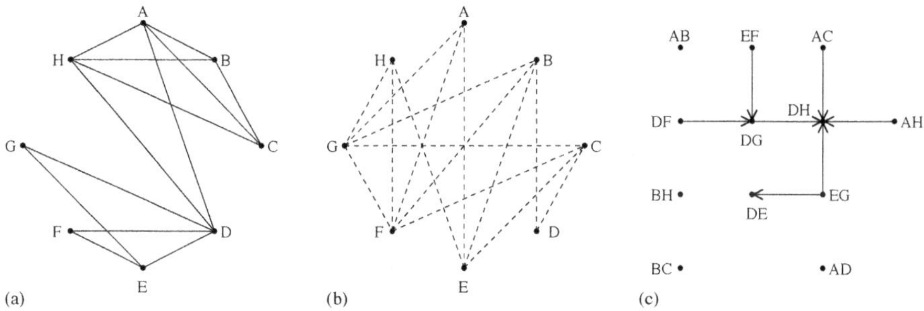

Figure 1.157. *a) Simplified merger graph for compatible states; b) simplified merger graph for incompatible states; c) compatibility graph*

NOTE 1.8.– Referring to the implication table shown in Table 1.148(b), we can also determine the maximal compatibility or incompatibility classes as follows:

– maximal compatibility classes:

Column G:	(G)			
Column F:	(F)	(G)		
Column E:	(EF)	(EG)		
Column D:	(DEF)	(DEG)	(DH)	
Column C:	(DEF)	(DEG)	(CH)	(DH)
Column B:	(BCH)	(DEF)	(DEG)	(DH)
Column A:	(ABCH)	(ADH)	(DEF)	(DEG)

– maximal incompatibility classes:

Column G:	(GH)									
Column F:	(FGH)									
Column E:	(FGH)	(EH)								
Column D:	(FGH)	(EH)	(D)							
Column C:	(CFG)	(FGH)	(CD)	(CE)	(EH)					
Column B:	(BFG)	(CFG)	(FGH)	(BE)	(BD)	(CD)	(CE)	(EH)		
Column A:	(AFG)	(BFG)	(CFG)	(FGH)	(AE)	(BE)	(BD)	(CD)	(CE)	(EH)

For the state machine 4, it is possible to construct the implication tables as shown in Tables 1.149(a)–(c). We can then determine the following pairs of compatible states: (AG), (AH), (BF), (BG), (CE), (CH) and (DE). Figures 1.158(a)–(b) show the simplified merger graphs for compatible and incompatible states, respectively. Assuming that $S_0 = (AG)$, $S_1 = (BF)$, $S_2 = (CH)$ and $S_3 = (DE)$, it is possible to reduce the state table as shown in Table 1.151.

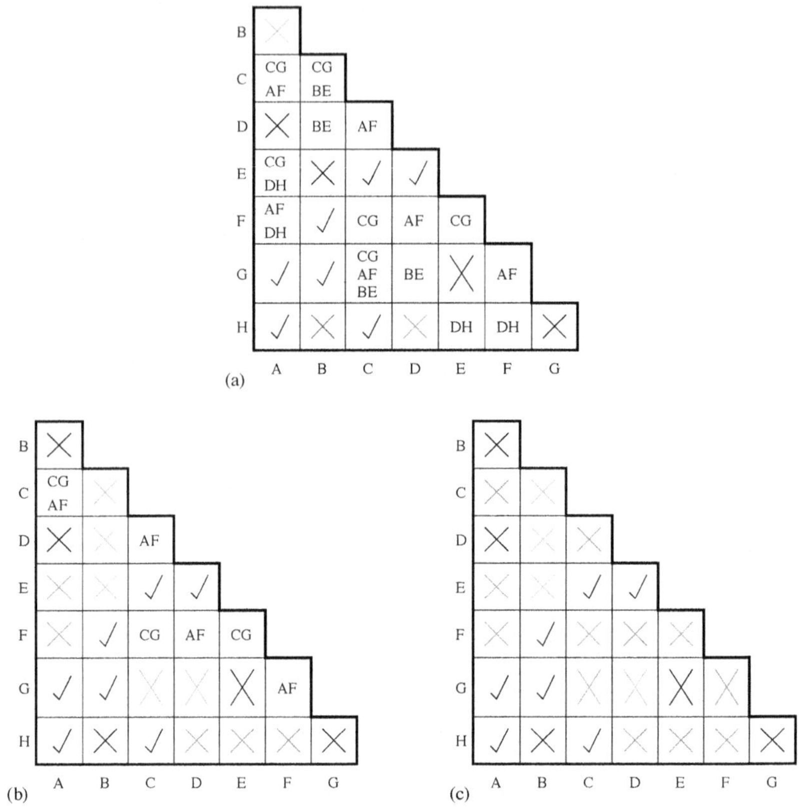

Table 1.149. *Implication table a) based on the state table, b) after one marking pass and c) after two marking pass*

PS	NS				Output Z
	$XY = 00$	01	10	11	
S_0	S_0	S_0	S_0	S_3	0
S_1	S_1	S_1	S_2	S_2	1
S_2	S_2	S_1	S_2	S_3	1
S_3	S_0	–	S_2	S_3	0

Table 1.150. *Reduced state table (state machine 3)*

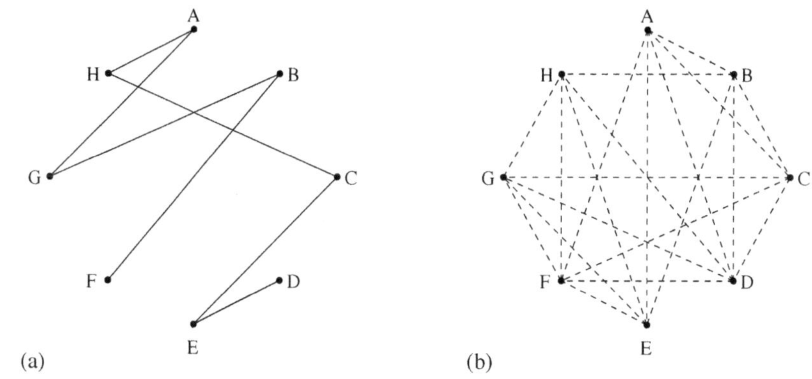

Figure 1.158. *a) Subgraph of the merger graph for compatible states; b) subgraph for merger graph of incompatible states*

PS	NS				Output Z			
	$XY = 00$	01	10	11	$XY = 00$	01	10	11
S_0	S_0	S_0	S_1	S_2	0	1	0	1
S_1	S_0	S_1	S_1	S_3	0	1	0	0
S_2	S_2	S_1	S_3	S_2	0	–	1	1
S_3	S_2	S_0	S_3	S_3	–	–	1	0

Table 1.151. *Reduced state table (state machine 4)*

NOTE 1.9.– All the maximal compatibility classes are only pairs of states and the maximal incompatibility classes can be determined using the implication table shown in Table 1.149(c), as follows:

Column G:	(GH)					
Column F:	(FGH)					
Column E:	(EFGH)					
Column D:	(DFGH)	(EFGH)				
Column C:	(DFGH)	(EFGH)	(CFG)	(CD)		
Column B:	(DFGH)	(EFGH)	(BEH)	(BCD)	(CFG)	
Column A:	(ABCD)	(DFGH)	(EFGH)	(AEF)	(BEH)	(CFG)

SOLUTION 1.22.– (Transformation of Moore Model Based State Machine to Equivalent Mealy Model Based State Machine).

Figure 1.159 shows the Moore state machines, while the equivalent Mealy state machines are represented in Figure 1.160.

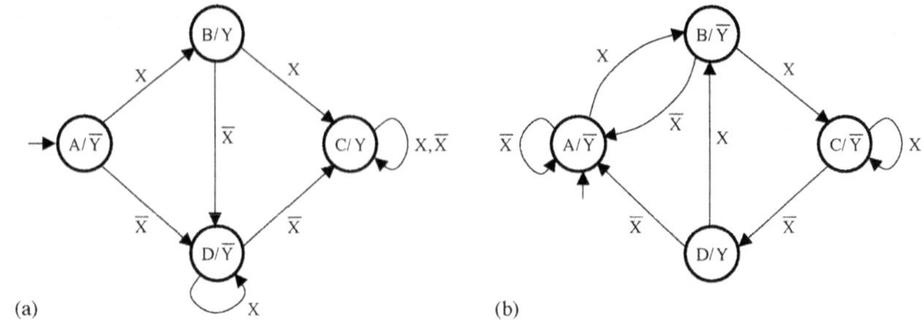

Figure 1.159. *Moore model: a) state machine 1; b) state machine 2*

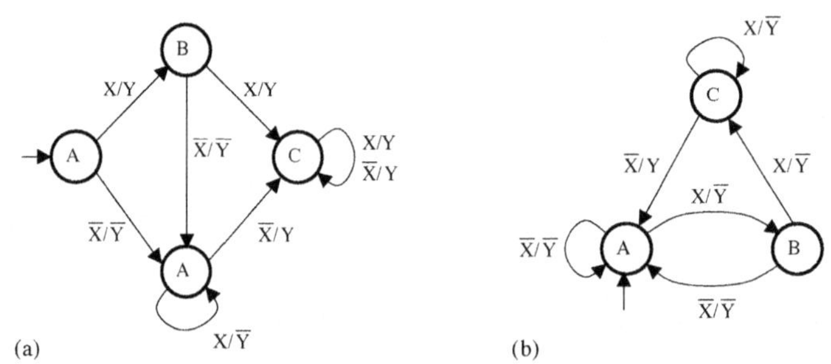

Figure 1.160. *Equivalent Mealy model: a) state machine 1; b) state machine 2*

SOLUTION 1.23.– (Transformation of Mealy Model Based State Machine to Moore Model Based State Machine).

Each of the state machines represented in Figure 1.161 can be transformed as shown in Figure 1.162.

Similarly, each of the state machines depicted in Figure 1.163 can be transformed as shown in Figure 1.164.

SOLUTION 1.24.– (Splitting Finite State Machines).

Applying the rules for splitting, each finite state machine (modulo 6 counter and 010 and 1001 sequence detector) shown in Figures 1.165 and 1.166 can be transformed as shown in Figures 1.167 and 1.168. Supposing that the states are represented by a

1-out-of-n code, each signal, Z_j, corresponds to a unique flip-flop output, Q_j, which is set to 1 whenever the state machine enters the corresponding state.

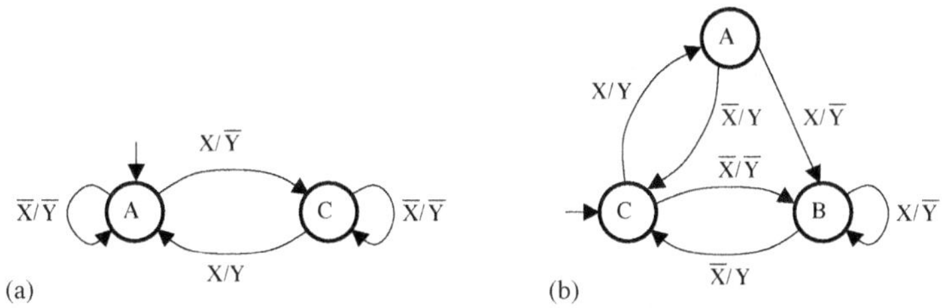

Figure 1.161. *Mealy model: a) state machine 1; b) state machine 2*

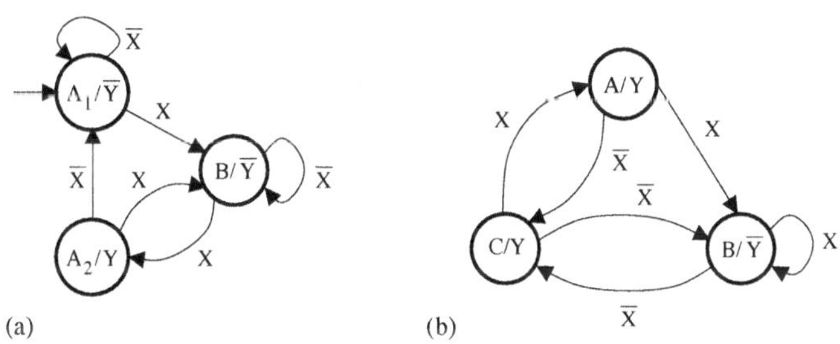

Figure 1.162. *Moore model: a) state machine 1; b) state machine 2*

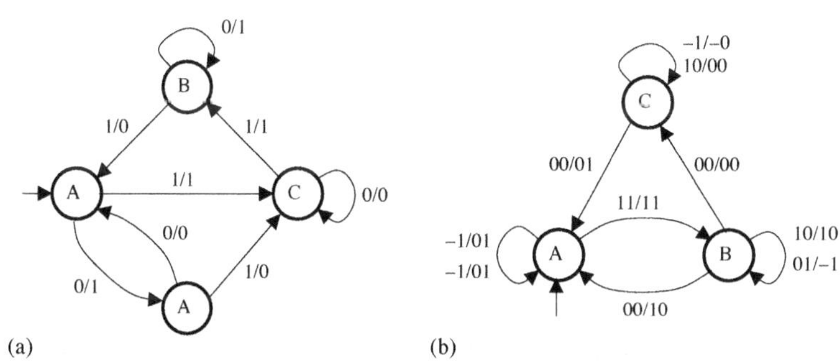

Figure 1.163. *Mealy model: a) state machine 1; b) state machine 2*

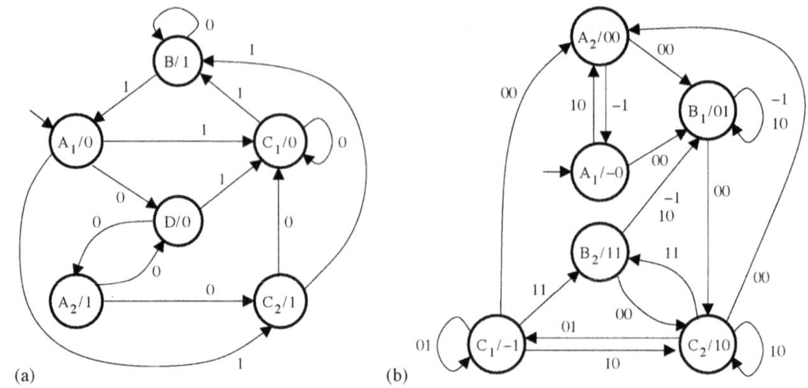

Figure 1.164. *Moore model: a) state machine 1; b) state machine 2*

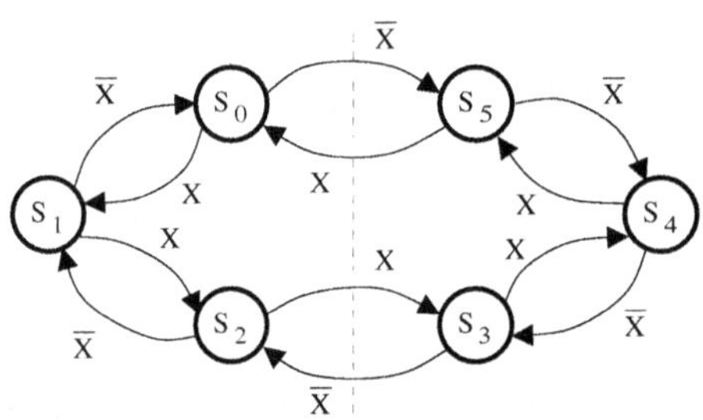

Figure 1.165. *State diagram of the modulo 6 counter*

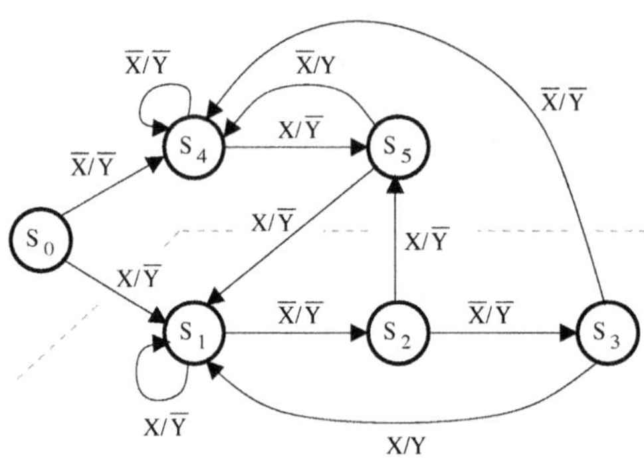

Figure 1.166. *State diagram of the 010 and 1001 sequence detector*

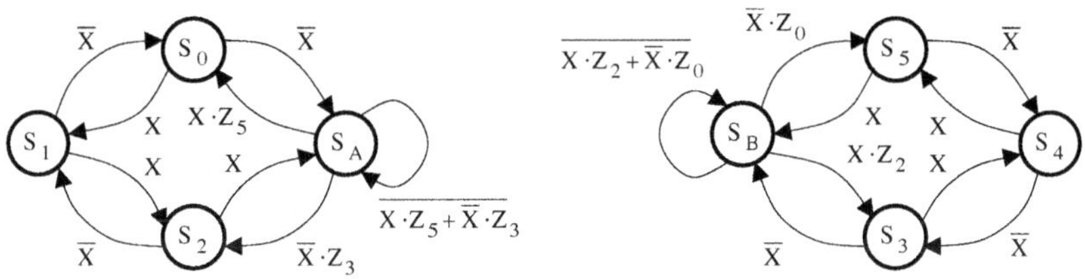

Figure 1.167. *State diagram of the modulo 6 counter (after splitting)*

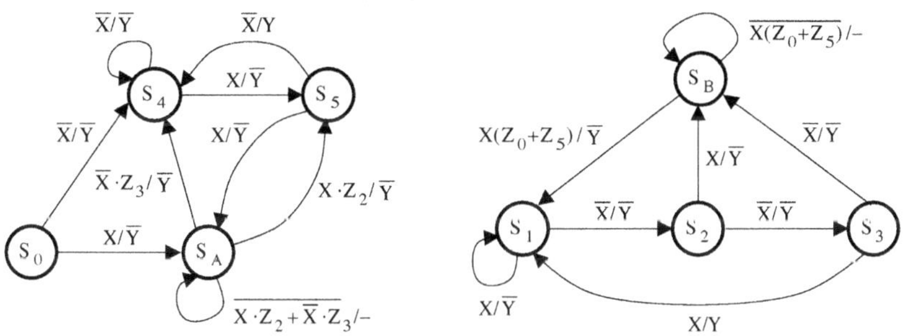

Figure 1.168. *State diagram of the 010 and 1001 sequence detector (after splitting)*

SOLUTION 1.25.– To complete the timing diagram of the proposed machine, the logic equations for the D input must first be determined. Analyzing the logic circuit, we can obtain:

$$D_1 - A^+ - A \cdot B + X \cdot \overline{A} \quad [1.131]$$

$$D_2 = B^+ = X \cdot A + \overline{A} \cdot \overline{B} \quad [1.132]$$

The content of each flip-flop is only defined from the moment the signal $\overline{CLR}$ takes the logic state 0.

Figure 1.169 shows the complete timing diagram of the finite state machine.

SOLUTION 1.26.– The logic equations for the inputs J and K can be written as follows:

$$J_1 = \overline{X} \cdot \overline{A} \quad [1.133]$$

$$K_1 = \overline{B} \quad [1.134]$$

and:

$$J_2 = K_2 = X + \overline{A} \qquad [1.135]$$

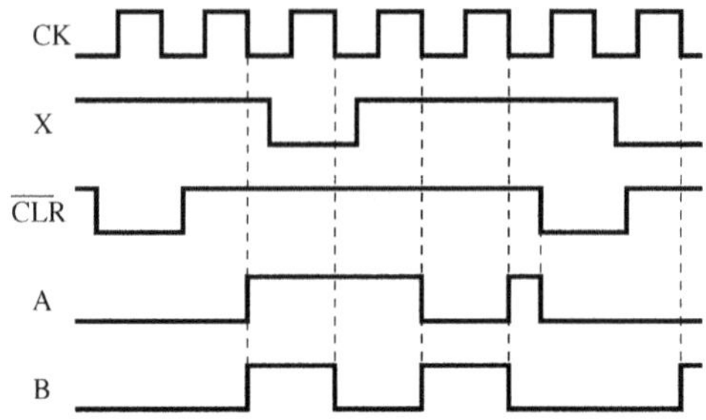

Figure 1.169. *Timing diagram*

The output logic equation is given by:

$$Y = A \cdot B \qquad [1.136]$$

The $\overline{CLR}$ signal is used to initialize each flip-flop. The timing diagram for the state machine can be completed using the truth table or the characteristic equation of the JK flip-flop. Figure 1.170 shows the timing diagram of the state machine.

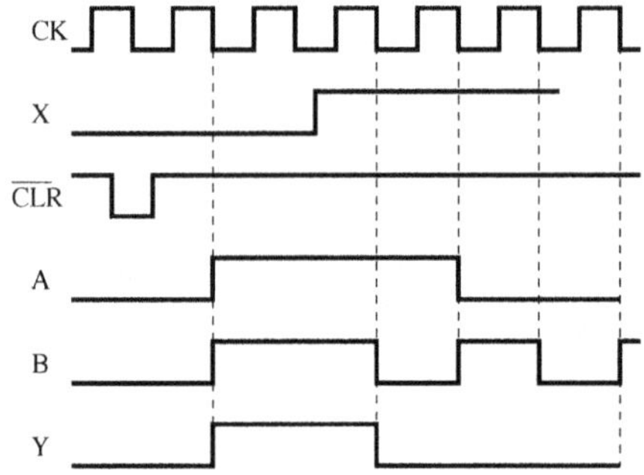

Figure 1.170. *Timing diagram*

SOLUTION 1.27.– The operation of the finite state machine can be described by the state diagram shown in Figure 1.171.

Table 1.152 presents the state table of the state machine.

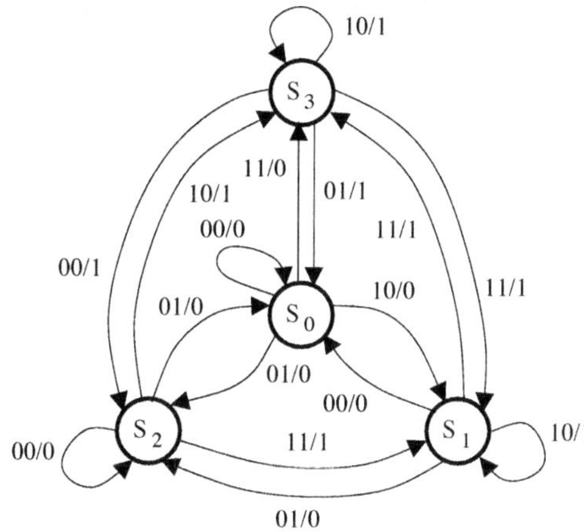

Figure 1.171. *State diagram*

PS	NS				Output Z			
	$XY = 00$	01	10	11	$XY = 00$	01	10	11
S_0	S_0	S_2	S_1	S_3	0	0	0	0
S_1	S_0	S_2	S_1	S_3	0	0	1	1
S_2	S_2	S_0	S_3	S_1	0	0	1	1
S_3	S_2	S_0	S_3	S_1	1	1	1	1

Table 1.152. *State table*

PS	NS A^+B^+				Output Z			
	$XY = 00$	01	10	11	$XY = 00$	01	10	11
00	00	10	01	11	0	0	0	0
01	00	10	01	11	0	0	1	1
10	10	00	11	01	0	0	1	1
11	10	00	11	01	1	1	1	1

Table 1.153. *Transition table*

The transition table shown in Table 1.153 is obtained by assigning the binary codes 00, 01, 10 and 11 to the states S_0, S_1, S_2 and S_3, respectively.

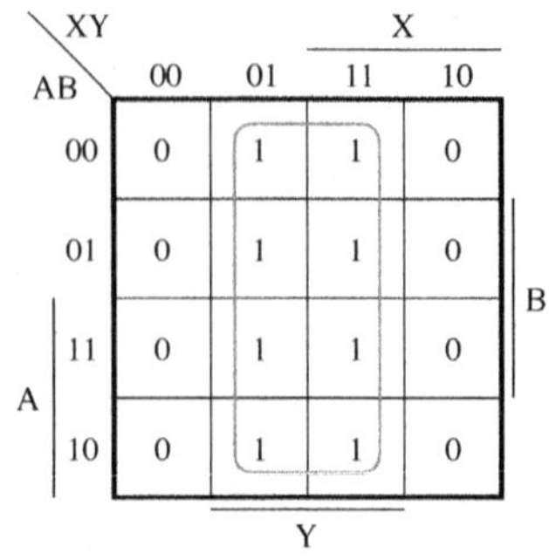

Figure 1.172. *Input* T_1
$T_1 = Y$

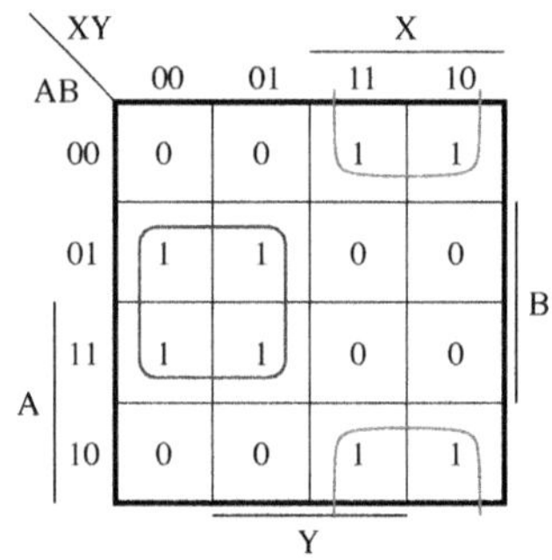

Figure 1.173. *Input* T_2
$T_2 = B \cdot \overline{X} + \overline{B} \cdot X$

The excitation table for the T flip-flop can be used to construct the Karnaugh maps that are shown in Figures 1.172 and 1.173, and are required for the determination of the following logic equations for the T inputs:

$$T_1 = Y \qquad [1.137]$$

$$T_2 = B \cdot \overline{X} + \overline{B} \cdot X = B \oplus X \qquad [1.138]$$

For the output of the state machine, the logic equation, deduced from the Karnaugh map in Figure 1.174, is given by:

$$Z = A \cdot B + X \cdot A + X \cdot B \qquad [1.139]$$

Figure 1.175 shows the logic circuit of the state machine, where the T flip-flop is implemented using a JK flip-flop with $J = K$.

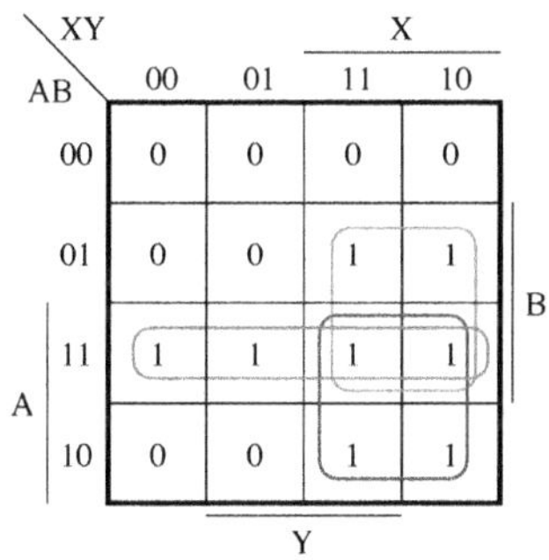

Figure 1.174. *Output Z*
$Z = A \cdot B + X \cdot A + X \cdot B$

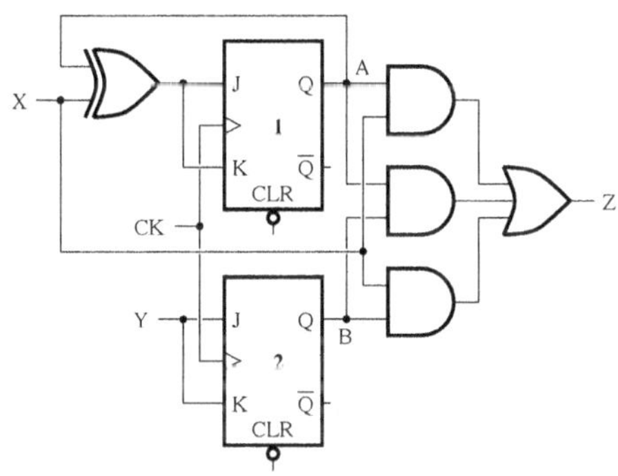

Figure 1.175. *Logic circuit*

SOLUTION 1.28.– (Median Filter).

A median filter can be described as a Moore state machine, the state diagram for which is represented in Figure 1.176. To detect each input bit with a value of 0 located between two bits with a value of 1, and which must be set to 1 at the output, the machine stores three bits, $Q_1Q_2Q_3$, at each clock pulse.

The state 101 is unused. But if the state machine enters this state, it will go to the initial state 000.

Table 1.154 presents the transition table of the median filter.

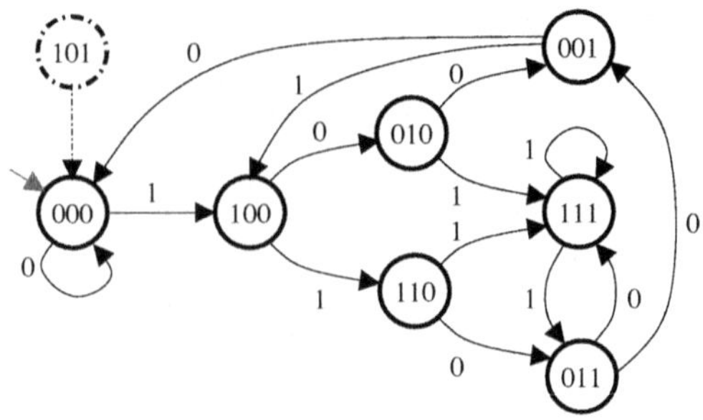

Figure 1.176. *State diagram*

Input	PS			NS		
X	Q_1	Q_2	Q_3	Q_1^+	Q_2^+	Q_3^+
0	0	0	0	0	0	0
0	0	0	1	0	0	0
0	0	1	0	0	0	1
0	0	1	1	0	0	1
0	1	0	0	0	1	0
0	1	0	1	0	0	0
0	1	1	0	0	1	1
0	1	1	1	0	1	1
1	0	0	0	1	0	0
1	0	0	1	1	0	0
1	0	1	0	1	1	1
1	0	1	1	1	1	1
1	1	0	0	1	1	0
1	1	0	1	0	0	0
1	1	1	0	1	1	1
1	1	1	1	1	1	1

Table 1.154. *Transition table*

The Karnaugh maps shown in Figures 1.177–1.179 can be constructed using the excitation table for the *D* flip-flop. The equations for the *D* inputs are given by:

$$D_1 = Q_1^+ = X \cdot \overline{Q_1} + X \cdot \overline{Q_3} + X \cdot Q_2 \quad [1.140]$$

$$D_2 = Q_2^+ = X \cdot Q_2 + Q_1 \cdot Q_2 + Q_1 \cdot \overline{Q_3} \quad [1.141]$$

and:

$$D_3 = Q_3^+ = Q_2 \qquad [1.142]$$

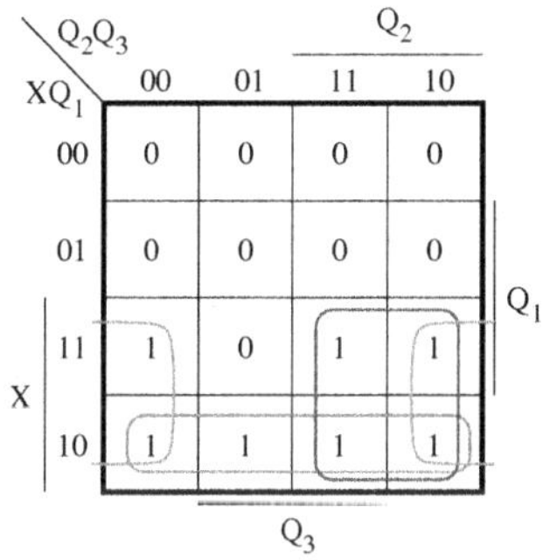

Figure 1.177. *Input* D_1
$D_1 = X \cdot \overline{Q_1} + X \cdot \overline{Q_3} + X \cdot Q_2$

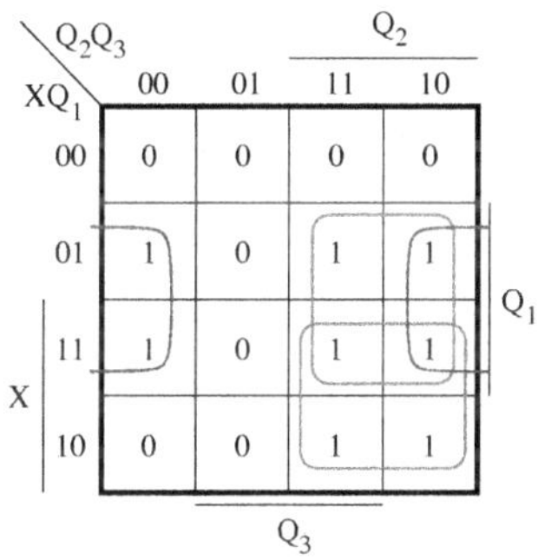

Figure 1.178. *Input* D_2
$D_2 = X \cdot Q_2 + Q_1 \cdot Q_2 + Q_1 \cdot \overline{Q_3}$

The logic circuit of the median filter is depicted in Figure 1.180, where X denotes the input and Y is the output.

SOLUTION 1.29.– (Bus Arbiter).

The operation of the bus arbiter can be described by the state diagram shown in Figure 1.181.

Using Gray code to represent the states, the transition table can be constructed as shown in Table 1.155, where S_3 corresponds to an unused state.

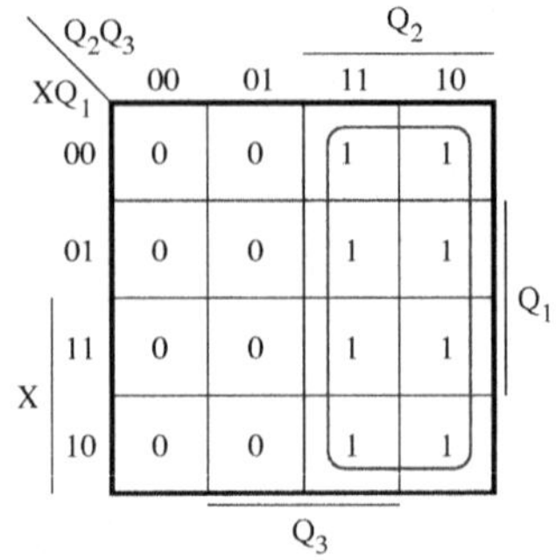

Figure 1.179. *Output D_3*

$D_3 = Q_2$

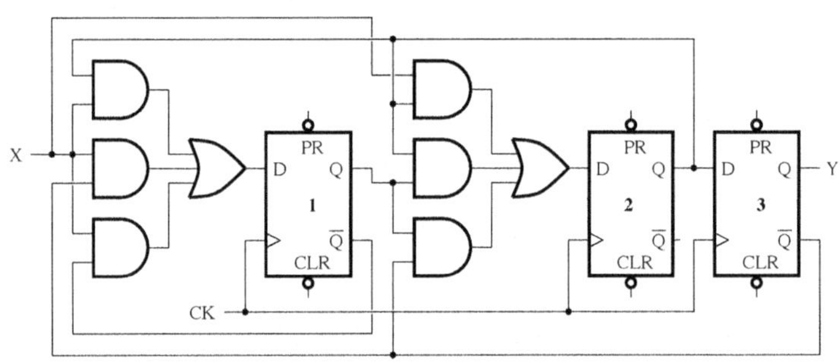

Figure 1.180. *Logic circuit of the median filter*

We can obtain the Karnaugh maps represented in Figures 1.182 and 1.183 using the excitation table of the *D* flip-flop. The logic equations for the flip-flop inputs can be written as follows:

$$D_1 = Q_1^+ = \overline{R_A} \cdot R_B + R_B \cdot Q_1 \qquad [1.143]$$

and:

$$D_2 = Q_2^+ = R_A + R_B \qquad [1.144]$$

The outputs are not dependent on the input signals, R_A and R_B. Based on the transition table, the following logic equations can be obtained:

$$G_A = \overline{Q_1} \cdot Q_2 \qquad [1.145]$$

and:

$$G_B = Q_1 \cdot Q_2 + Q_1 \cdot \overline{Q_2} = Q_1 \qquad [1.146]$$

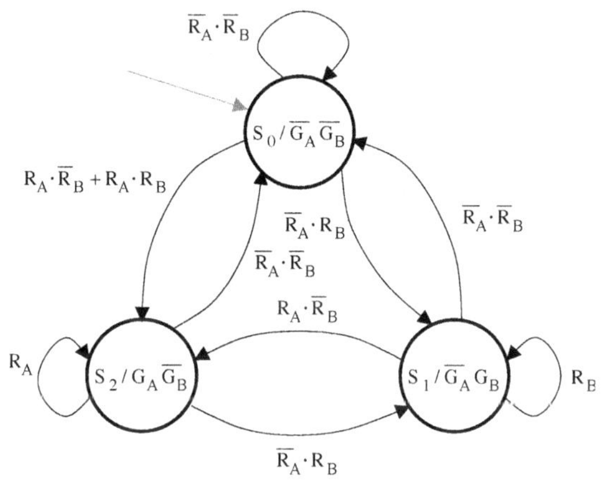

Figure 1.181. *State diagram*

PS		NS				Outputs	
	Q_1Q_2	$Q_1^+Q_2^+$				G_AG_B	
		$R_AR_B = 00$	01	10	11		
S_0	00	00	11	01	01	0	0
S_1	01	00	11	01	01	1	0
S_2	11	00	11	01	11	0	1
S_3	10	–	–	–	–	x	x

Table 1.155. *Transition table*

Figure 1.184 shows the logic circuit of the bus arbiter. Each flip-flop has an asynchronous reset input.

SOLUTION 1.30.– (Robot Ant).

The robot ant is equipped with two antennae, L and R, and servo motors that are controlled by a finite state machine. In order to come out of the labyrinth, the ant must move trying to keep the wall to its right after each control pulse.

The operation principle of the robot ant is illustrated in Figure 1.185.

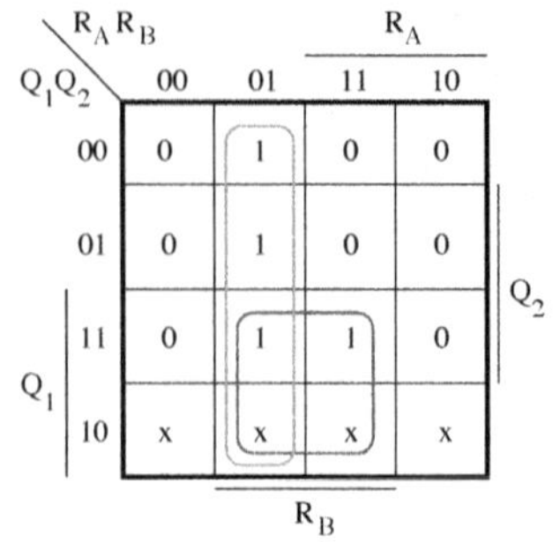

Figure 1.182. *Input D_1*

$D_1 = Q_1^+ = \overline{R_A} \cdot R_B + R_B \cdot Q_1$

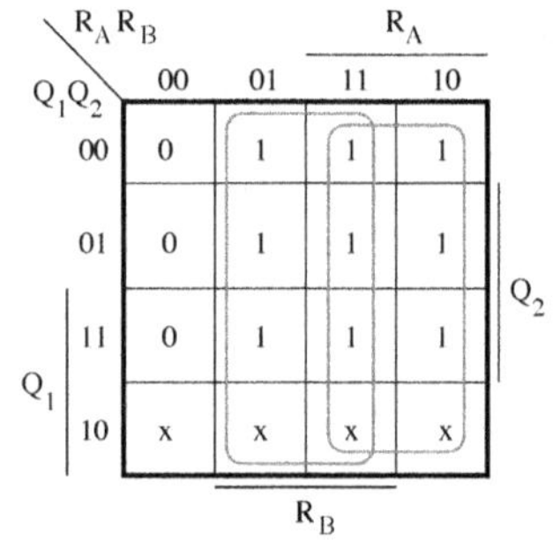

Figure 1.183. *Input D_2*

$D_2 = Q_2^+ = R_A + R_B$

The signals sent by the antennae correspond to the following situations:

– $LR = 00$: no contact with the wall;

– $LR = 01$: contact with the right wall;

– $LR = 10$: contact with the left wall;

– $LR = 11$: frontal contact with the wall.

The state table of the finite state machine is given in Table 1.156. It can be reduced to the form in Table 1.157 by noting that the states C and E are equivalent.

Assigning a binary code to each state, we can obtain the transition table as shown in Table 1.158.

Figure 1.186 presents the state diagram of the finite state machine.

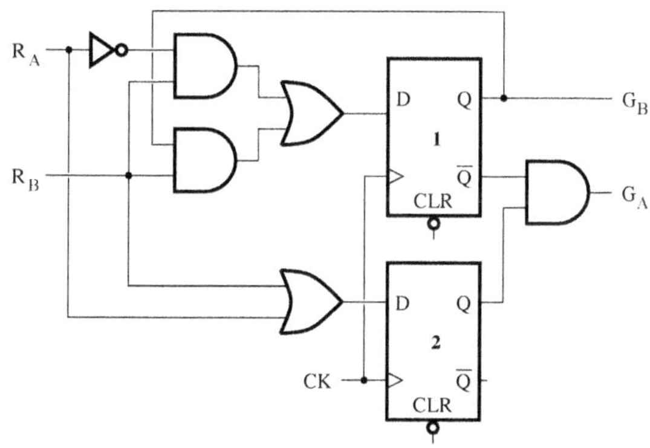

Figure 1.184. *Bus arbiter*

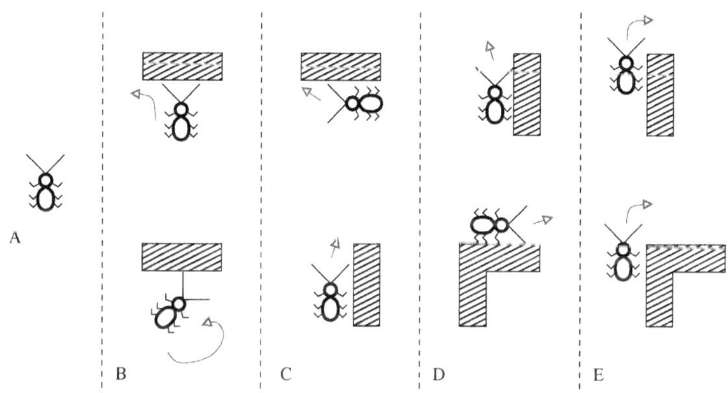

Figure 1.185. *Operation principle of the robot ant*

PS	NS				Outputs		
					TL	TR	F
	$LR = 00$	01	10	11			
A	A	B	B	B	0	0	1
B	C	B	B	B	1	0	0
C	E	D	E	D	0	1	1
D	C	D	B	B	1	0	1
E	E	D	E	D	0	1	1

Table 1.156. *State table*

PS	NS				Outputs		
					TL	TR	F
	$LR = 00$	01	10	11			
A	A	B	B	B	0	0	1
B	C	B	B	B	1	0	0
C	C	D	C	D	0	1	1
D	C	D	B	B	1	0	1

Table 1.157. *Reduced state table*

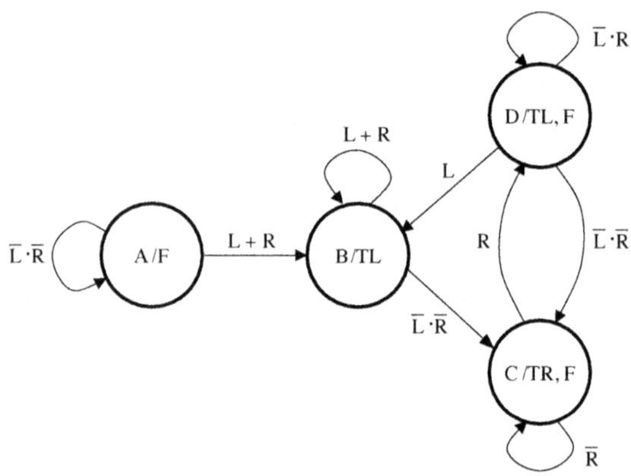

Figure 1.186. *State diagram*

PS		NS				Outputs		
	Q_1Q_2	$Q_1^+Q_2^+$				TL	TR	F
		$LR = 00$	01	10	11			
A	00	00	01	01	01	0	0	1
B	01	10	01	01	01	1	0	0
C	10	10	11	10	11	0	1	1
D	11	10	11	01	01	1	0	1

Table 1.158. *Transition table*

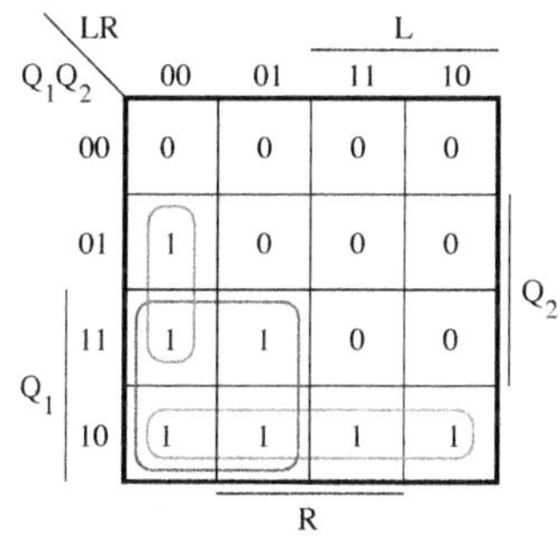

Figure 1.187. *Input D_1*
$D_1 = \overline{L} \cdot \overline{R} \cdot Q_2 + \overline{L} \cdot Q_1 + Q_1 \cdot \overline{Q_2}$

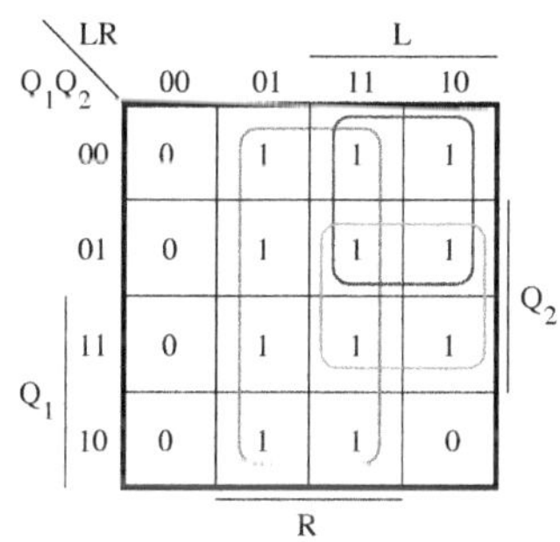

Figure 1.188. *Input D_2*
$D_2 = R + L \cdot Q_2 + L \cdot \overline{Q_1}$

The transition table and the excitation table of the D flip-flop can be used to construct the Karnaugh maps that are shown in Figures 1.187 and 1.188, and are required for the determination of the following logic equations for the flip-flop inputs:

$$D_1 = \overline{L} \cdot \overline{R} \cdot Q_2 + \overline{L} \cdot Q_1 + Q_1 \cdot \overline{Q_2} \quad [1.147]$$

and:

$$D_2 = R + L \cdot Q_2 + L \cdot \overline{Q_1} \quad [1.148]$$

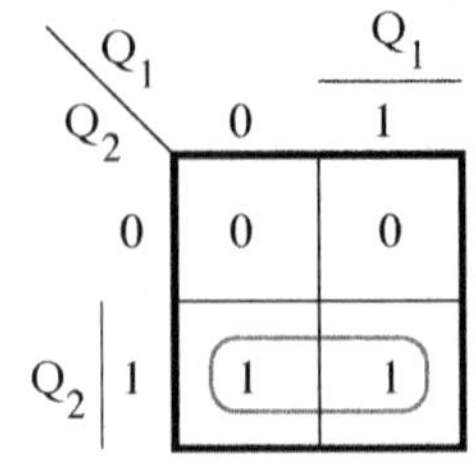

Figure 1.189. *Output TR*
$TR = Q_2$

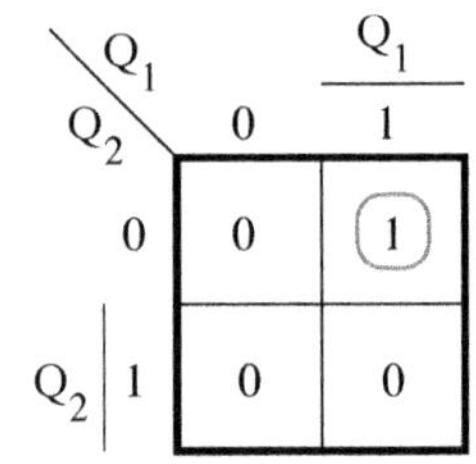

Figure 1.190. *Output TL*
$TL = Q_1 \cdot \overline{Q_2}$

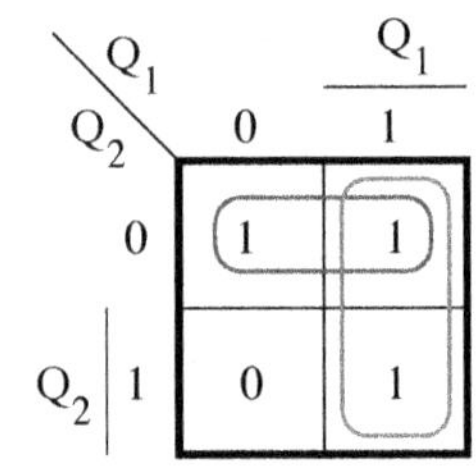

Figure 1.191. *Output F*
$F = Q_1 + \overline{Q_2}$

Figures 1.189–1.191 present the Karnaugh maps that are obtained from the transition tables and can be used to derive the following output equations:

$$\text{TR} = Q_2 \quad [1.149]$$

$$\text{TL} = Q_1 \cdot \overline{Q_2} \quad [1.150]$$

and:

$$F = Q_1 + \overline{Q_2} \qquad [1.151]$$

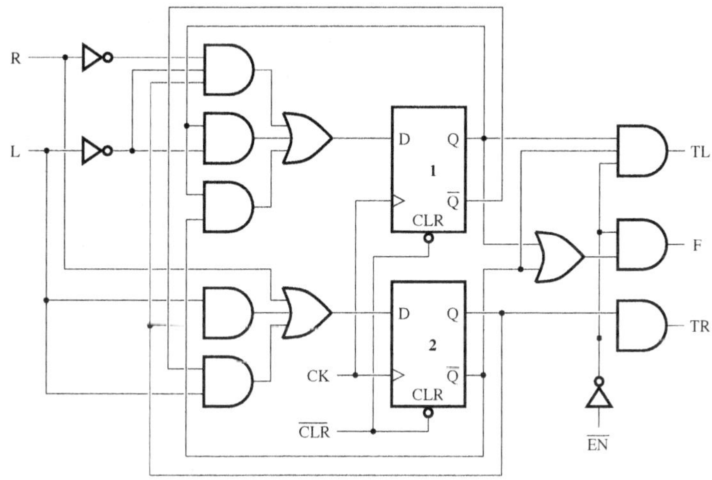

Figure 1.192. *Logic circuit (robot ant)*

Taking into account the signal $\overline{EN}$, the output equations can be put into the form:

$$TR = \overline{EN} \cdot Q_2 \qquad [1.152]$$

$$TL = \overline{EN} \cdot Q_1 \cdot \overline{Q_2} \qquad [1.153]$$

and:

$$F = \overline{EN}(Q_1 + \overline{Q_2}) \qquad [1.154]$$

The logic circuit of the finite state machine is represented in Figure 1.192.

2

Algorithmic State Machines

2.1. Introduction

In general, circuits that comprise combinational and sequential logic modules may be described as finite state machines.

Using a state diagram or state table may prove inadequate for complex digital systems. It is preferable to adopt algorithmic state machines (ASMs) when the number of inputs and outputs becomes large. The ASM can be used to describe the operation of both state machines based on Moore and Mealy models, as well as of systems that have the output characteristics of both Mealy and Moore models. It is also directly related to hardware implementation when the machine states are represented using a one-hot (or 1-out-of-n) code.

2.2. Structure of an ASM

An ASM is a finite state machine based on a flowchart that can be used to represent the transitions between states and outputs. Compared to a state diagram, this flowchart is based on the sequence of operations to be carried out rather than the sequence of states. It offers the advantage of not requiring the listing of all input conditions and the possible output combinations.

In general, a digital system can be subdivided into a data processing unit and control unit, as shown in Figure 2.1.

The control of combinational and sequential logic components such as adder, comparator, multiplexer, decoder, counter and register, which form the processing unit, ensures the synchronization of data operations. The control unit can be implemented as a finite state machine.

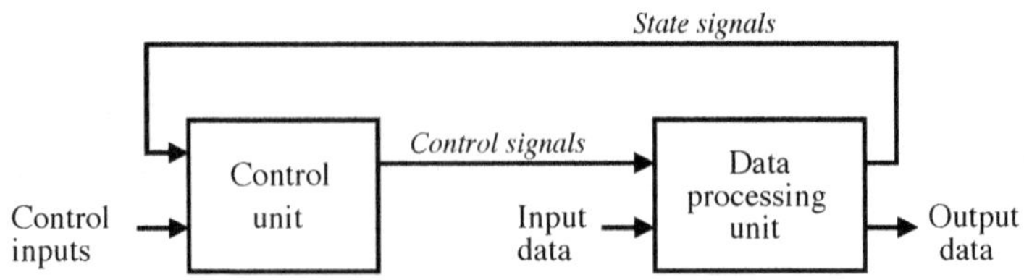

Figure 2.1. *Structure of a digital system*

2.3. ASM chart

An ASM chart is a graphical representation of the functional and hierarchical links that exist between the different operations carried out by the elements of a system. It consists of symbols representing the different types of operations as well as lines and arrows that indicate the relationships that exist between these operations. Figure 2.2 presents three types of symbols used in the construction of ASM charts:

1) a rectangle is required to represent the outputs that are not dependent on the input conditions, such as the flip-flop outputs;

2) a diamond or hexagon is associated with a condition that governs one or several inputs and that modifies the execution of the operations depending on whether or not this condition is satisfied. It should be noted that the different output possibilities, a maximum of 2^N possibilities if there are N input variables, must be mutually exclusive;

3) a rectangle with rounded corners is used to yield the conditional outputs or the outputs that are dependent on input combinations. Most often, this rectangle precedes a diamond or a hexagon, which specifies the conditions that are required to generate these outputs.

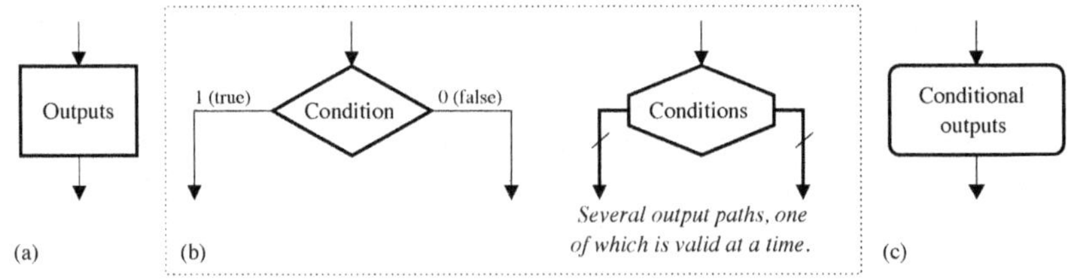

Figure 2.2. *Symbols representing a) the outputs, b) the decisions and c) the conditional outputs*

It should be noted that only the output variables whose logic state changes appear in the rectangles of an ASM chart.

ASMs are most often implemented using one-hot (or 1-out-of-n) encoding to represent states. Assuming that only one of the state variables takes the value 1 and that all the others are set to 0 at each instant, it follows that one flip-flop is required for each state. The implementation of a machine with N states then requires N flip-flops. Consequently, the logic circuit for an ASM can include several flip-flops, but only a small number of logic gates, and is easy to implement. It can be directly derived from an ASM chart. Figure 2.3 depicts some symbols and the corresponding logic circuits. A D flip-flop can be used to represent a state and a demultiplexer can be used to realize a condition. The intersection of the ASM chart paths corresponds to an OR logic gate. Two consecutive states are implemented by connecting two D flip-flops in series, as shown in Figure 2.4.

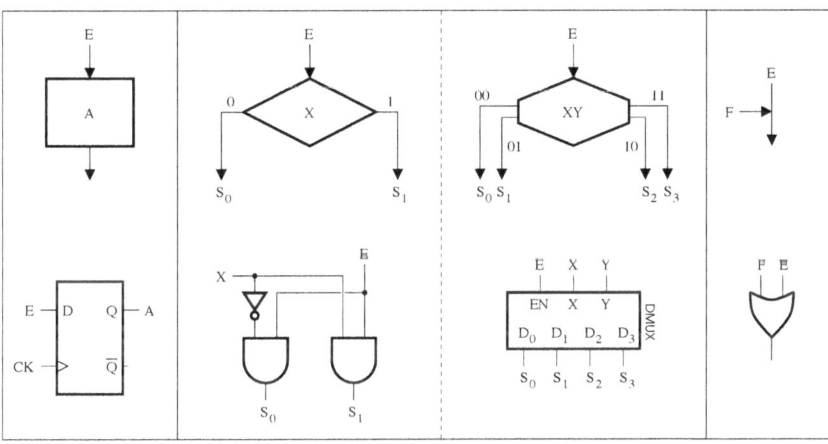

Figure 2.3. *Symbols and corresponding logic circuits*

The two methods used to represent finite state machines, ASM charts and state diagrams, are equivalent and interchangeable. An ASM chart can, thus, be converted to a state diagram and vice versa.

Each rectangle in an ASM chart contains only those outputs or operations that can be carried out in the same clock period and thus corresponds to a state of the state diagram.

The input conditions that allow for transitions from one state to another in a state diagram can be determined from an ASM chart by following the possible paths through the decision symbols located between two rectangles.

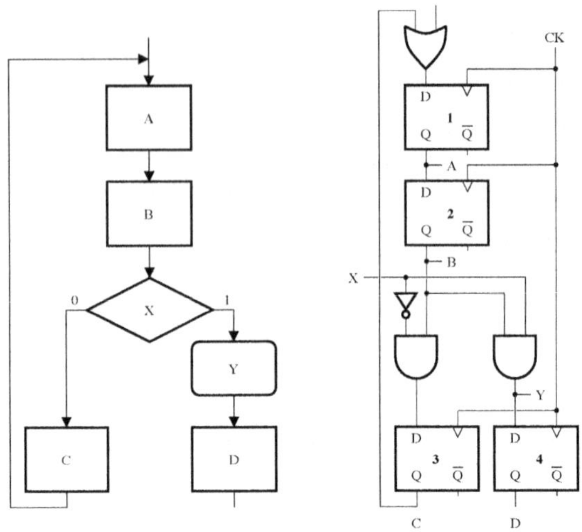

Figure 2.4. *Section of an ASM chart and the corresponding logic circuit*

Each conditional output that is found between two rectangles is enabled upon reaching the state associated with the rectangle that precedes this output symbol, and the necessary conditions are satisfied.

The conversion of a state diagram to an ASM chart is illustrated in Figure 2.5 for a Mealy machine and Figure 2.6 for a Moore machine.

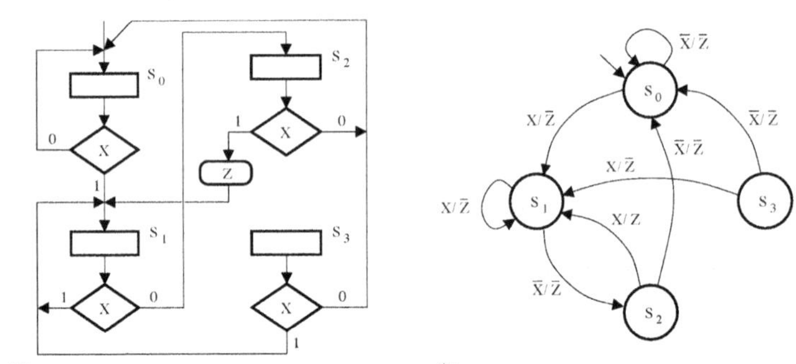

Figure 2.5. *a) ASM chart for a Mealy machine; b) state diagram*

In the case of a Mealy machine, the ASM chart contains symbols for the conditional outputs, while for a Moore machine, the ASM chart has no symbols for conditional outputs.

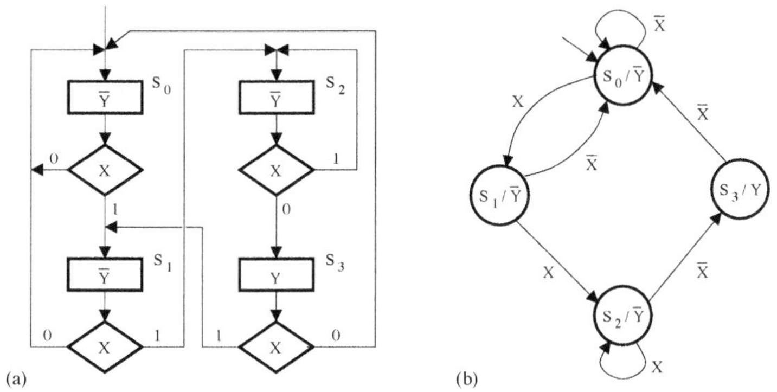

Figure 2.6. *a) ASM chart for a Moore machine; b) state diagram*

NOTE.– The ASM chart does not allow a feedback loop that encompasses only one symbol, as shown in Figure 2.7(a). The correct representation shown in Figure 2.7(b) is obtained by inserting a state S in the loop.

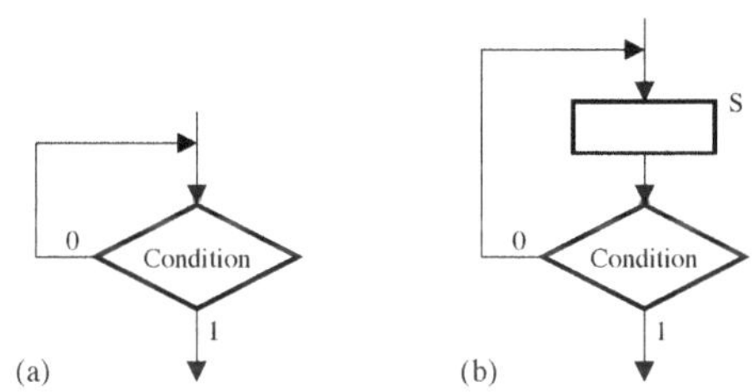

Figure 2.7. *a) Incorrect and b) correct representations of a loop in an ASM chart*

The ASM chart section in Figure 2.8(a) does not comply with the rule stating that only one output path must be valid at a time. When the inputs X and Y take the same state, the two valid paths can lead to the next states, which are distinct.

Multiple parallel paths can be valid only if they lead to the same output node (or next state) as shown in Figure 2.8(b), where the outputs P and Q are assumed to be independent.

For the version of the ASM chart section represented in Figure 2.8(c), only one output path is valid for each combination of inputs X and Y.

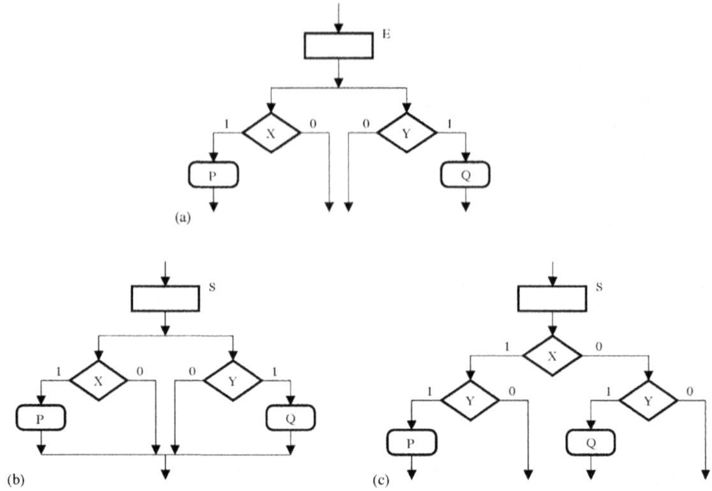

Figure 2.8. *Examples of ASM chart sections*

The ASM chart sections shown in Figure 2.9(a) are equivalent, as are those in Figure 2.9(b). The order in which the variables appear between two states may be inverted.

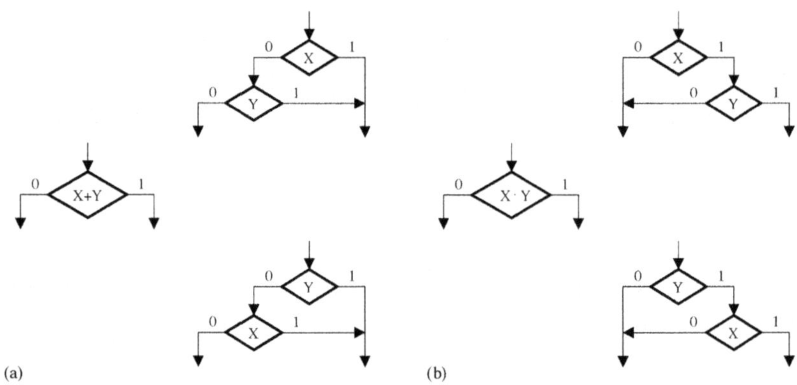

Figure 2.9. *Representations of equivalent ASM chart sections*

In practice, each ASM must have an input that can be used to initialize it to a known state, and a mechanism to prevent its operation from being disturbed by the unused states.

2.4. Applications

ASMs find applications in various fields such as the design of circuits and systems, and process modeling and control.

2.4.1. *Serial adder/subtracter*

Serial arithmetic operations are generally performed on data stored in shift registers. This approach offers the advantage of requiring only a small chip area, but it is limited by low speed as the execution of an operation requires the same number of clock signal pulses as the number of bits in the representation of each operand.

A serial adder is a sequential circuit that generates one bit of the sum at a time, while the logic state of the carry out bit is memorized in a flip-flop for the addition of the next bit. The operands A and B are initially stored in shift registers and, beginning with the least significant bit, are shifted by one position on each clock pulse.

A bit-serial adder can be designed using the full adder whose truth table is given in Table 2.1.

A	B	C_i	S	C_{i+1}
0	0	0	0	0
0	0	1	1	0
0	1	0	1	0
0	1	1	0	1
1	0	0	1	0
1	0	1	0	1
1	1	0	0	1
1	1	1	1	1

Table 2.1. *Truth table of a full adder*

Analysis shows that it is possible to divide the truth table into two parts, one with 0 being the carry out and the other with 1 being the carry out. The bit-serial adder can then be described using a state table, as shown in Table 2.2, where each state is represented by a binary code. Figures 2.10 and 2.11 present the Karnaugh maps constructed based on the state table. The logic equations can, thus, be obtained as follows:

$$Q^+ = A_i \cdot B_i + A_i \cdot Q + B_i \cdot Q \qquad [2.1]$$

and:

$$S_i = A_i \oplus B_i \oplus Q \qquad [2.2]$$

where $Q = C_i$ and $Q^+ = C_{i+1}$. Another version of the state table can be drawn up by denoting the state with 0 as the carry out by S_0, and the state with 1 as the carry out by S_1. This table is represented in Table 2.3. The state diagram of the bit-serial adder is shown in Figure 2.12(a). Using a D flip-flop to represent the two states of a bit-serial adder, we can obtain the logic circuit depicted in Figure 2.12(b).

PS	NS				Output S_i			
	$A_iB_i = 00$	01	10	11	$A_iB_i = 00$	01	10	11
0	0	0	0	1	0	1	1	0
1	0	1	1	1	1	0	0	1

Table 2.2. *State table of the serial adder with encoded states*

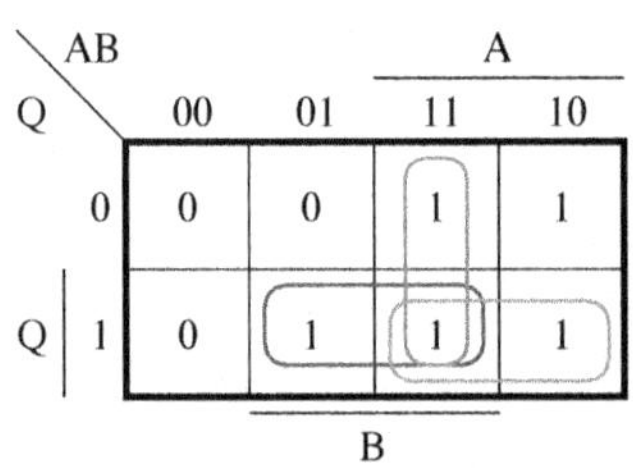

Figure 2.10. *Karnaugh map for the next state Q^+*

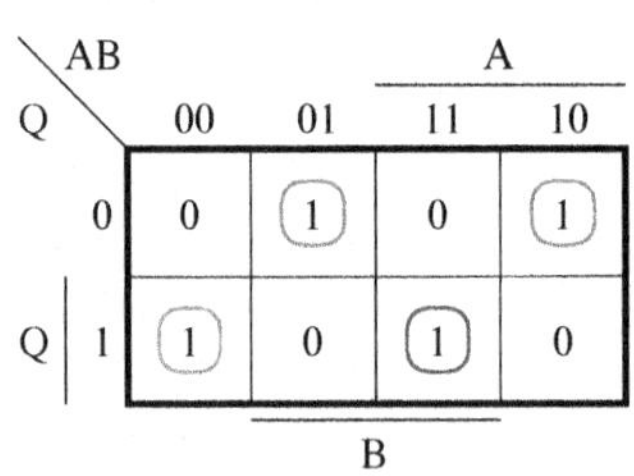

Figure 2.11. *Karnaugh map for the output S*

A serial adder/subtracter, as shown in Figure 2.13, includes a datapath and a control unit. The result for a given input data sequence is generated by applying the appropriate control signals to the different components of the datapath. The Start signal takes the logic level 1 to indicate the beginning of an operation. The $\overline{\text{Add}}$/Sub signal is set either to 0, for an addition, or to 1, for subtraction. The Reset input can be used to asynchronously reset the control circuit.

PS	NS				Output S_i			
	$A_iB_i = 00$	01	10	11	$A_iB_i = 00$	01	10	11
S_0	S_0	S_0	S_0	S_1	0	1	1	0
S_1	S_0	S_1	S_1	S_1	1	0	0	1

Table 2.3. *State table of the serial adder*

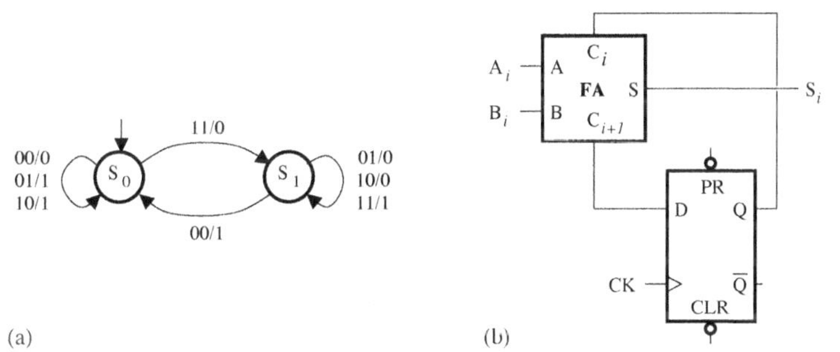

Figure 2.12. *Bit-serial adder: a) state diagram; b) logic circuit*

The datapath is made up of two shift registers, a full adder, a *D* flip-flop and an XOR logic gate (or programmable inverter). The control unit consists of a down counter and a control circuit that operates as a finite state machine.

The ASM chart shown in Figure 2.14, where $N = 4$, describes the operation of the control unit. The control unit is initially at the state S_0 and the shift registers the down counter, and the *D* flip-flop are deactivated by the logic level 0 of the Enable signal.

When the Start signal takes the logic state 1, the LoadR signal is set to 1 and the operands can be loaded into the shift registers. Meanwhile, all the flip-flops of the down counter are initialized to 1 when the $\overline{\text{PrCNT}}$ is set to 0, and the control unit then moves to the state S_1. It should be noted that the two signals, LoadR and $\overline{\text{PrCNT}}$, are logical complements. The carry in and the Compl signal are then specified depending on the logic state of the input signal, $\overline{\text{Add}}$/sub. The control unit only exits the state S_1 when the Start signal is reset.

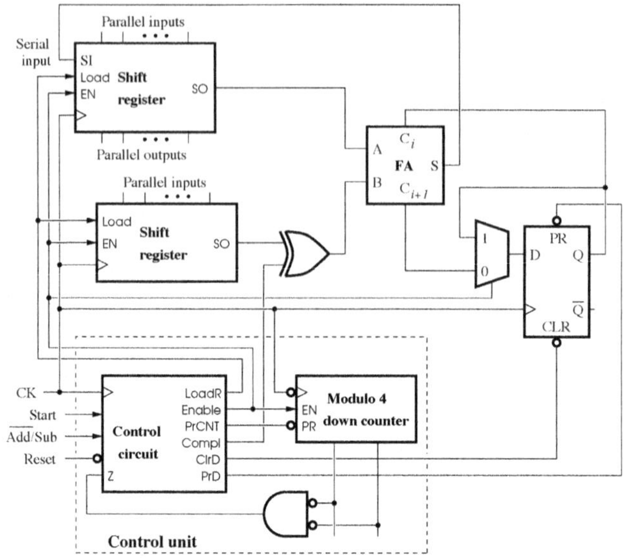

Figure 2.13. *Datapath and control unit for a serial adder/subtracter*

After the control unit goes to the state S_2, the shift registers, the down counter and the *D* flip-flop are activated by setting the Enable signal to 1. The operands can, thus, be applied to the full adder/subtracter bit by bit, beginning with the least significant bit; the down counting phase then begins from $N - 1$, where N equals 4 and corresponds to the number of bits in each operand.

One bit of the result is generated and transferred to register A on each clock signal pulse, while the carry out bit is stored in the *D* flip-flop. The execution of these different operations continues in the same way for the other bits of the operands and is only stopped when the signal Z takes the logic state 1 to indicate the detection of a zero generated by the down counter.

The control unit goes back to the state S_0 and waits for the Start signal to be set to 1 again, indicating the beginning of a new arithmetic operation.

The state table of the control unit is shown in Table 2.4, where the binary codes 00, 01 and 11 are assigned to the states S_0, S_1 and S_2, respectively. As the control unit has three states, a minimum of two flip-flops are required to encode these states. Using the excitation table for the JK flip-flop, the transition table can be constructed as shown in Table 2.5. When the state represented by the binary code 10 is considered as a don't-care state, the Karnaugh maps depicted in Figure 2.15 can be used to simplify the logic functions required for the implementation of the control circuit. The logic expressions for the J and K inputs are given by:

$$J_A = B \cdot \overline{Start} \qquad [2.3]$$

$$K_A = Z \qquad [2.4]$$

$$J_B = Start \qquad [2.5]$$

and:

$$K_B = A \cdot Z \qquad [2.6]$$

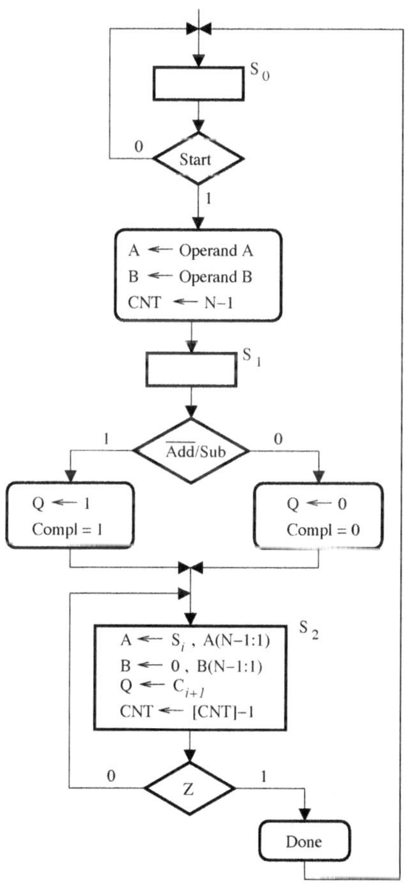

Figure 2.14. *ASM chart of the control unit* ($N = 4$)

PS	Inputs		NS	Outputs					
$A\,B$	Start	Z	A^+B^+	LoadR	$ClrD$	PrD	Compl	Enable	Done
0 0	0		0 0	0	0	0	0	0	0
0 0	1		0 1	0	0	0	0	0	0
0 1	0		1 1	1	$\overline{\overline{Add}/Sub}$	$\overline{Add}/Sub$	$\overline{Add}/Sub$	0	0
0 1	1		0 1	1	$\overline{\overline{Add}/Sub}$	$\overline{Add}/Sub$	$\overline{Add}/Sub$	0	0
1 1		0	1 1	0	0	0	0	1	0
1 1		1	0 0	0	0	0	0	1	1

Table 2.4. *State table of the control unit*

PS	Inputs		NS	Inputs J and K			
$A\,B$	Start	Z	A^+B^+	J_A	K_A	J_B	K_B
0 0	0		0 0	0	x	0	x
0 0	1		0 1	0	x	1	x
0 1	0		1 1	1	x	x	0
0 1	1		0 1	0	x	x	0
1 1		0	1 1	x	0	x	0
1 1		1	0 0	x	1	x	1

Table 2.5. *Transition table*

The outputs of the control circuit can be obtained as follows:

$$LoadR = \overline{A} \cdot B \qquad [2.7]$$

$$PrCNT = \overline{LoadR} \qquad [2.8]$$

$$Enable = A \qquad [2.9]$$

$$ClrD = \overline{A} \cdot B \cdot \overline{\overline{Add}/Sub} \qquad [2.10]$$

$$PrD = \overline{A} \cdot B \cdot \overline{Add}/Sub \qquad [2.11]$$

and:

$$Compl = A \cdot \overline{Add}/Sub \qquad [2.12]$$

The control circuit is represented in Figure 2.16. Some outputs are active low, while others are active high.

Figure 2.17 presents the timing diagram of the serial adder/subtracter when the $\overline{Add}/Sub$ signal is set to 1. It should be noted that if the shift registers and D flip-flop are activated by the rising edge of the clock signal, and the down counter by the falling edge of the clock signal, the count sequence only starts after each bit-serial arithmetic operation.

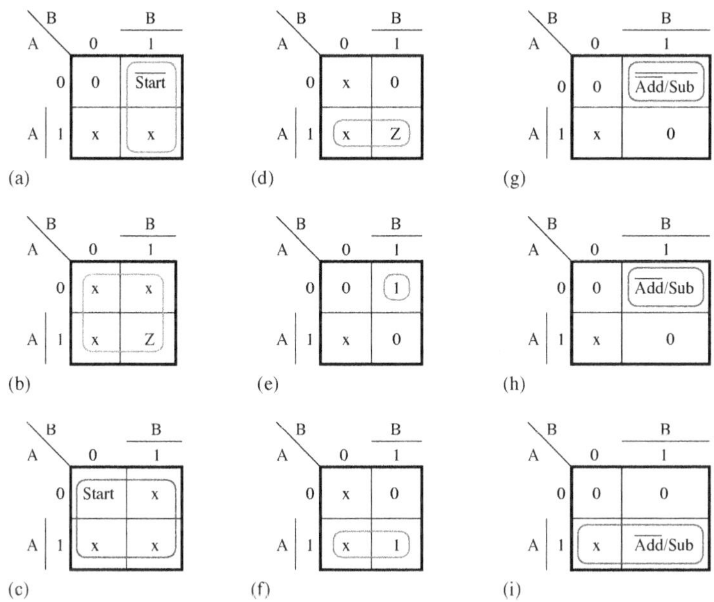

Figure 2.15. *Karnaugh maps: a) J_A; b) K_A; c) J_B; d) $K_B = Done$; e) $LoadR$ $(PrCNT = \overline{LoadR})$; f) Enable; g) $ClrD$; h) PrD; i) Compl*

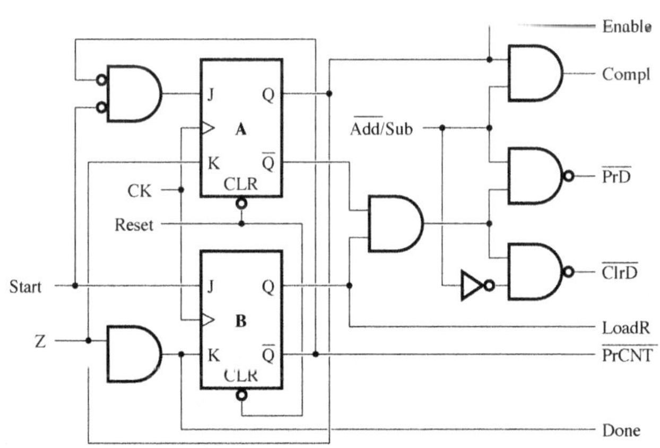

Figure 2.16. *Control circuit*

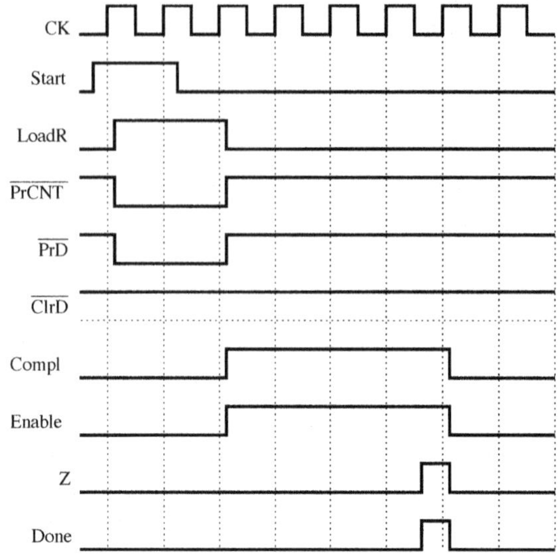

Figure 2.17. *Timing diagram of the serial adder/subtracter ($\overline{Add}/sub = 1$)*

Table 2.6 summarizes the control signals with the corresponding operations and the components involved.

Control signal	Operation	Component
LoadR	$A \leftarrow$ Operand A	Register A
	$B \leftarrow$ Operand B	Register B
$\overline{\text{PrCNT}}$	$CNT \leftarrow N - 1$	CNT down counter
$\overline{\text{ClrD}}$	$Q \leftarrow 0$	Flip-flop D
$\overline{\text{PrD}}$	$Q \leftarrow 1$	Flip-flop D
Compl	Complement signal	
Enable	$A \leftarrow S_i, A(N-1:1)$	Register A
	$B \leftarrow 0, B(N-1:1)$	Register B
	$CNT \leftarrow [CNT] - 1$	CNT down counter
Done	End of addition/subtraction	–

Table 2.6. *Serial adder/subtracter: summary of the control signals with the corresponding operations and the components involved*

The design approach based on a one-hot (or 1-out-of-n) code and the use of D flip-flops leads to another architecture of the control circuit, which is directly linked to the ASM chart. In this case, the logic equations for the *D* inputs of the flip-flops are given by:

$$D_A = A \cdot \overline{Start} + C \cdot Z \quad [2.13]$$

$$D_B = A \cdot Start + B \cdot Start \quad [2.14]$$

$$D_C = B \cdot \overline{Start} + C \cdot \overline{Z} \quad [2.15]$$

The logic equations of the outputs can be written as:

$$LoadR = A \quad [2.16]$$

$$PrCNT = \overline{LoadR} \quad [2.17]$$

$$ClrD = B \cdot \overline{\overline{Add}/Sub} \quad [2.18]$$

$$PrD = B \cdot \overline{Add}/Sub \quad [2.19]$$

$$Compl = B \cdot \overline{Add}/Sub \quad [2.20]$$

$$Enable = C \quad [2.21]$$

and:

$$Done = C \cdot Z \quad [2.22]$$

As one flip-flop is allocated to each state, an increase in the size of the final circuit is generally expected. However, the operation of the final circuit is not affected by any critical race condition.

2.4.2. *Multiplier based on addition and shift operations*

A multiplier is an important component in the arithmetic unit of a microprocessor. It can be implemented either as a combinational logic circuit or as a sequential logic circuit, which generally has the advantage of being smaller in size, but is slower.

A multiplier for unsigned four-bit numbers is shown in Figure 2.18. It is based on addition and shift operations.

The processing unit includes a *D* flip-flop, registers (Q, A and B) and an adder. Initially, the *Init* signal allows the storage of the multiplier and multiplicand in the registers Q and B, respectively, while the *D* flip-flop is reset by the *ClearC* signal. For each addition, the sum, S, is placed in register A following a parallel transfer, and the carry, C, is placed in the *D* flip-flop using the inputs enabled by the *AddA* signal. The *RshiftDec* signal initiates the transfer of the carry, C, stored in the *D* flip-flop to register A, whose content is shifted to the right in order to be transferred to register Q. Each shift results in the loss of the least significant bit of the register Q. Thus, the control unit can successively access each multiplier bit starting from the position Q_0 of the least significant bit of the register Q.

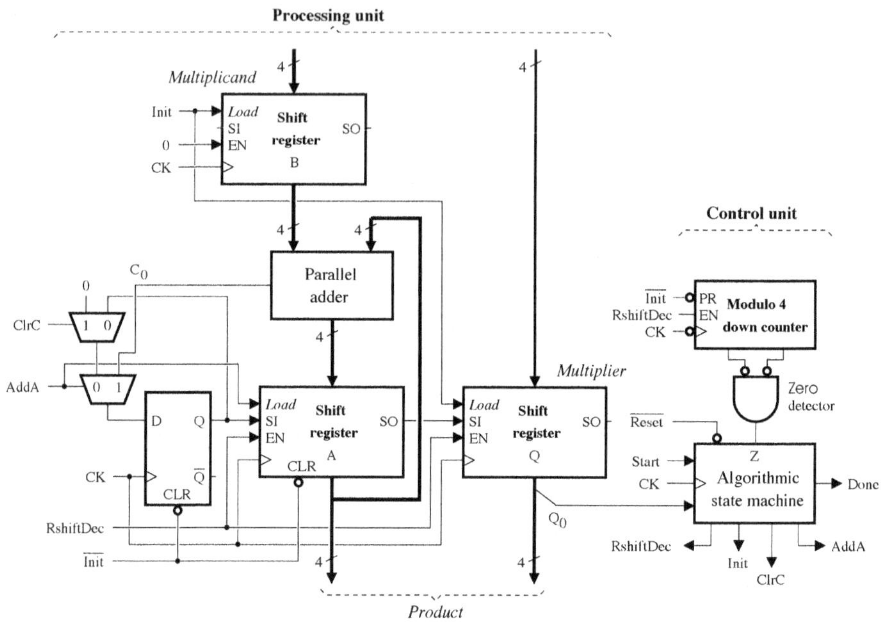

Figure 2.18. *Four-bit multiplier based on addition and shift operations*

A modulo N down counter, CNT (where N is the number of bits of the multiplier or multiplicand that is equal to 4), is used to monitor the change in the number of iterations. After being initialized by the *Init* signal, it is successively triggered by the *RshiftDec* signal to count in a cyclic manner from 3 to 0. The signal Z takes the logic state 1 as soon as the counter reaches 0.

The control unit uses the logic state of the *Start* signal, the bit Z, and the least significant bit, Q_0, of the register Q to determine the control signals for the different components that make up the datapath. For each bit of the multiplier, it determines, at each iteration or step, whether to execute an addition followed by a shift operation or only a shift operation. The product is contained in registers A and Q (the most significant bits being in register A), and all the multiplier bits are lost at the end of the process.

The control unit comprises an ASM, the ASM chart of which is shown in Figure 2.19. The multiplication process is initiated by setting the *Start* signal to 1, and three states, S_0, S_1 and S_2, are needed to generate all the signals required to control the different operations. Once the multiplication is completed, the *Done* signal is set to 1. Table 2.7 summarizes the control signals with the corresponding operations and the components involved. It should be noted that the *Done* signal, which indicates the end of a multiplication operation, does not affect the state of any component.

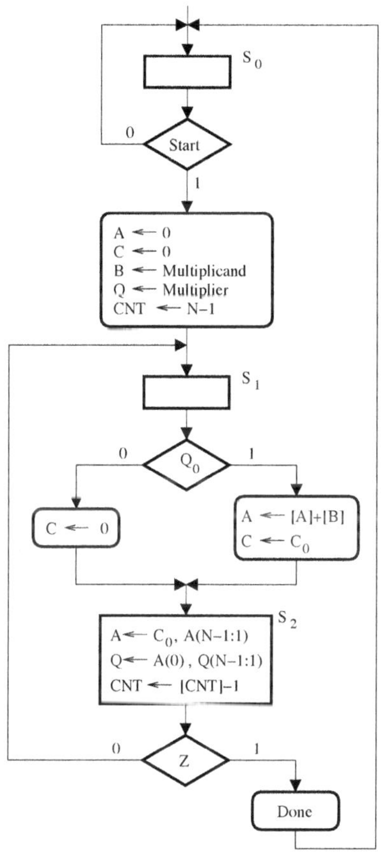

Figure 2.19. *ASM chart of the control unit* ($N = 4$)

An example of a multiplication of two unsigned binary numbers is illustrated in Figure 2.20(a). Upon a multiplication by 0 or 1, each partial product is either a copy of the multiplicand that is shifted by an appropriate number of bits, or zero. As the product of two unsigned N-bit numbers (with $N = 4$) has $2N$ bits, it can only be generated using a $2N$-bit adder.

1100	Multiplicand
x 1011	Multiplier
1100	First partial product
1100	Second partial product
0000	Third partial product
+ 1100	Fourth partial product
10000100	Product

(a)

1100	Initially, register B contains the multiplicand.
x 1011	Initially, register Q contains the multiplier.
0000	Initially, register A is reset.
+ 1100	First bit of the multiplier = 1; A ← [A] + [B]
1100	Content of the register A = Sum
0110**0**	Right shift of the sum by one bit, [A] = 0110, [Q] = 0101
+ 1100	Second bit of the multiplier = 1; A ← [A] + [B]
1 0010**0**	Content of the register A = Sum; Carry C = 1
1001 **00**	Right shift of the carry and sum by one bit, [A] = 1001, [Q] = 0010
0100 **100**	Third bit of the multiplier = 0; right shift by one bit, [A] = 0100, [Q] = 1001
+ 1100	Fourth bit of the multiplier = 1; A ← [A] + [B]
1 0000 **100**	Content of the register A = Sum; Carry C = 1
1000 **0100**	Right shift of the carry and sum by one bit, [A] = 1000, [Q] = 0100
	The multiplier is loss and the product is contained in registers A and Q.

(b)

Figure 2.20. *a) Multiplication of two unsigned binary numbers; b) operation principle of the multiplier*

Control signal	Operation	Component
Init	$B \leftarrow$ Multiplicand	Register B
	$Q \leftarrow$ Multiplier	Register Q
$\overline{Init}$	$A \leftarrow 0$	Register A
	$CNT \leftarrow N - 1$	CNT down counter
$\overline{ClrC}$	$C \leftarrow 0$	Flip-flop *D*
AddA	$A \leftarrow [A] + [B]$	Register A
	$C \leftarrow C_0$	Flip-flop *D*
RshiftDec	$A \leftarrow C_0, A(N-1:1)$	Register A
	$Q \leftarrow A(0), Q(N-1:1)$	Register Q
	$CNT \leftarrow [CNT] - 1$	CNT counter
Done	End of the multiplication	–

Table 2.7. *Multiplier: summary of the control signals with the corresponding operations and the components involved*

The working principle of a 4-bit multiplier based on addition and shift operations is illustrated in Figure 2.20(b). Considering each bit of the multiplier from right to left, the repetitive execution of the addition operation on the most significant bits of the partial products and the right-shift operation can provide the product of two unsigned binary numbers. An addition followed by a right-shift operation is required for a bit at the logic state 1, while only a right-shift operation is required for a bit at the logic state 0. Thus, the implementation of a multiplier for unsigned N-bit numbers ($N = 4$) only requires an N-bit adder.

2.4.3. *Divider based on subtraction and shift operations*

The division of unsigned binary numbers can be carried out using several types of structures. For applications that require a small circuit size, it is most often implemented as a sequential circuit based on subtraction and shift operations.

As an example, Figure 2.21 presents a logic circuit that, from an eight bit dividend and a four-bit divider, provides a quotient of four bits. The divider is based on a configuration that reduces the number of registers required to store data. Initially, the dividend is found in the five-bit register A and the four-bit register Q, and the divisor is placed in the four-bit register B.

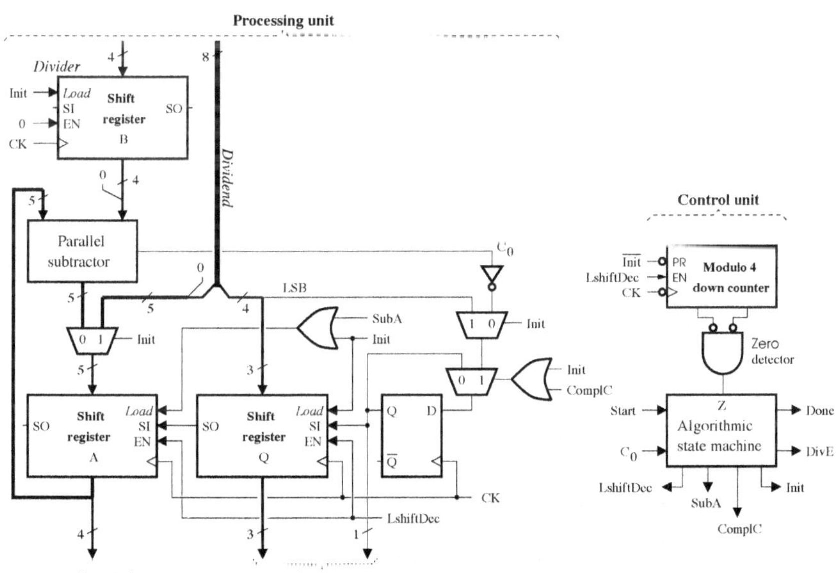

Figure 2.21. *A divider based on subtraction and shift operations*

During the division process, instead of shifting the divisor to the right before each subtraction, the dividend is rather shifted to the left. An additional bit is, therefore, required at the extreme left of register A to prevent the anticipated loss of the most significant bit of the dividend. As the dividend is shifted to the left, it is replaced by the quotient, which is stored bit by bit from the right end of the register Q.

The control unit consists of a binary counter, CNT, and an ASM.

The CNT counter has a modulo equal to the number of clock cycles required for the execution of a division, that is N, $2N$ being the number of the bits of the dividend and N being the number of bits of the divisor. It is followed by a zero detector whose output bit, Z, is connected to the ASM.

Based on the logic state of the *Start* signal, and the bits Z and C_0, the ASM produces control signals that can be used to sequence the different cycles required for a division. The operation of this machine is based on the ASM chart shown in Figure 2.22. Each division begins by setting the signal *Start* to 1. The dividend is then placed in the registers A and Q, the divisor in the register B and the number, N, in the CNT counter.

To determine whether the division is possible, the content of the register A is subtracted from that of the register B. If the carry out is equal to 0, the *DivE* signal takes the logic state 1 to indicate that the quotient is undefined or has a number of bits greater to the word length of the register Q. On the other hand, if the carry out is equal to 1, we proceed to the step of computing the quotient and the remainder.

Before subtracting the divisor from the dividend, the dividend is shifted by one bit to the left. If the result is greater than or equal to 0, the carry out takes the value 0 and the value of the dividend is updated. The corresponding quotient bit is set to 1. If, on the other hand, the result is less than 0, the carry out takes the value 1 and the value of the dividend is not modified. The corresponding quotient bit is set to 0. The process of shifting and subtracting is reiterated to determine the other quotient bits and ends when the signal Z assumes the logic state 1. It is based on a division algorithm with restoration of the dividend.

Table 2.8 provides a summary of the control signals with the corresponding operations and the components involved. The *Load* signal is used to initialize the registers. The result of the subtraction is stored only when the *SubA* signal, which enables the parallel inputs of the register A, is set to 1. Each quotient bit is obtained as the logical complement of the carry out and is loaded in the D flip-flop by setting the *ComplC* signal to 1. It is then applied to the serial input of the register Q. The logic state of the *LShiftDec* signal determines whether the contents of the registers A and Q are to be shifted left and the down counter, CNT, must be decremented.

The division of two binary numbers is performed using successive subtraction and shift operations, as in the example shown in Figure 2.23(a). The operating principle of the divider is illustrated in Figure 2.23(b). It should be noted that each shift results in the loss of the left-most bit of the register A : Q. At the end of the process, the quotient is in the register Q, and the remainder in register A.

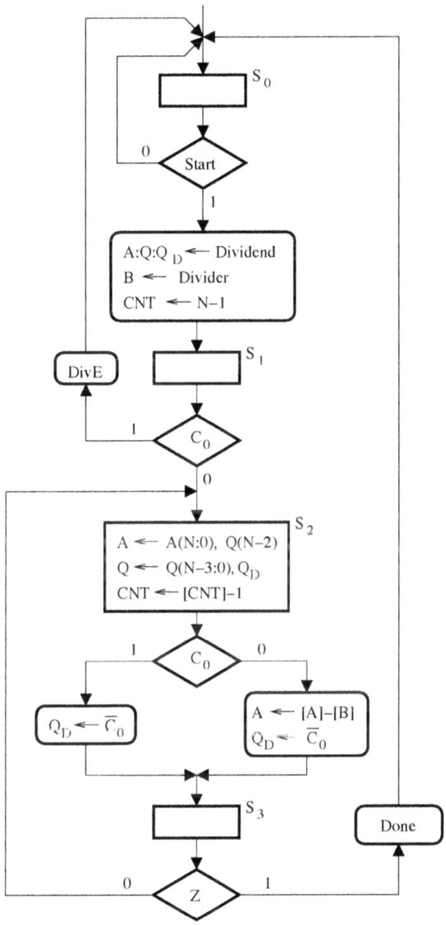

Figure 2.22. *ASM chart of the control unit* ($N = 4$)

2.4.4. *Controller for an automatic vending machine*

An automatic vending machine is a machine that can provide different products (drinks, sandwiches, biscuits, tissues, etc.) in exchange for coins.

Control signal	Operation	Component
Init	$A : Q : Q_D \leftarrow$ Dividend	Registers A and Q, *D* flip-flop
	$B \leftarrow$ Divider	Register B
$\overline{Init}$	$CNT \leftarrow N - 1$	CNT counter
SubA	$A \leftarrow [A] - [B]$	Register A
ComplC	$Q_D \leftarrow \overline{C_0}$	*D* flip-flop
LshiftDec	$A \leftarrow A(N : 0), Q(N - 2)$	Register A
	$Q \leftarrow Q(N - 3 : 0), Q_D$	Register Q
	$CNT \leftarrow [CNT] - 1$	CNT down counter
DivE	Error message	–
Done	End of division	–

Table 2.8. *Divider: summary of the control signals with the corresponding operations and the components involved*

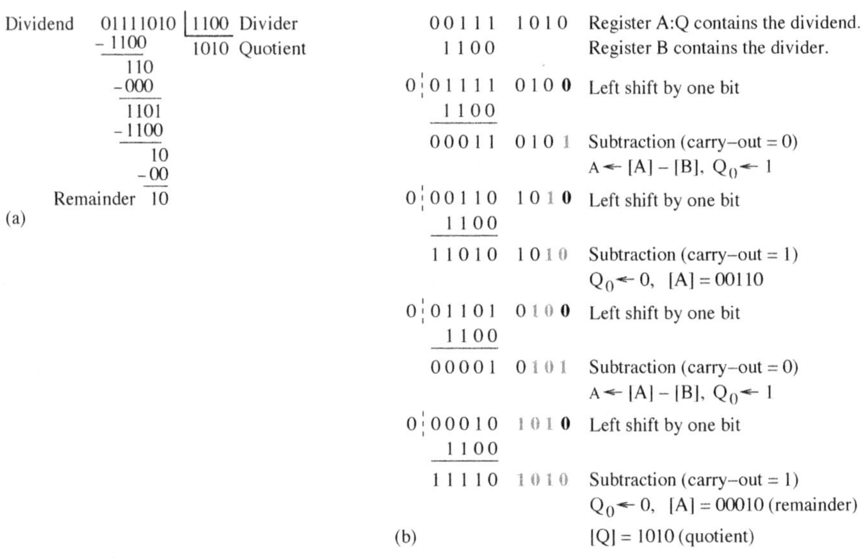

Figure 2.23. *a) Division of two unsigned binary numbers; b) operation principle of the divider*

The vending machine shown in Figure 2.24 consists of a slot for the 5, 10 and 25 cent coins, a push button to cancel any incomplete transaction, a compartment for returned coins and a digital keyboard to choose a product. At the start, the keyboard is disabled and the slot is open. To identify the coins, the signals produced by the sensors are sampled at each clock pulse to generate the inputs X and Y of the controller, as illustrated in Table 2.9. The input R is set to 1 for one clock period, every time the push button is pressed. This sets into motion the mechanism to return the coin and, consequently, the cancellation of a transaction.

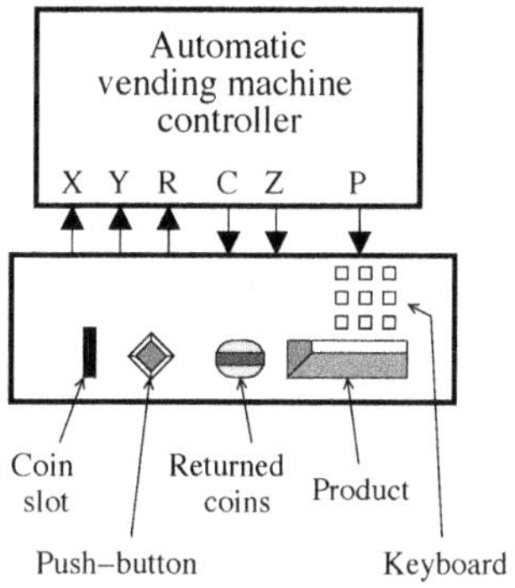

Figure 2.24. *Automatic vending machine*

X	Y	Received coins
0	0	None
0	1	25 cents
1	0	5 cents
1	1	10 cents

Table 2.9. *Binary code associated with each type of coin*

Each product costs 20 cents. After a sufficient number of coins is inserted, the keyboard is enabled to allow the selection of a product, while the slot is closed to prevent the insertion of any additional coins. When the total amount of coins is equal to 25 cents, the controller output, C, is also set to 1 so that coins can be returned. The vending machine does not accept more than 25 cents per transaction, and returns all the coins inserted if the total is 30, 35 or 40 cents. This is done by setting the output Z to 1.

The controller of the automatic vending machine is implemented as a Moore state machine, which offers the advantage of being unaffected by transient signals that can appear at the inputs. Its operation is described by the ASM chart shown in Figure 2.25.

The controller, which is initially in the state S_0, can move to S_5, S_{10} or S_{25} when the corresponding coin is detected. Starting from the state S_5 or S_{10}, only one coin of 5 or 10 cents can be used to cause the controller to go to one of the states S_{10}, S_{15} or S_{20}. Similarly, the transition from the state S_{15} to the state S_{20} or S_{25} only occurs after the detection of a 5 or 10 cent coin. The transition from one of the states S_5, S_{10} or S_{15} to the state S_Z requires the insertion of a 25 cent coin. A transaction can be cancelled using the R signal, which is set to 1 by pressing the push button, when the controller is in one of the following states: S_5, S_{10} or S_{15}. Each time the R signal is set to 1, the controller returns to the state S_0, and the coins are returned. Setting the R signal to 1 when the controller is in the state S_0 provides an opportunity to verify that all the returned coins were actually recovered.

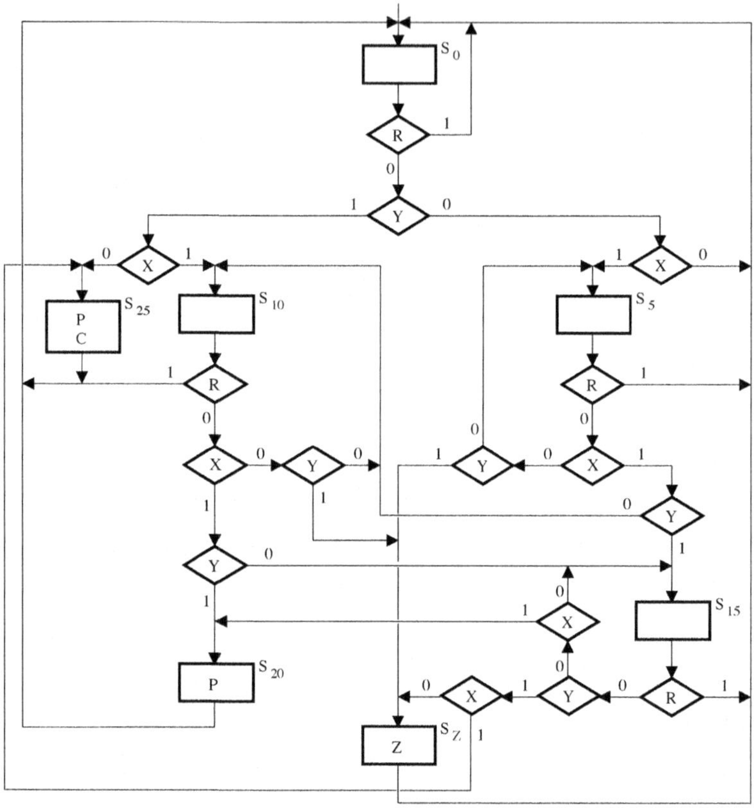

Figure 2.25. *ASM chart for the controller of the automatic vending machine*

The state table of the controller for the automatic vending machine is represented in Table 2.10. The controller has seven states. The slot used to insert the coins is open during the states S_0, S_5, S_{10} and S_{15}, and closed during the states S_{20}, S_{25} and S_Z. All the input combinations of the form 1xx, where x can take the value 0 or 1, bring the controller back to the state S_0.

PS	NS					Outputs		
						P	C	Z
	$RXY = 000$	001	010	011	1xx			
S_0	S_0	S_{25}	S_5	S_{10}	S_0	0	0	0
S_5	S_5	S_Z	S_{10}	S_{15}	S_0	0	0	0
S_{10}	S_{10}	S_Z	S_{15}	S_{20}	S_0	0	0	0
S_{15}	S_{15}	S_Z	S_{20}	S_{25}	S_0	0	0	0
S_{20}	S_0	–	–	–	S_0	1	0	0
S_{25}	S_0	–	–	–	S_0	1	1	0
S_Z	S_0	–	–	–	S_0	0	0	1

Table 2.10. *State table of the controller*

2.4.5. *Traffic light controller*

We wish to implement a controller for traffic lights that can regulate the movement of vehicles at an intersection of a main road (north-south) and a secondary road (east-west), as shown in Figure 2.26. The inputs and outputs are specified as follows:

– inputs:

R: places the controller in the initial state;

C: sensor used to detect the presence of a vehicle in both directions of the secondary road;

S: signal indicating the end of the short counting sequence;

L: signal indicating the end of the long counting sequence.

– outputs:

IC: initialization signal of the counter;

NR: red light in the north-south direction;

NG: green light in the north-south direction;

NY: yellow light in the north-south direction;

ER: red light in the east-west direction;

EG: green light in the east-west direction;

EY: yellow light in the east-west direction.

Figure 2.27 presents the structure of the traffic light controller, and the logic circuit for the counter is represented in Figure 2.28.

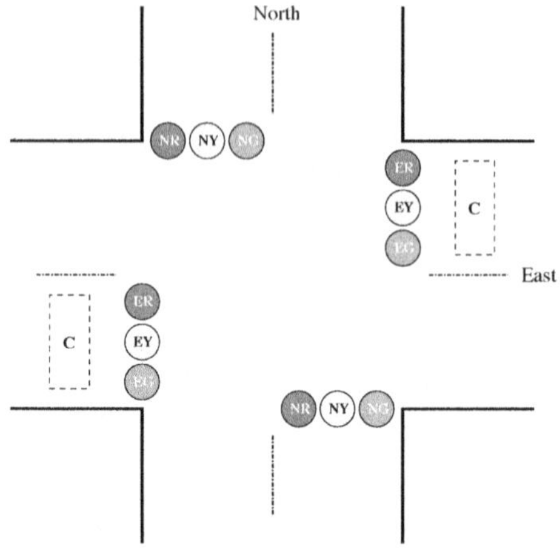

Figure 2.26. *Intersection with traffic lights. for a color version of this figure, www.iste.co.uk/ndjountche/electronics3.zip*

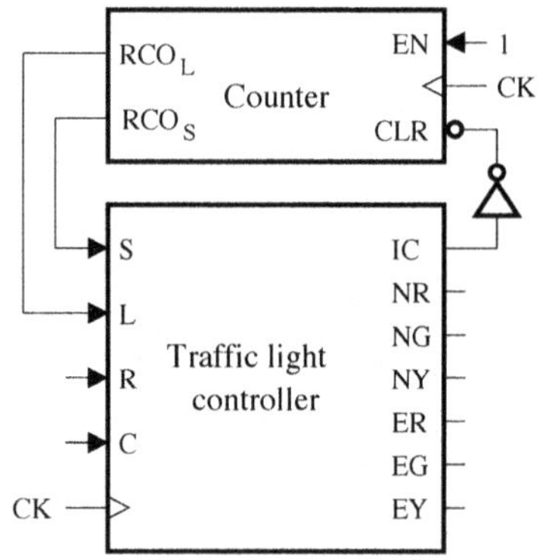

Figure 2.27. *Traffic light controller*

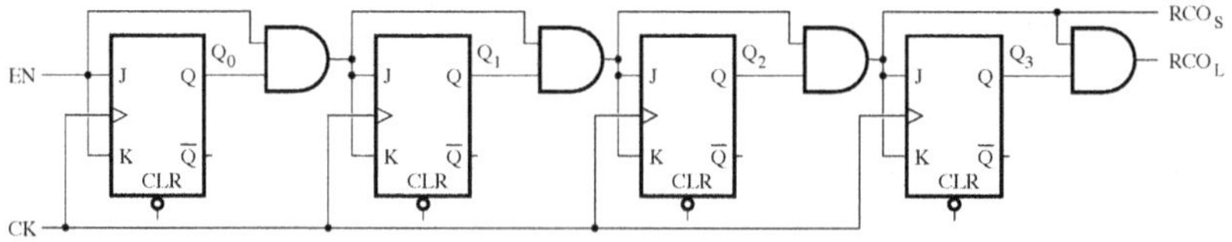

Figure 2.28. *Logic circuit of the counter*

The lights must be switched on in the following cyclic sequence:

NR-ER, NG-ER, NY-ER, NR-ER, NR-EG, NR-EY

The states in which the red lights are switched on in both directions are used to provide a safety margin. Traffic may be manually directed using the push-button R that holds the controller in one of the states where the red lights are on in both directions. A sensor, C, placed on the secondary road allows the detection of a vehicle at the stop. After the vehicle is detected, when the long counting sequence ends (L set to 1), the light switches from green to yellow, then to red on the main road, allowing the light on the secondary road to turn green. The light remains green on the secondary road only as long as a vehicle is detected and never longer than the duration of the long counting sequence. The duration for the yellow light is identical in both directions and is determined by the short counting sequence (S set to 1).

The ASM chart of the traffic light controller is represented in Figure 2.29.

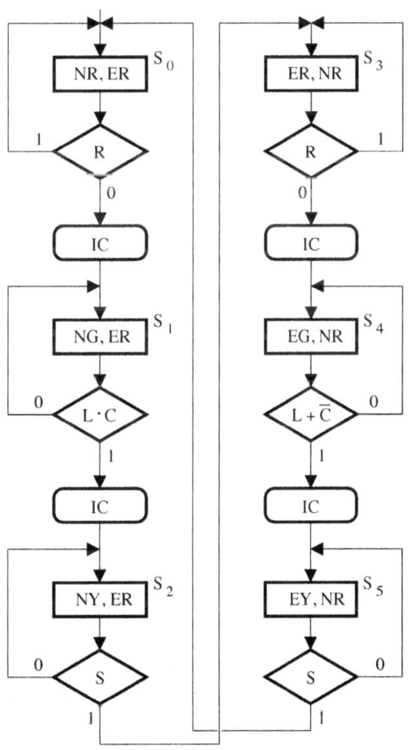

Figure 2.29. *ASM chart of the traffic light controller*

	PS	Inputs				NS	Outputs						
	$Q_1Q_2Q_3$	R	C	L	S	$Q_1^+Q_2^+Q_3^+$	IC	NR	NG	NY	ER	EG	EY
S_0	0 0 0	1				0 0 0	0	1	0	0	1	0	0
S_0	0 0 0	0				0 0 1	$\overline{R}$	1	0	0	1	0	0
S_1	0 0 1		0	0		0 0 1	0	0	1	0	1	0	0
S_1	0 0 1		0	1		0 0 1	0	0	1	0	1	0	0
S_1	0 0 1		1	0		0 0 1	0	0	1	0	1	0	0
S_1	0 0 1		1	1		0 1 1	$L \cdot C$	0	1	0	1	0	0
S_2	0 1 1				0	0 1 1	0	0	0	1	1	0	0
S_2	0 1 1				1	0 1 0	0	0	0	1	1	0	0
S_3	0 1 0	1				0 1 0	0	1	0	0	1	0	0
S_3	0 1 0	0				1 1 0	$\overline{R}$	1	0	0	1	0	0
S_4	1 1 0		0	0		1 1 1	$L + \overline{C}$	1	0	0	0	1	0
S_4	1 1 0		0	1		1 1 1	$L + \overline{C}$	1	0	0	0	1	0
S_4	1 1 0		1	0		1 1 0	0	1	0	0	0	1	0
S_4	1 1 0		1	1		1 1 1	$L + \overline{C}$	1	0	0	0	1	0
S_5	1 1 1				0	1 1 1	0	1	0	0	0	0	1
S_5	1 1 1				1	0 0 0	0	1	0	0	0	0	1

Table 2.11. *Transition table*

Three bits are required to encode the states of the traffic light controller. Table 2.11 presents the transition table obtained by representing the states using Gray code.

The D flip-flop is chosen as the memory element for the implementation of the traffic light controller. The transition table and the excitation table for the D flip-flop can used to construct the Karnaugh maps shown in Figures 2.30–2.32. The logic equations for the D inputs of flip-flops can then be obtained as follows:

$$D_1 = Q_1^+ = \overline{R} \cdot Q_2 \cdot \overline{Q_3} + \overline{S} \cdot Q_1 \cdot Q_2 + Q_1 \cdot Q_2 \cdot \overline{Q_3} \quad [2.23]$$

$$D_2 = Q_2^+ = \overline{S} \cdot Q_2 + L \cdot C \cdot \overline{Q_1} \cdot Q_3 + \overline{Q_1} \cdot Q_2 + Q_2 \cdot \overline{Q_3} \quad [2.24]$$

$$D_3 = Q_3^+ = \overline{R} \cdot \overline{Q_1} \cdot \overline{Q_2} + \overline{S} \cdot Q_2 \cdot Q_3 + (L + \overline{C})Q_2 \cdot \overline{Q_3} + \overline{Q_1} \cdot \overline{Q_2} \cdot Q_3 + \overline{Q_1} \cdot Q_2 \cdot \overline{Q_3} \quad [2.25]$$

The Karnaugh maps for the outputs of the traffic light controller are depicted in Figures 2.33–2.39. Hence, the logic equations are given by:

$$IC = \overline{R} \cdot \overline{Q_1} \cdot \overline{Q_3} + L \cdot C \cdot \overline{Q_1} \cdot \overline{Q_2} \cdot Q_3 + (L + \overline{C})Q_1 \cdot Q_2 \cdot \overline{Q_3} \quad [2.26]$$

$$= \overline{R} \cdot \overline{Q_1} \cdot \overline{Q_3} + L \cdot C \cdot NG + (L + \overline{C})EG \quad [2.27]$$

$$NR = Q_1 + \overline{Q_3} \qquad [2.28]$$

$$NG = \overline{Q_1} \cdot \overline{Q_2} \cdot Q_3 \qquad [2.29]$$

$$NY = \overline{Q_1} \cdot Q_2 \cdot Q_3 \qquad [2.30]$$

$$ER = \overline{Q_1} \qquad [2.31]$$

$$EG = Q_1 \cdot Q_2 \cdot \overline{Q_3} \qquad [2.32]$$

$$EY = Q_1 \cdot Q_2 \cdot Q_3 \qquad [2.33]$$

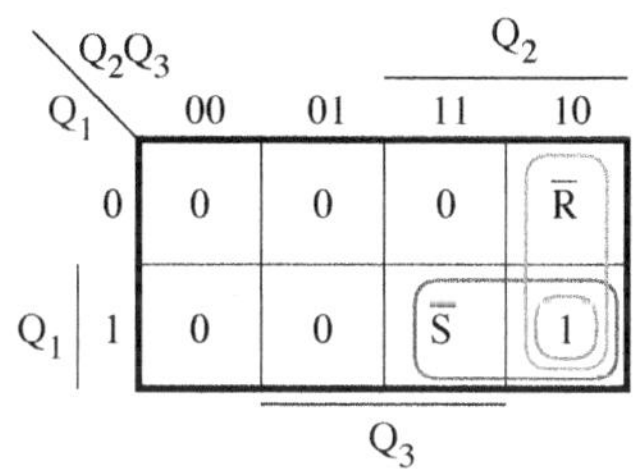

Figure 2.30. *Input* D_1

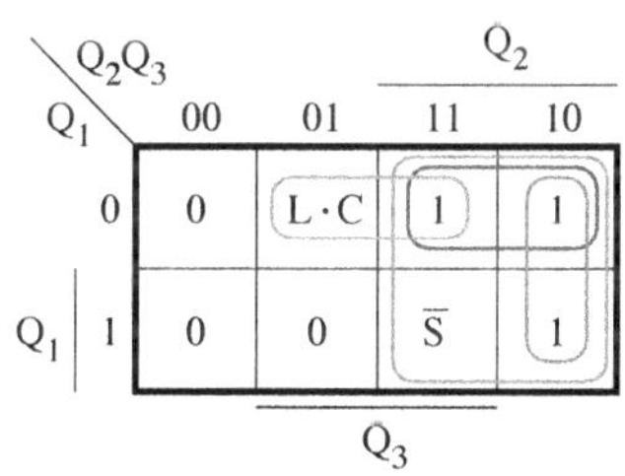

Figure 2.31. *Input* D_2

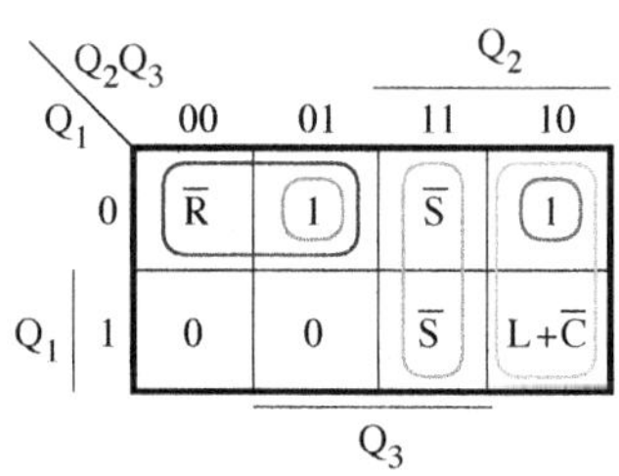

Figure 2.32. *Input* D_3

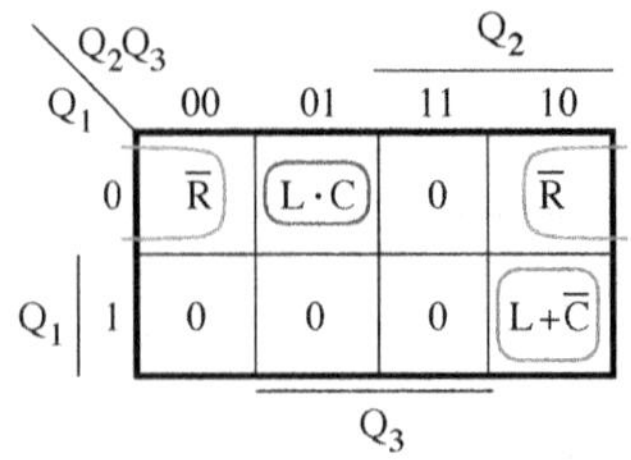

Figure 2.33. *Output IC*

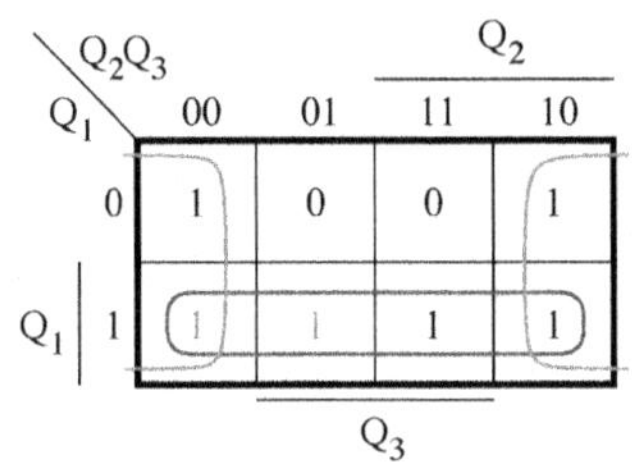

Figure 2.34. *Output NR*

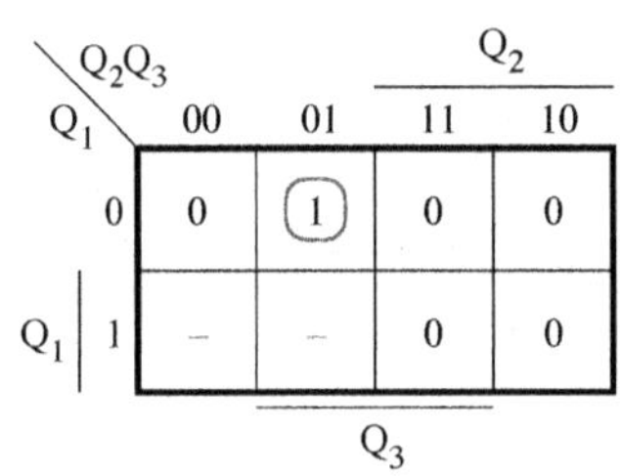

Figure 2.35. *Output NG*

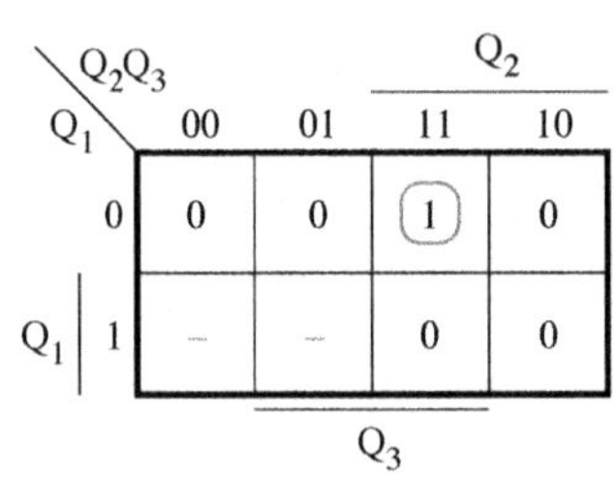

Figure 2.36. *Output NY*

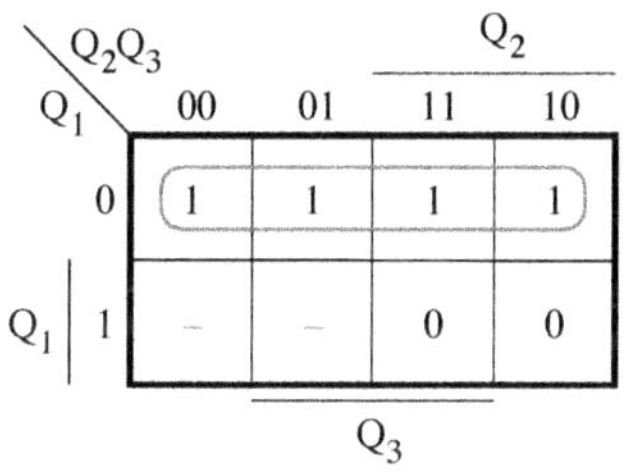

Figure 2.37. *Output* ER

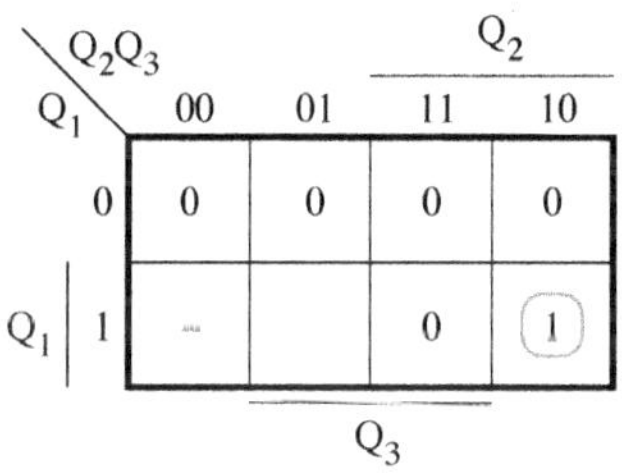

Figure 2.38. *Output* EG

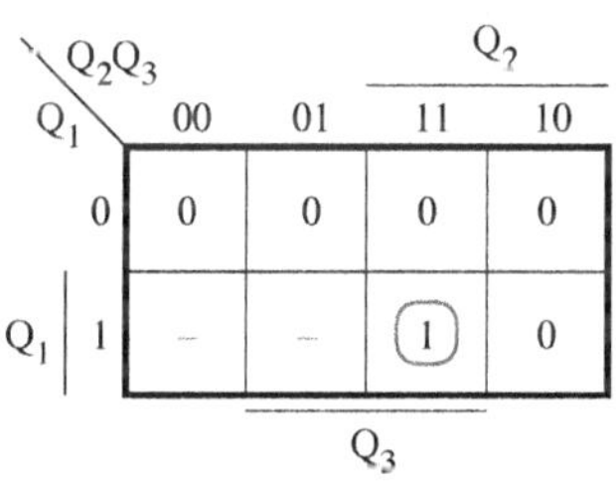

Figure 2.39. *Output* EY

It should be noted that the outputs corresponding to the two unused codes, 100 and 101, were defined so as to minimize the logic equation of the output NR.

Figure 2.40 depicts the logic circuit of the traffic light controller.

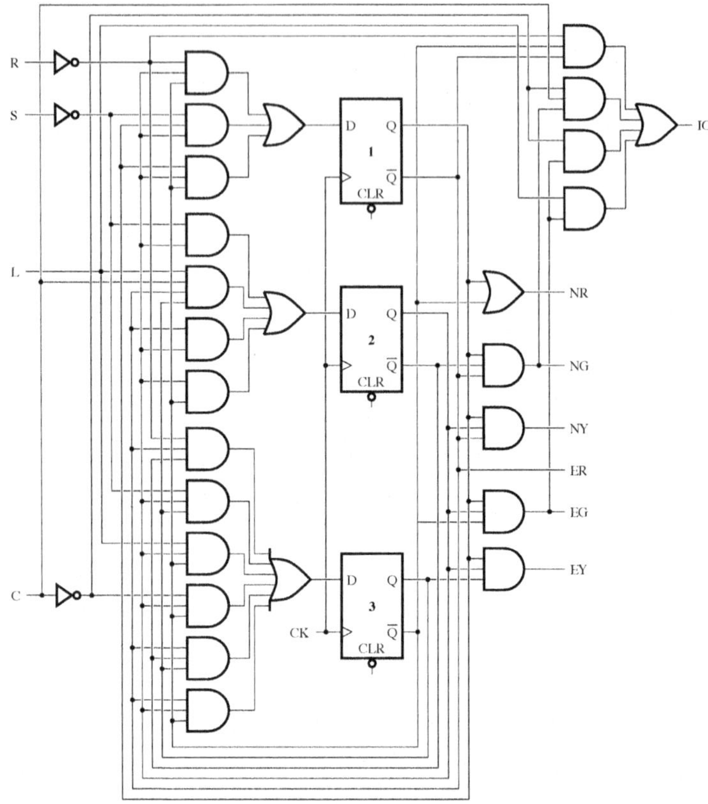

Figure 2.40. *Logic circuit for the controller of the traffic lights*

2.5. Exercises

EXERCISE 2.1.– The operation of a finite state machine can be described by one of the ASM charts shown in Figure 2.41:

a) Assume that the finite state machine is implemented based on the ASM chart in Figure 2.41(a):

– construct the state table;

– determine the logic equations for the D inputs and the output, Y, when the states are represented using a one-hot (or 1-out-of-n) code.

b) To take into account the fact that some D flip-flops do not have an asynchronous reset input, $\overline{PR}$, a reset state is added to the ASM chart of the finite state machine, as shown in Figure 2.41(b) and the one-hot code with zero is adopted to represent the states.

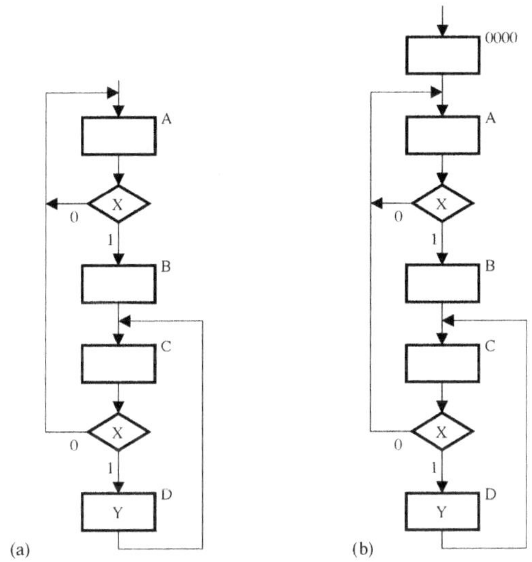

Figure 2.41. *Two ASM chart versions for a finite state machine*

Determine the logic equations for the D input and the Y output.

EXERCISE 2.2.– Consider the finite state machine whose ASM chart is represented in Figure 2.42:

– construct the state table of this state machine;

– assuming that a one-hot (or 1-out-of-n) code is used to represent the states, determine the logic equation for the D inputs of flip-flops and for the outputs U and V.

EXERCISE 2.3.– A finite state machine with two inputs, X and Y, and one output, Z, is characterized by the state table shown in Table 2.12.

Construct the corresponding ASM chart for this state machine.

EXERCISE 2.4.– Suggest an ASM chart to describe the finite state machine whose state table is shown in Table 2.13, where the inputs are denoted by X and Y, and the output by Z.

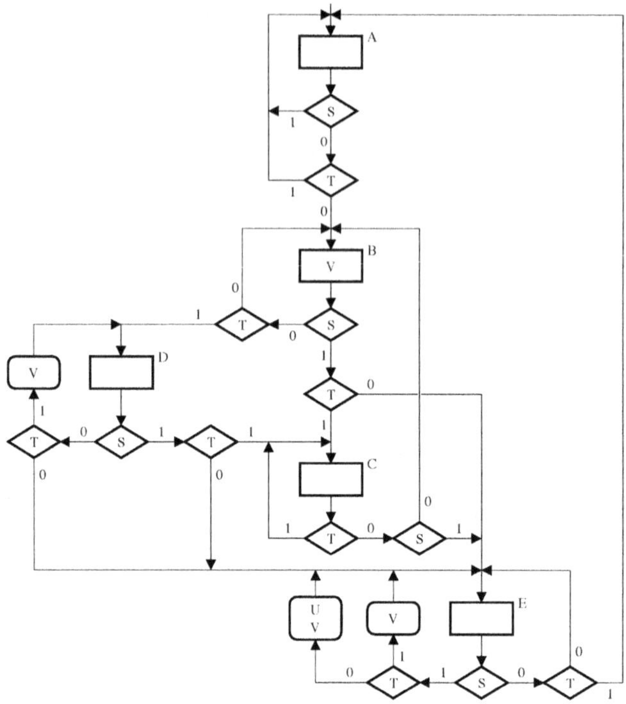

Figure 2.42. *ASM chart of a finite state machine*

PS	NS $XY = 00$	01	10	11	Output Z
S_0	S_3	S_1	S_0	S_0	0
S_1	S_3	S_1	S_1	S_2	1
S_2	S_3	S_2	S_0	S_2	$X \cdot \overline{Y}$
S_3	S_3	S_1	S_3	S_2	Y

Table 2.12. *State table of the state machine*

EXERCISE 2.5.– We wish to implement the finite state machines, whose operation is described by each of the ASM charts shown in Figure 2.43, using *D* flip-flops and logic gates.

Assuming that the states are represented using a one-hot (or 1-out-of-n) code, determine the logic equations for the *D* inputs of flip-flops and the outputs.

PS	NSI				Output Z
	$XY = 00$	01	10	11	
S_0	S_0	S_1	S_0	S_2	0
S_1	S_1	S_1	S_0	S_3	0
S_2	S_1	S_2	S_3	S_2	1
S_3	S_0	S_2	S_1	S_3	$\overline{X} \cdot \overline{Y}$

Table 2.13. *State table*

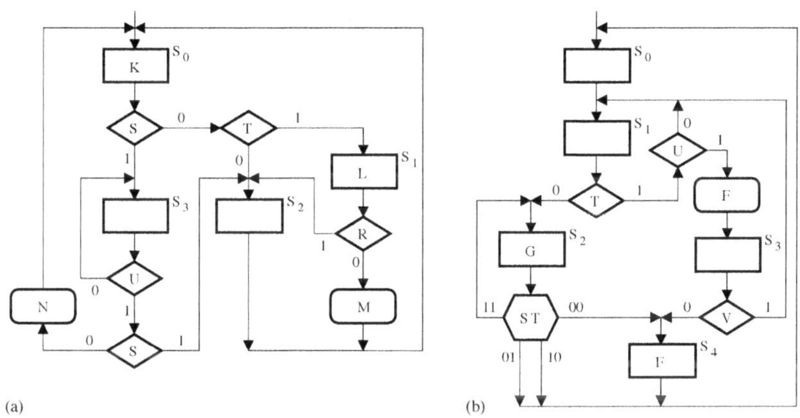

Figure 2.43. *ASM charts: machine a); machine b)*

EXERCISE 2.6.– (Programmable Signal Generator).

We wish to design a programmable signal generator based on the ASM chart shown in Figure 2.44 using *D* flip-flops and logic gates. For each combination of the inputs X and Y, a new signal type is available at the output Z of this generator:

– construct the state table of the generator;

– assigning the binary codes 00, 01, 11 and 10 to the states S_0, S_1, S_2 and S_3, respectively, determine the logic equations for the *D* inputs of flip-flops and the output Z;

– complete the timing diagram shown in Figure 2.45.

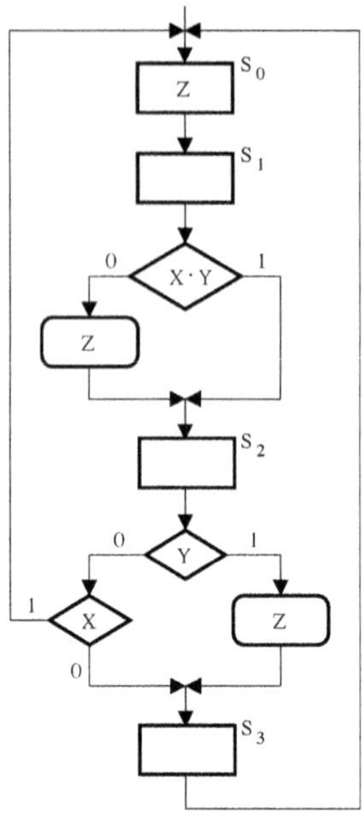

Figure 2.44. *ASM chart of the programmable signal generator*

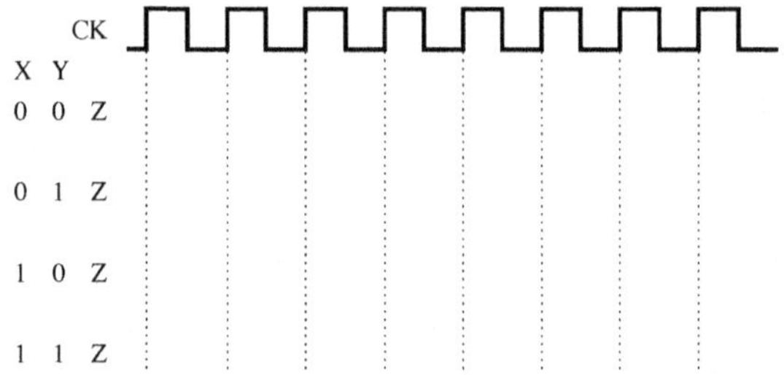

Figure 2.45. *Timing diagram of the programmable signal generator*

EXERCISE 2.7.– (Controller for an Elevator).

The movements of an elevator between three floors are controlled by the finite state machine whose operation, based on the Mealy model, is described by the state diagram in Figure 2.46. The input signals A_1, A_2 and A_3 are used to call the elevator car to floor 1, 2 and 3, respectively. The outputs D_1, D_2 and D_3 cause the elevator car to move down one, two and three floors, while the outputs M_1, M_2 and M_3 cause it to move up one, two and three floors, respectively. The output signal, R, allows the elevator to be held in one state.

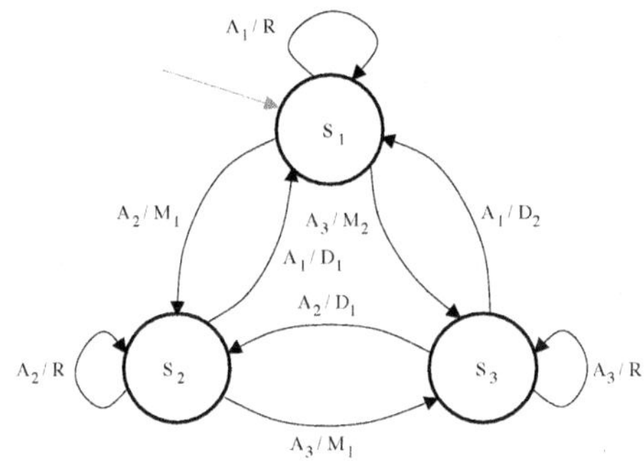

Figure 2.46. *State diagram (Mealy machine) of the controller*

Propose the state diagram for the equivalent Moore machine.

2.6. Solutions

SOLUTION 2.1.– (Analysis of ASMs).

a) Table 2.14 presents the state table obtained from the state diagram.

PS	NS		Output Y
	$X = 0$	1	
A	A	B	0
B	C	C	0
C	A	D	0
D	C	C	1

Table 2.14. *State table*

The logic equations for the D inputs of flip-flops and for the output can be written as follows:

$$D_A = Q_A \cdot \overline{X} + Q_C \cdot \overline{X} \quad [2.34]$$

$$D_B = Q_A \cdot X \quad [2.35]$$

$$D_C = Q_B + Q_D \quad [2.36]$$

$$D_D = Q_C \cdot X \quad [2.37]$$

$$Y = D \quad [2.38]$$

b) The logic equations for the D inputs of flip-flops and the output are given by:

$$D_A = Q_A \cdot \overline{X} + Q_C \cdot \overline{X} + \overline{Q_A} \cdot \overline{Q_B} \cdot \overline{Q_C} \cdot \overline{Q_D} \quad [2.39]$$

$$D_B = Q_A \cdot X \quad [2.40]$$

$$D_C = Q_B + Q_D \quad [2.41]$$

$$D_D = Q_C \cdot X \quad [2.42]$$

$$Y = D \quad [2.43]$$

SOLUTION 2.2.– The operation of the finite state machine can be described by the state table shown in Table 2.15, where S and T are the inputs and U and V represent the outputs.

PS	NS				Outputs	
					U	V
	$ST = 00$	01	10	11		
A	B	A	A	A	0	0
B	B	D	E	C	0	1
C	B	C	E	C	0	0
D	E	D	E	C	0	$\overline{S} \cdot T$
E	E	A	E	E	$S \cdot \overline{T}$	S

Table 2.15. *State table*

The logic equations for the D inputs of flip-flops and for the outputs can be obtained from the ASM chart, as follows:

$$D_A = Q_A \cdot S + Q_A \cdot T + Q_E \cdot \overline{S} \cdot T \quad [2.44]$$

$$D_B = Q_A \cdot \overline{S} \cdot \overline{T} + Q_B \cdot \overline{S} \cdot \overline{T} + Q_C \cdot \overline{S} \cdot \overline{T} \quad [2.45]$$

$$D_C = Q_B \cdot S \cdot T + Q_C \cdot T + Q_D \cdot S \cdot T \quad [2.46]$$

$$D_D = Q_B \cdot \overline{S} \cdot T + Q_D \cdot \overline{S} \cdot T \quad [2.47]$$

$$D_E = Q_B \cdot S \cdot \overline{T} + Q_C \cdot S \cdot \overline{T} + Q_D \cdot \overline{T} + Q_E \cdot S + Q_E \cdot \overline{T} \quad [2.48]$$

$$U = Q_E \cdot S \cdot \overline{T} \quad [2.49]$$

and:

$$V = Q_B + Q_D \cdot \overline{S} + Q_E \cdot S \quad [2.50]$$

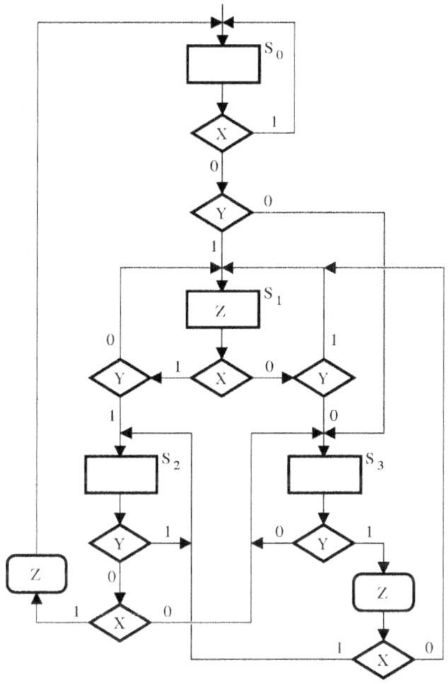

Figure 2.47. *ASM chart of the finite state machine*

SOLUTION 2.3.– The ASM chart obtained from the state table is represented in Figure 2.47.

SOLUTION 2.4.– The ASM chart shown in Figure 2.48 is obtained from the state table.

SOLUTION 2.5.– The logic equations of the flip-flop inputs are directly related to the conditions for transitions between states when one-hot encoding and D flip-flops are used.

Upon analysis of the ASM chart of the machine (a), the logic equations for the inputs of D flip-flops can be written as follows:

$$D_0 = \overline{R} \cdot Q_1 + Q_2 + \overline{S} \cdot U \cdot Q_3 \quad [2.51]$$

$$D_1 = \overline{S} \cdot T \cdot Q_0 \quad [2.52]$$

$$D_2 = \overline{S} \cdot \overline{T} \cdot Q_0 + R \cdot Q_1 + S \cdot U \cdot Q_3 \quad [2.53]$$

and:

$$D_3 = S \cdot Q_0 + \overline{U} \cdot Q_3 \quad [2.54]$$

where the flip-flop outputs are designated by Q_i $(i = 0, 1, 2, 3)$. For the state machine outputs, we can obtain:

$$K = Q_0 \quad [2.55]$$

$$L = Q_1 \quad [2.56]$$

$$M = \overline{R} \cdot Q_1 \quad [2.57]$$

and:

$$N = \overline{S} \cdot U \cdot Q_3 \quad [2.58]$$

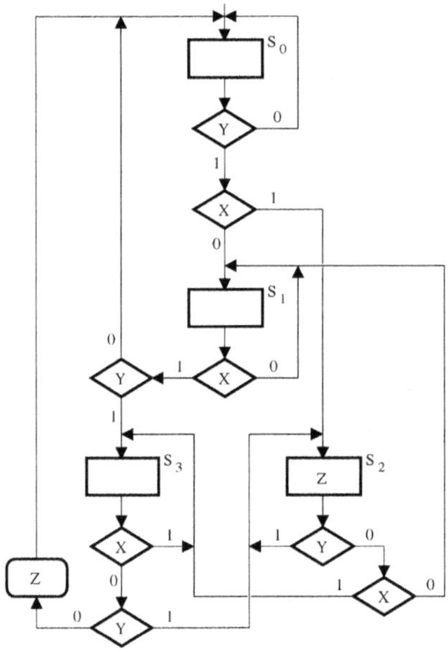

Figure 2.48. *ASM chart of the finite state machine*

In the case of the state machine (b), the logic equations for the *D* inputs of flip-flops can take the form:

$$D_0 = (\overline{S} \cdot T + S \cdot \overline{T})Q_2 + Q_4 \quad [2.59]$$

$$D_1 = Q_0 + T \cdot \overline{U} \cdot Q_1 + V \cdot Q_3 \quad [2.60]$$

$$D_2 = \overline{T} \cdot Q_1 + S \cdot T \cdot Q_2 \quad [2.61]$$

$$D_3 = T \cdot U \cdot Q_1 \quad [2.62]$$

and:

$$D_4 = \overline{S} \cdot \overline{T} \cdot Q_2 + \overline{V} \cdot Q_3 \quad [2.63]$$

The logic equations for the machine outputs are given by:

$$F = T \cdot U \cdot Q_1 + Q_4 \quad [2.64]$$

and:

$$G = Q_2 \quad [2.65]$$

SOLUTION 2.6.– (Programmable Signal Generator).

The ASM chart of the programmable signal generator can be used to construct the state table, as shown in Table 2.16.

PS	NS				Output Z			
	$XY = 00$	01	10	11	$XY = 00$	01	10	11
S_0	S_1	S_1	S_1	S_1	1	1	1	1
S_1	S_2	S_2	S_2	S_2	1	1	1	0
S_2	S_3	S_3	S_0	S_3	0	1	0	1
S_3	S_0	S_0	–	S_0	0	0	–	0

Table 2.16. *State table of the signal generator*

The characteristic equation of the *D* flip-flop is of the form $Q^+ = D$. Figure 2.49 presents the Karnaugh maps obtained from the state table. The logic equations for the *D* inputs of flip-flops and for the output can, thus, be written as follows:

$$Q_1^+ = \overline{Q_1} \cdot Q_0 + Q_0 \cdot \overline{X} + Q_0 \cdot Y \quad [2.66]$$

$$Q_0^+ = \overline{Q_1} \quad [2.67]$$

and:

$$Z = \overline{Q_1} \cdot \overline{Q_0} + \overline{Q_1} \cdot \overline{X} + \overline{Q_1} \cdot \overline{Y} + Q_1 \cdot Q_0 \cdot Y \quad [2.68]$$

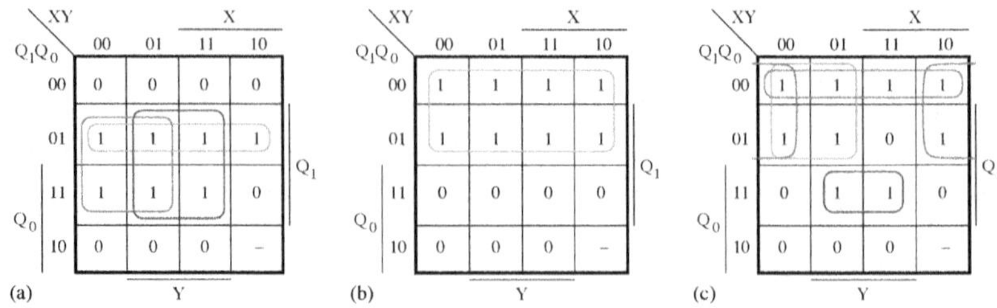

Figure 2.49. *Karnaugh maps for the determination of a) Q_1^+, b) Q_0^+ and c) Z*

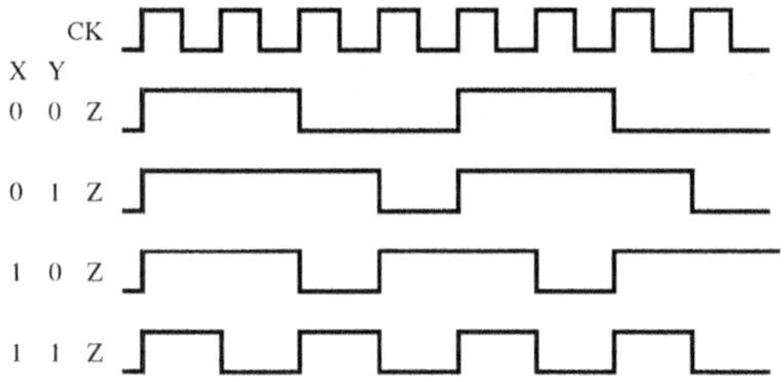

Figure 2.50. *Timing diagram of the programmable signal generator*

The timing diagram of the programmable signal generator is illustrated in Figure 2.50 for the different combinations of the inputs X and Y.

The logic circuit of the programmable signal generator is represented in Figure 2.51, while Table 2.52 shows the truth table specifying the decoder inputs and ROM outputs.

SOLUTION 2.7.– (Elevator Controller).

The state diagram (Moore model) of the controller is depicted in Figure 2.53. It comprises nine states and each state is associated with a particular output signal.

The Moore machine outputs only depend on the present state, while those of the Mealy machine are determined by the present state as well as the inputs. The Moore machine, thus, offers the advantage of being less sensitive to undesirable disturbances that can affect the inputs.

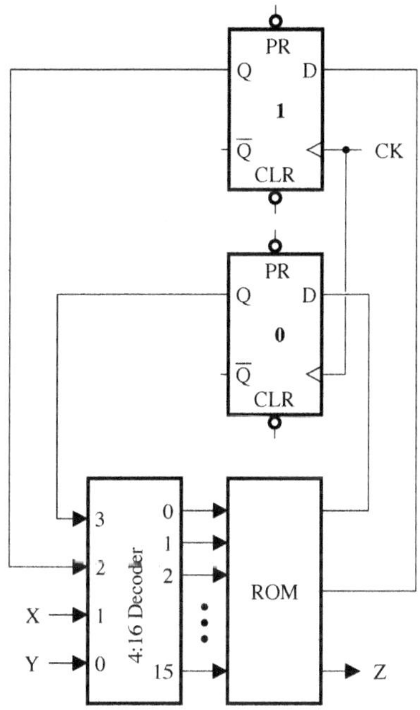

Figure 2.51. *Logic circuit of the ROM-based programmable signal generator*

Inputs				Outputs		
Q_0	Q_1	X	Y	Q_0^+	Q_1^+	Z
0	0	0	0	0	1	1
0	0	0	1	0	1	1
0	0	1	0	0	1	1
0	0	1	1	0	1	1
0	1	0	0	1	1	1
0	1	0	1	1	1	1
0	1	1	0	1	1	1
0	1	1	1	1	1	0
1	0	0	0	0	0	0
1	0	0	1	0	0	0
1	0	1	0	0	0	0
1	0	1	1	0	0	0
1	1	0	0	1	0	0
1	1	0	1	1	0	1
1	1	1	0	0	0	0
1	1	1	1	1	0	0

Figure 2.52. *Truth table (decoder inputs/ROM outputs)*

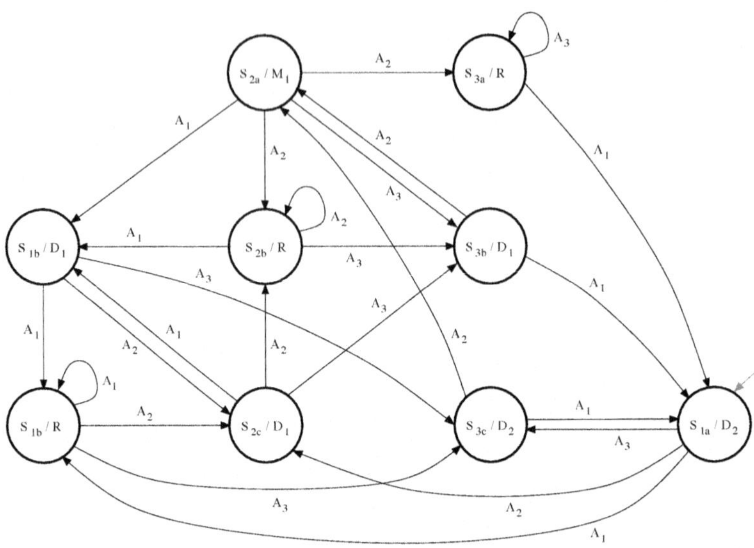

Figure 2.53. *State diagram (Moore model) of the elevator controller*

3

Asynchronous Finite State Machines

3.1. Introduction

Finite state machines can be synchronous or asynchronous. The operation of asynchronous state machines, unlike that of synchronous state machines, does not require a clock signal. Data transfer or the synchronization of asynchronous machines is carried out through the bidirectional exchange of request signals and acknowledge signals, also called *handshake communication*.

Asynchronous machines offer the advantage of being faster. However, they are more sensitive to synchronization errors (namely critical race conditions, propagation delay or hazard, oscillation). As a result, it is much more difficult to design reliable asynchronous state machines.

Asynchronous state machines can be classified based on their operating mode, such as the *fundamental mode*, *pulse mode* or *burst mode*.

Operation in the fundamental mode is possible only when a single input can change at any one time and the state machine is in a stable state.

A state machine designed to operate in the pulse mode uses latches or flip-flops triggered by data signals, and it is required that the input signal pulses do not overlap.

To operate in burst mode, the state machine must allow multiple inputs to change simultaneously.

3.2. Overview

An asynchronous state machine can have stable and transient (or unstable) states.

A state in which the asynchronous state machine can remain before and after a transition is said to be stable.

The operation of an asynchronous state machine can be described using a flow table.

A flow table is the tabular transposition of the possible transitions and outputs for each input combination of an asynchronous state machine. It highlights stable states, which are encircled, while the others are unstable.

A flow table for an asynchronous state machine is identical to a state table for a synchronous machine. It is said to be primitive if each row has only one stable state.

A transition table displays, for each input combination, the transitions that can take place between the states represented in terms of state variables.

An asynchronous state machine can exit a stable state only if an input changes. It can move from one stable state to another either directly, or transiting through several unstable states or states that do not satisfy the stability condition.

An asynchronous state machine operates in the fundamental mode, provided that only one input changes logic states at a time and then remains constant until a steady state is reached.

Changing a state variable forces the state machine to pass through a transient state.

For an asynchronous state machine, there is not such a marked difference between Moore and Mealy models as there is with synchronous state machines. In fact, the transitions between states in an asynchronous state machine are always initiated by the inputs, and the outputs are, either directly or through state variables, related to the inputs.

3.3. Gated D latch

A latch is a logic circuit that maintains a stable state even when the inputs become inactive. Figure 3.1(a) presents the logic circuit of a gated D latch, which consists of logic gates (an inverter, two AND gates, and an OR gate) configured as a 2:1 multiplexer.

Assuming that the gated D latch operates in fundamental mode, we obtain the equivalent circuit shown in Figure 3.1(b), where a delay element is inserted in the feedback path. The delay element can be considered to provide the sequential circuit with a short-term memory. It usually represents the delay caused along the feedback loop because of the propagation delay of the combinational logic section. The characteristic equation for the gated D latch can be written as follows:

$$Q^+ = D \cdot C + \overline{C} \cdot Q \qquad [3.1]$$

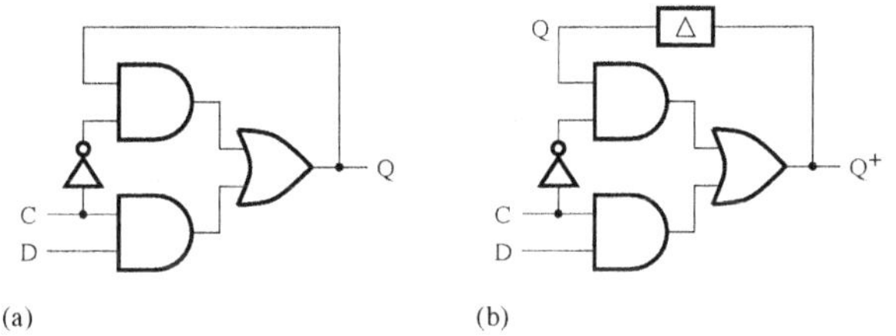

Figure 3.1. *a) Logic circuit for a gated D latch; b) equivalent circuit in fundamental mode*

If the signal C is set to 1, the output takes the logic state of the input D. Otherwise, the output remains in the previous state.

The Karnaugh maps shown in Figure 3.2 represent the values of Q^+ as given by the characteristic equation for all the combinations of the variables D, C and Q.

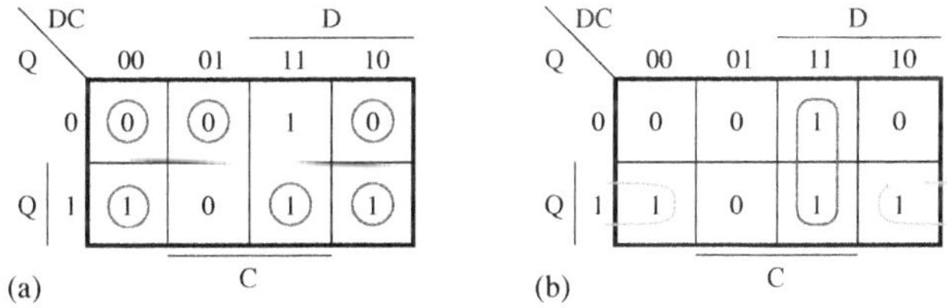

Figure 3.2. *Karnaugh maps: a) stable states; b) characteristic equation Q^+*

A transition leads to a stable state if and only if the present state is identical to the next state, that is $Q^+ = Q$. The stable states are encircled in Figure 3.2(a).

Figure 3.2(b) shows the loops of the Karnaugh map that correspond to the terms of the characteristic equation. As these loops are adjacent, the operation of the latch can be affected by propagation delays of the logic gates. Figure 3.3 presents a timing diagram that highlights the transient signals, which are transformed into oscillations at the latch output. To eliminate this problem, a redundant term can be added to the characteristic equation, as shown in the Karnaugh map in Figure 3.4(a). Thus:

$$Q^{+} = D \cdot C + \overline{C} \cdot Q + D \cdot Q \qquad [3.2]$$

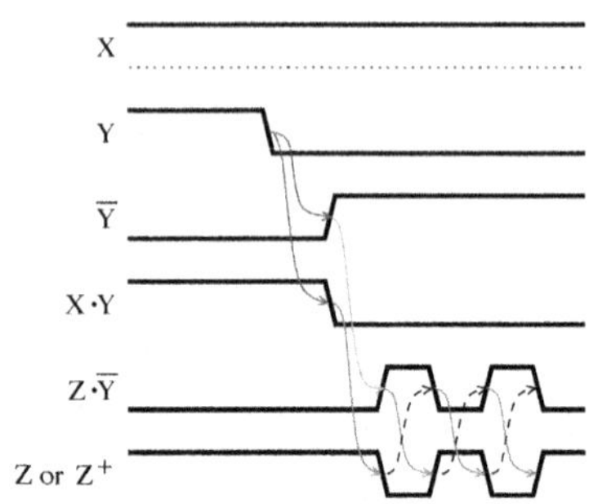

Figure 3.3. *Timing diagram illustrating the effect of propagation delays*

When the variables D and Q are set to 1, Q^{+} takes the logic state 1 and is no longer dependent on the term $C + \overline{C}$ as previously. An improved version of the logic circuit for the gated D latch is given in Figure 3.4(b).

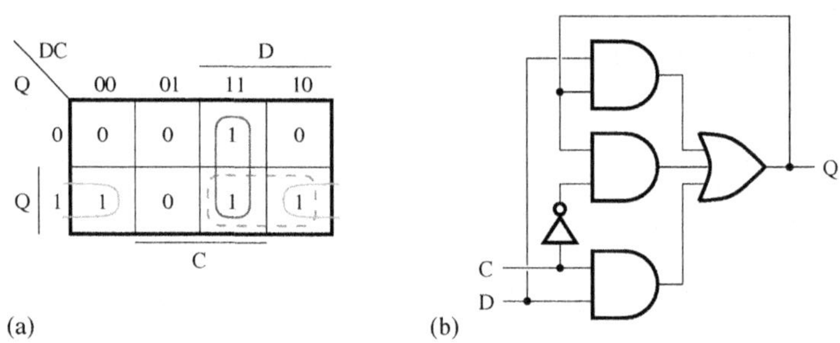

Figure 3.4. *a) Karnaugh map; b) improved version of the logic circuit for the gated D latch*

Another approach is the use of an SR latch and logic gates to implement a gated D latch based on the state table given in Table 3.1.

The characteristic equation for an SR latch that does not use a forbidden state is given by:

$$Q^{+} = S + \overline{R} \cdot Q \quad \text{and} \quad S \cdot R = 0 \qquad [3.3]$$

PS Q	NS				Output Q^+
	DC = 00	01	10	11	
0	0	0	0	1	0
1	1	0	1	1	1

Table 3.1. *State table of the gated D latch*

Table 3.2 presents the excitation table for the SR latch, which can be used along with the state table to determine the values required for the construction of the Karnaugh maps, as shown in Figure 3.5, for the inputs S and R. The corresponding logic equation can then be obtained as follows:

$$S = D \cdot C \qquad [3.4]$$

and

$$R = \overline{D} \cdot C \qquad [3.5]$$

Q	→	Q^+	S	R
0	→	0	0	x
0	→	1	1	0
1	→	0	0	1
1	→	1	x	0

Table 3.2. *Excitation table of the SR latch*

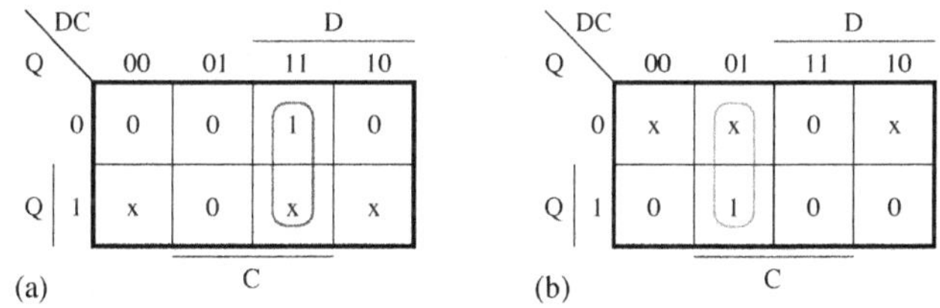

Figure 3.5. *Karnaugh maps: a) $S = D \cdot C$; b) $R = \overline{D} \cdot C$*

The gated D latch can, thus, be implemented by connecting an SR latch to AND gates and an inverter, as illustrated in Figure 3.6. This circuit is only composed of logic gates with a maximum of two inputs, and its operation is not affected by the propagation delays of logic gates.

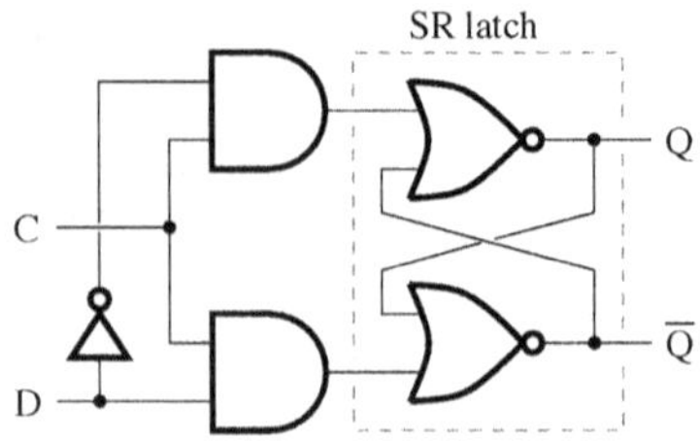

Figure 3.6. *Implementation of a gated D latch based on an SR latch*

3.4. Muller C-element

A Muller C-element can be considered as a building block of asynchronous circuits. It is used primarily to synchronize events because it can merge two requests into one.

The logic circuit for a Muller C-element with two inputs and one output is shown in Figure 3.7(a). It can be described as the implementation of a majority function with three inputs and whose output signal is fed back to one input. By analyzing this circuit, we can obtain the following logic equation:

$$Q^+ = A \cdot B + (A + B)Q \qquad [3.6]$$

where the present state and the next state of the output are represented by Q and Q^+, respectively. The output of the C-element takes the logic level 1 if the majority of the inputs A, B and Q are set to 1. Otherwise, it remains at the logic state 0. The symbol of the C-element is depicted in Figure 3.7(b) and an example of the timing diagram is shown in Figure 3.7(c).

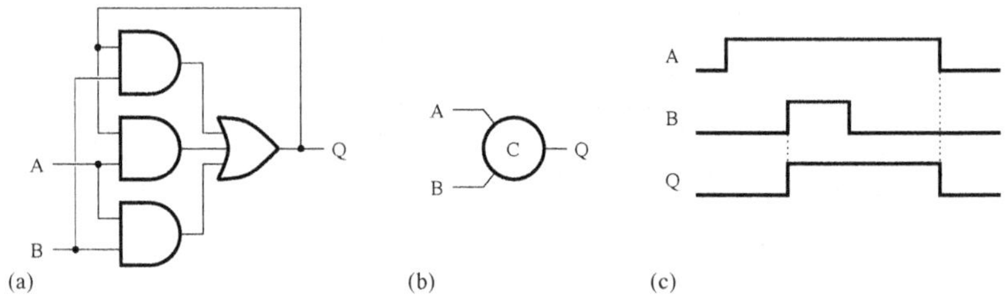

Figure 3.7. *a) Logic circuit b) symbol and c) timing diagram of the Muller C-element*

The truth table of the C-element is represented in Table 3.3. When the inputs A and B take the logic state 0, the output is set to 0. When the logic states of the inputs A and B are different, the output remains in the previous state. Finally, when both inputs A and B assume the logic state 1, the output is set to 1.

A	B	Q^+
0	0	0
0	1	Q
1	0	Q
1	1	1

Table 3.3. *Truth table of the Muller C-element*

An active-low reset input can be added to the C-element, as shown in the logic circuit and symbol in Figures 3.8(a) and (b).

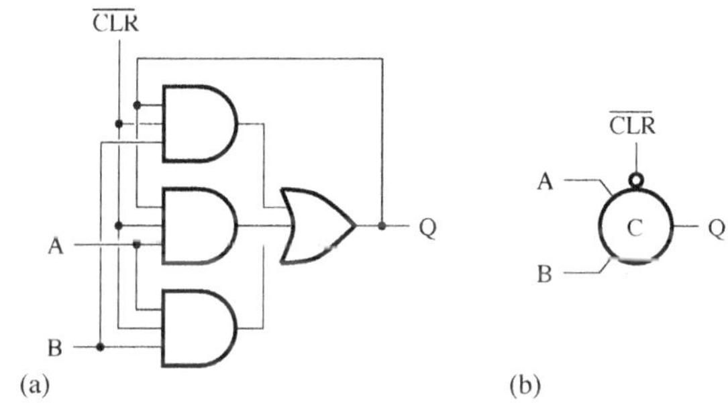

Figure 3.8. *a) Logic circuit and b) symbol of the C-element with a reset input*

The excitation table of the C-element is represented in Table 3.4. Two combinations of the inputs A and B can cause each of the transitions $0 \rightarrow 0$ and $1 \rightarrow 1$.

A C-element can also be implemented by combining logic gates and the SR latch. The transition table given in Table 3.5 is constructed based on the excitation table for the SR latch and the truth table for the C-element. The Karnaugh maps shown in Figure 3.9 can be used to derive the minimum logic expressions for the inputs S and R of the latch. That is:

$$S = A \cdot B \qquad [3.7]$$

and

$$S = \overline{A} \cdot \overline{B} \qquad [3.8]$$

Q	→	Q^+	A	B
0	→	0	0	x
			x	0
0	→	1	1	1
1	→	0	0	0
1	→	1	1	x
			x	1

Table 3.4. *Excitation table of the C-element*

A	B	Q	Q^+	S	R
0	0	0	0	0	x
0	0	1	0	0	1
0	1	0	0	0	x
0	1	1	1	x	0
1	0	0	0	0	x
1	0	1	1	x	0
1	1	0	1	1	0
1	1	1	1	x	0

Table 3.5. *Transition table*

The resulting logic circuit of the C-element is depicted in Figure 3.10.

3.5. Self-timed circuit

Self-timed circuits consist of modules that communicate with each other using handshake protocol. They can be implemented by exploiting the operating principle of C-elements.

There are two handshake communication protocols: one with two phases and the other with four phases.

The two-phase protocol can be implemented as shown in Figure 3.11, where continuous (dashed) lines represent the operations carried out by the sender (receiver). Each transition is interpreted as an event carrying information. Thus, a transition of the request (REQ) signal corresponds to a binary word being placed on the data bus by the sender, while a transition of the acknowledgment (ACK) signal is sent by the receiver to indicate the end of the transfer operation.

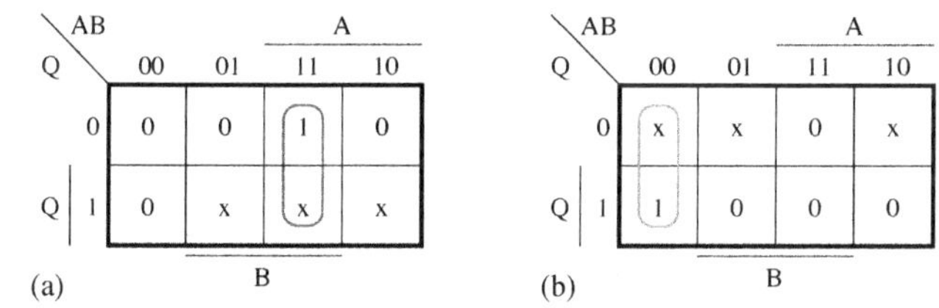

Figure 3.9. *Karnaugh maps: a)* $S = A \cdot B$*; b)* $R = \overline{A} \cdot \overline{B}$

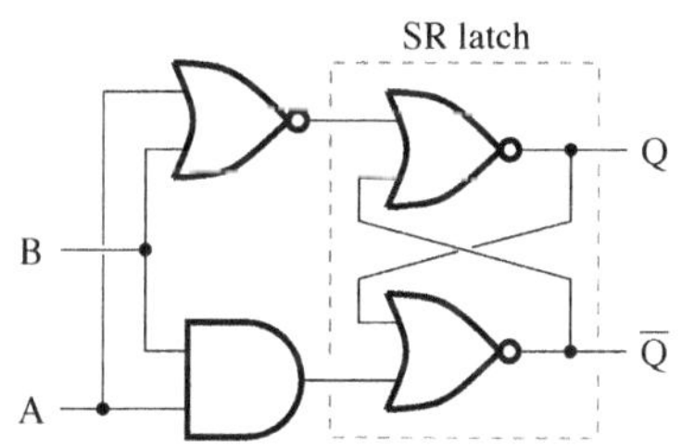

Figure 3.10. *Implementation of the C-element using an SR latch*

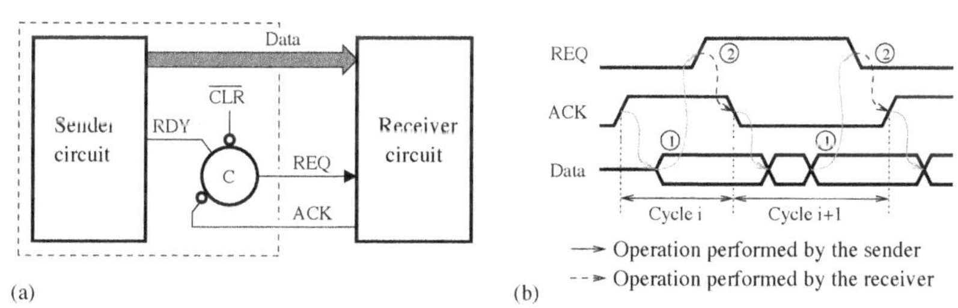

Figure 3.11. *a) Handshake communication system based on a two-phase protocol; b) timing diagram*

Initially, all signals are set to 0. The sender usually generates a *data ready* (RDY) pulse to mark the beginning of the data transfer and the activation of the C-element is initiated as a result of its two inputs taking the logic state 1. The transition of the REQ signal that follows enables the receiver to take control, thereby preventing any other placement of data on the bus. When the data transfer ends, the receiver produces a transition of the ACK signal and the C-element may again change the state of the REQ signal if another RDY pulse is detected.

This protocol has the advantage of minimizing the number of transitions required. However, the need to store data on flip-flops with a complex triggering mechanism – or that can be triggered by both the rising and falling edges of the signal – is only usually possible at the cost of increasing the power consumption and size of the circuit.

In the case of the four-phase protocol, as illustrated in Figure 3.12, a single transition type (rising edge or falling edge) is considered for each event. As soon as a binary word is placed on the data bus by the sender, a transition of the REQ signal is initiated and the control is passed to the receiver. When the receiver is available, the data transfer can be completed. This causes a transition of the ACK signal. The next operations consist of returning first the REQ signal and then the ACK signal to their initial states. Data transfer is performed during each cycle, which comprises four phases, two of which are carried out by the sender and two of which are performed by the receiver. The detection of these four phases requires the use of two C-elements, if the RDY and ACK signals are considered as pulses.

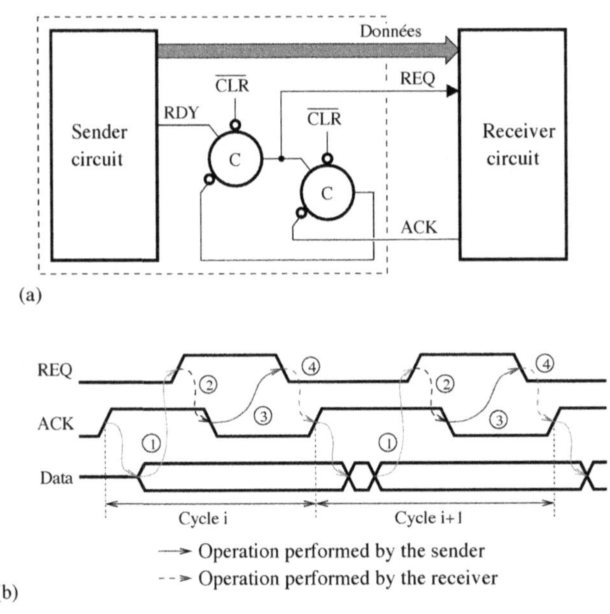

Figure 3.12. *a) Handshake communication system based on a four-phase protocol; b) timing diagram*

The four-phase protocol is slower and more complex. But its operation is not completely disturbed by faulty initialization, and only flip-flops triggered by either the rising or the falling edge of the signal are required.

Pipeline architecture is used in microprocessors to introduce parallelism in the execution of instructions, thereby reducing the response time.

The implementation of a pipeline datapath based on the two-phase protocol is represented in Figure 3.13, where each logic circuit executes a predefined operation. The use of double-edge triggered D flip-flops helps reduce the complexity of the control section. To ensure the proper functionality of the pipeline, the propagation delay introduced by the buffer circuit connecting the output of one C-element to the input of another C-element must be long enough to satisfy the data setup time requirement, while the propagation delay of each C element must long enough to meet the data hold time constraint. The minimum duration of a pipeline cycle is equal to the smallest time interval between two successive pulses of the request signal.

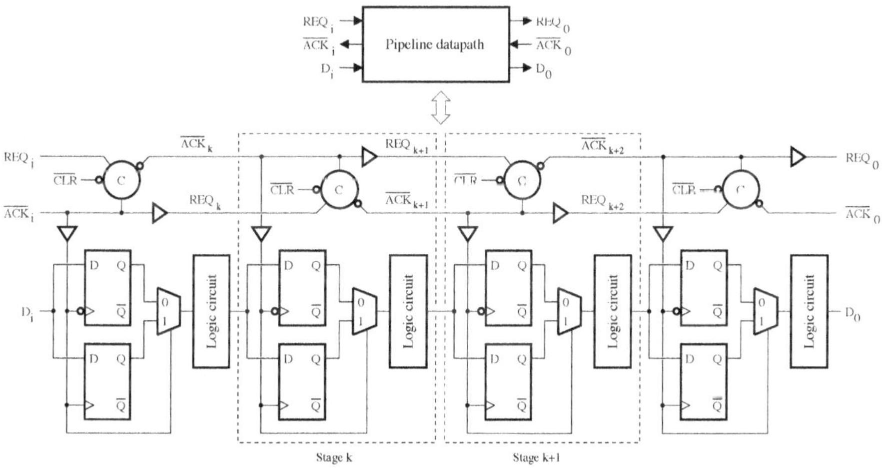

Figure 3.13. *Implementation of a pipeline datapath based on a two-phase protocol*

For the kth stage of the pipeline, a transition of the request signal, R_k, indicates that the input data are valid. This implies that the transfer of the previous data has been completed and the next stage has set the $\overline{ACK}_k$ signal to 0. The output of the C-element is used as the signal for the new data arriving at the input of the stage and as the R_{k+1} signal after a certain period of time, corresponding to the propagation delay introduced by the buffer circuit.

It is assumed by this that the transfer of the preceding data is complete and the next stage has set the $\overline{ACK_k}$ signal to 0. The output of the C-element is used as the acknowledgment signal for the new data arriving at the input of the stage and as the R_{k+1} signal after a certain period of time corresponding to the propagation delay introduced by the buffer circuit.

One advantage of the asynchronous control mode is that it facilitates the insertion of logic or arithmetic functions between the pipeline stages.

3.6. Encoding the states of an asynchronous state machine

In general, in order to avoid introducing critical race conditions in an asynchronous machine, the states must be encoded such that only one variable can change during each transition. But if certain critical race conditions still persist, they will be eliminated by introducing cycles through existing undefined or unstable states, or by making use of additional states or state variables to allow insertion of the appropriate cycles. However, sequencing through additional states can lead to a reduction of the final circuit speed.

Consider the asynchronous state machine described by the flow table shown in Figure 3.6. It is assumed that both inputs do not change simultaneously. The transition table, which is used to find the appropriate code for the representation of the states, can be constructed as shown in Figure 3.14. It highlights the transitions between the states as specified by the state table and is also called the adjacency diagram. As each state must be adjacent to three other states, encoding states with two state variables cannot solve the critical race problem. One possible solution consists, therefore, of using three state variables and considering a *shared-row state assignment* or a *multiple-row state assignment*.

PS	NS				Output			
	$XY = 00$	01	10	11	$XY = 00$	01	10	11
A	(A)	(A)	D	D	0	0	–	–
B	(B)	A	(B)	C	1	–	0	–
C	A	(C)	B	(C)	–	1	–	0
D	B	C	(D)	(D)	–	–	1	1

Table 3.6. *Flow table*

With the shared-row state assignment, the availability of additional state variables is exploited to define the intermediate states, so that some transitions can take place without being affected by critical race conditions.

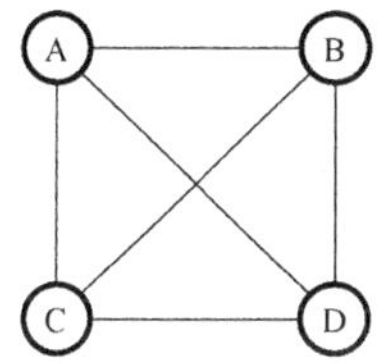

Figure 3.14. *Transition diagram*

The flow table shown in Table 3.7 is obtained by assigning binary codes to the states as follows: A (000), B (001), C (100), D (010), E (011), F (101) and G (110). It should be noted that the code 111 is not used. Using three state variables, it is possible to add lines to the flow table, and transitions between states can be controlled using cycles.

PS	NS				Output			
	$XY = 00$	01	10	11	$XY = 00$	01	10	11
A	(A)	(A)	D	D	0	0	–	–
B	(B)	A	(B)	F	1	–	0	0
C	A	(C)	F	(C)	–	1	0	0
D	E	G	(D)	(D)	1	1	1	1
E	B	–	–	–	1	–	–	–
F	–	–	B	C	–	–	0	0
G	–	C	–	–	–	1	–	–

Table 3.7. *Flow table for a state machine using shared-row state assignment*

Multiple-row state assignment consists of allocating more than one binary code to each state so that each transition from one state to another requires a change in only one state variable.

A systematic procedure to implement multiple-row state assignment independent of the configuration of the flow table is based on the use of an encoding that is said to be universal. Figures 3.15(a) and (b) illustrate the universal encoding that can be applied to any case where the minimum number of states does not exceed 4 and 8, respectively. For universal encodings, it should be noted that the states are pairwise equivalent and that the binary codes for equivalent states are logical complements. The synthesis of a state machine based on multiple-row state assignment can be carried out by replacing each state in a reduced flow table by two equivalent states.

The universal encoding in Figure 3.15(a) must be considered, as an example, to implement the multiple-row state assignment for the state machine whose flow table is represented in Table 3.6. The correspondence between the states and the binary codes is established as follows: A_1 (000), A_2 (111), B_1 (001), B_2 (110), C_1 (010), C_2 (101), D_1 (011) and D_2 (100). The extended flow table can then be represented as shown in Table 3.8. As only one state variable can change during the transition from one state to another, each of the states in a pair of equivalent states is adjacent to one of the states of each of the other pairs of equivalent states.

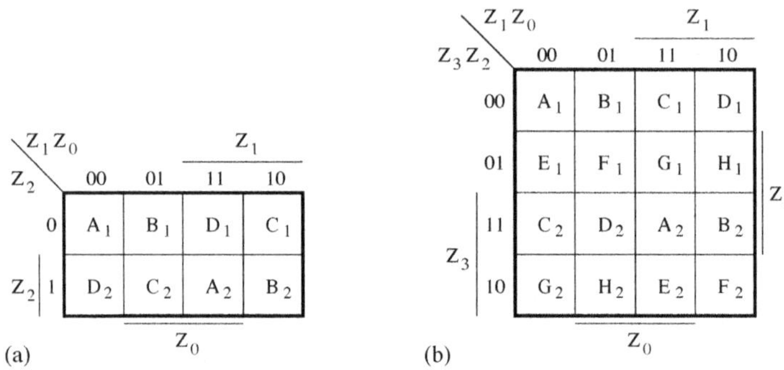

Figure 3.15. *Universal encoding when the minimum number of states does not exceed a) 4, and b) 8*

PS	NS				Output			
	$XY = 00$	01	10	11	$XY = 00$	01	10	11
A_1	(A_1)	(A_1)	D_2	D_2	0	0	-	-
A_2	(A_2)	(A_2)	D_1	D_1	0	0	-	-
B_1	(B_1)	A_1	(B_1)	C_2	1	-	0	0
B_2	(B_2)	A_2	(B_2)	C_1	1	-	0	0
C_1	A_1	(C_1)	B_2	(C_1)	-	1	0	0
C_2	A_2	(C_2)	B_1	(C_2)	-	1	0	0
D_1	B_1	C_1	(D_1)	(D_1)	1	1	1	1
D_2	B_2	C_2	(D_1)	(D_1)	1	1	1	1

Table 3.8. *Flow table of a state machine using multiple-row state assignment*

Multiple-row state assignment is simpler to implement than shared-row state assignment. However, it is most often characterized by a greater increase in the rows in the flow table and, therefore, by an increase in the complexity of the final circuit.

3.7. Synthesis of asynchronous circuits

During the design of an asynchronous state machine, the initial flow table obtained from the specifications most often has several unspecified inputs. This is due to the fact that the different possible transitions are listed assuming that a single variable changes state each time. The unspecified inputs offer a certain flexibility that can be used to minimize the number of states. Finally, the resulting logic equations depend on the binary code used to represent the machine states.

In addition to the critical race conditions and static hazards, the operation of an asynchronous state machine in the fundamental mode can be affected by other synchronization errors that result in oscillatory cycles, *essential* hazards or *delay-trio* (*d-trio*) hazards.

It should be noted that dynamic hazards occur mainly in logic circuits with more than two levels of AND and OR logic gates. As we can avoid using this type of circuit in the implementation of asynchronous state machines, dynamic hazards are not generally considered.

3.7.1. *Oscillatory cycle*

The transition of a machine from one stable state to another stable state, passing through one or more unstable states, constitutes a cycle.

When, under certain conditions, a machine enters and remains in a section of a cycle that involves only unstable states, an oscillatory cycle or oscillations can be observed.

Cycles are useful for the operation of asynchronous machines, while the oscillatory cycles must be eliminated.

Figure 3.16(a) depicts the state diagram of a machine with two inputs X and Y. The codes 00, 01, 11 and 10 being assigned to the states S_a, S_b, S_c and S_d, respectively; we obtain the Karnaugh map shown in Figure 3.16(b) that illustrates the transitions involving the states S_b and S_c.

An oscillatory cycle exists between the states S_b and S_c because the condition $\overline{X} \cdot \overline{Y}$ can be verified at the same time as the two transition conditions $X \odot Y$ and $\overline{Y}$.

Once a machine enters one of the states S_b or S_c, it oscillates between these two states as long as the condition $\overline{X} \cdot \overline{Y}$ is maintained at the inputs.

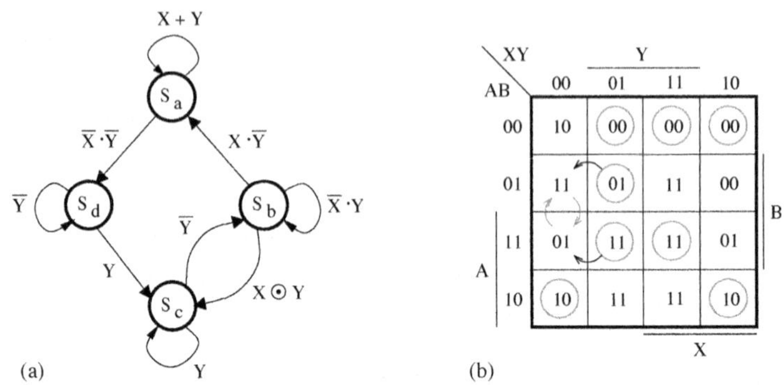

Figure 3.16. *a) State diagram; b) Karnaugh map illustrating an oscillatory cycle*

In general, when the conditions for the transition from the state S_I to the state S_J, T_{IJ}, and from the state S_J to the state S_I, T_{JI}, can be simultaneously true for the same combination of inputs, then:

$$T_{IJ} \cdot T_{JI} \neq 0 \qquad [3.9]$$

and there exists a logic expression that is common to both transitions and that can cause oscillations.

One solution to eliminate oscillatory cycles consists of modifying the transition conditions to suppress or reassign the term $\overline{X} \cdot \overline{Y}$, as illustrated in Figure 3.17(a). For the state S_b, the holding condition becomes $\overline{X} \cdot Y + \overline{X} \cdot \overline{Y}$, which is equal to $\overline{X}$. This implies, as shown in the Karnaugh map in Figure 3.17(b), inserting the 01 from the state S_b instead of the code 11 from the state S_c in the cell corresponding to the combinations $XY = 00$ and $AB = 01$.

Reassigning the term $\overline{X} \cdot \overline{Y}$ to eliminate the oscillatory cycle, the machine that is in the state S_c now goes to the state S_b, where it can be held under the condition $\overline{X} \cdot \overline{Y}$.

3.7.2. *Essential and d-trio hazards*

Essential and d-trio hazards are inherent to asynchronous sequential circuits with at least three states and operating in the fundamental mode. They arise because of the propagation delay that can be introduced by the parasitic components in a logic gate or an interconnect wire, so as to create a critical race condition along two paths (direct and indirect) stretching from the same input node up to the same logic gate.

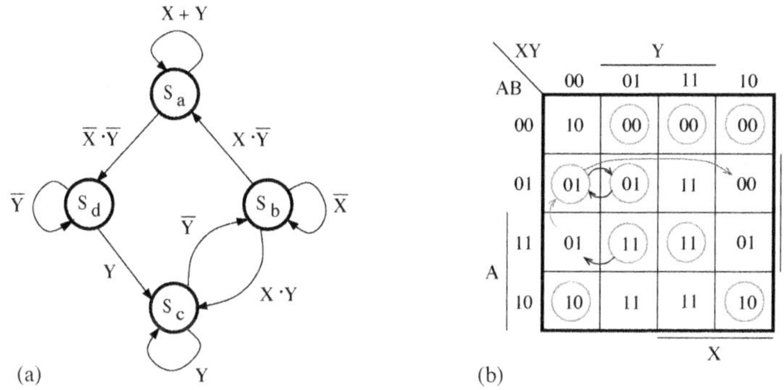

Figure 3.17. *a) State diagram after transformation; b) Karnaugh map illustrating the elimination of the oscillatory cycle*

The effect of each of these hazards can be eliminated by introducing delay elements in the feedback loop of the indirect path to prevent the execution of incorrect transitions.

3.7.2.1. *Essential hazard*

An essential hazard is a possible timing error related to a propagation delay introduced in one of the two paths, starting from an input node and converging toward the same logic gate. It results in a critical race condition between an input and a state variable activated by this input to reach a logic gate, and the state machine is found in an undesirable final state after at least two successive transitions.

Let us consider, as an example, the asynchronous state machine whose working is described by the state diagram shown in Figure 3.18.

The flow table of this state machine is represented in Table 3.9. Assigning the binary codes 00, 01, 11 and 10 to the states S_a, S_b, S_c and S_d, respectively, it can be used to construct the Karnaugh maps shown in Figure 3.19. The logic equations for the next states and the output can then be written as follows:

$$Z_1^+ = \overline{X} \cdot \overline{Y} \cdot Z_0 + \overline{X} \cdot Z_1 \cdot Z_0 + Y \cdot Z_1 \quad [3.10]$$

$$Z_0^+ = \overline{X} \cdot Z_0 + \overline{X} \cdot Y + Y \cdot Z_1 \quad [3.11]$$

and:

$$Z = \overline{Z}_1 \cdot Z_0 \qquad [3.12]$$

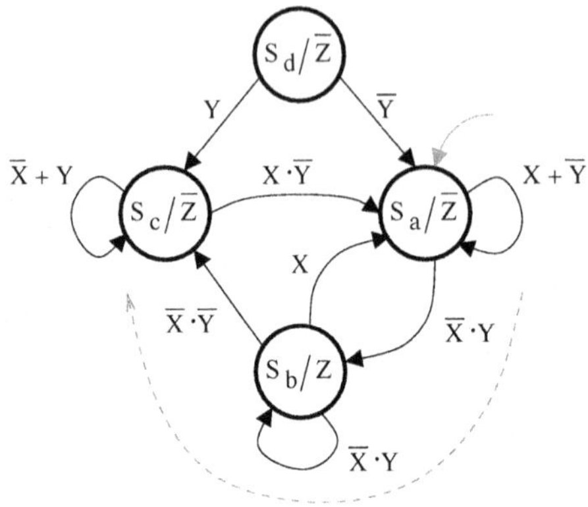

Figure 3.18. *State diagram*

PS	NS				Output, Z
	$XY = 00$	01	10	11	
S_a	(S_a)	S_b	(S_a)	(S_a)	0
S_b	S_c	(S_b)	S_a	S_a	1
S_c	(S_c)	(S_c)	S_a	(S_c)	0
S_d	S_a	S_c	S_a	S_c	0

Table 3.9. *Flow table*

Adding the redundant term $\overline{X} \cdot Z \cdot Z_0$ is required to obtain an implementation free from static hazards. Figure 3.20 depicts the logic circuit for the state machine.

When the state machine is in the state S_a with both inputs X and Y set to 0, the condition $\overline{X} \cdot Y$ drives the state machine to the state S_b. If, on the other hand, a certain delay, $\triangle t_e$, is explicitly located in the direct path of the input Y, the effect of the state change for Y will first reach the lowermost AND gate and cause the state variable, Z_0, to take to the logic state 1. This causes the state machine to move to the state S_b. When the logic state 1 of the state variable Z_0 is then applied to the uppermost AND gate (see Figure 3.20), the state variable Z_1 is set to 1 and the machine enters the state S_c. The machine is held in the state S_c even when the effect of the state change for Y finally reaches the uppermost AND gate (labeled with $*$ on the logic circuit in Figure 3.20) because of the logic state 1 taken by the outputs of the other AND gates connected to the OR gate whose output is Z_1.

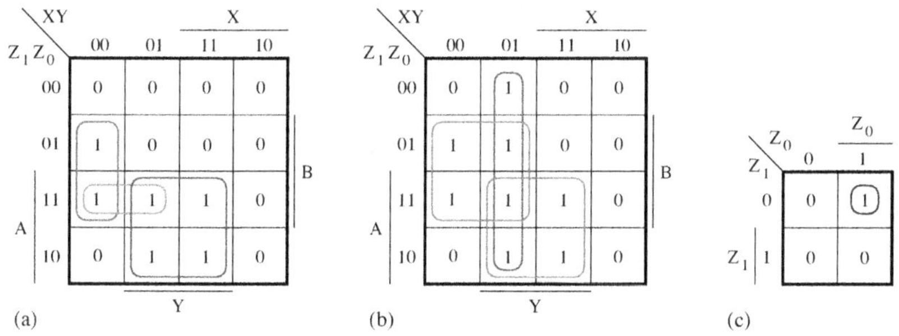

Figure 3.19. *Karnaugh maps: a) Z_1^+; b) Z_0^+; c) Z*

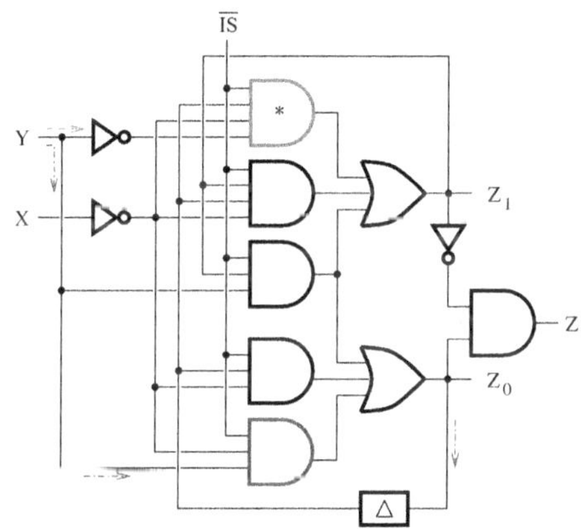

Figure 3.20. *Logic circuit*

Due to the effect of the essential hazard, the transition $S_a \rightarrow S_b$ or $00 \rightarrow 01$, under the condition $\overline{X} \cdot Y$, is carried out as a sequence of transitions, namely $S_a \rightarrow S_b \rightarrow S_c$ or $00 \rightarrow 01 \rightarrow 11$.

To avoid the formation of the above-mentioned essential hazard, delay elements must be inserted on the feedback path of the state variable Z_0, as illustrated in Figure 3.20.

Another implementation approach is based on the use of SR latches. In this case, the excitation table for the SR latch can be used to obtain the Karnaugh maps as represented in Figure 3.21, based on the flow table where each state is replaced by its binary code. The logic equations for the inputs of latches and the output of the state machine are given by:

$$S_1 = \overline{X} \cdot \overline{Y} \cdot Z_0 \qquad [3.13]$$

$$R_1 = X \cdot \overline{Y} + \overline{Y} \cdot \overline{Z}_0 \qquad [3.14]$$

$$S_0 = \overline{X} \cdot Y + Y \cdot Z_1 \qquad [3.15]$$

$$R_0 = X \cdot \overline{Y} + X \cdot \overline{Z}_1 \qquad [3.16]$$

and:

$$Z = \overline{Z}_1 \cdot Z_0 \qquad [3.17]$$

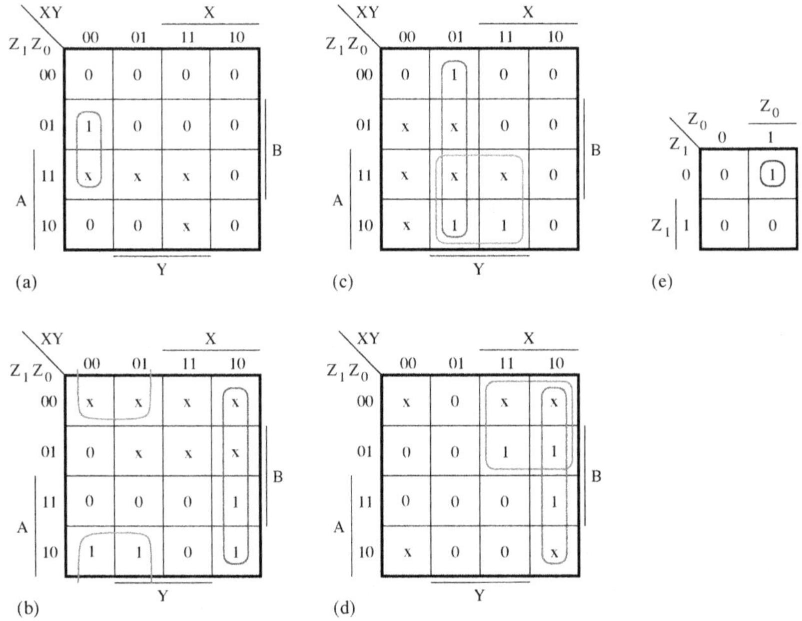

Figure 3.21. *Karnaugh maps: a) S_1; b) R_1; c) S_0; d) R_0; e) Z*

The logic circuit of the state machine is depicted in Figure 3.22. The $\overline{IS}$ signal is active low and is used to initialize the machine.

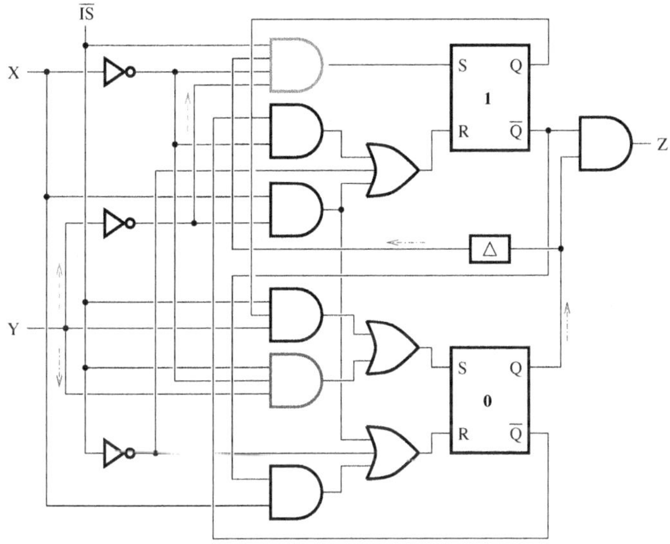

Figure 3.22. *Logic circuit*

As suggested previously, delay elements are introduced in the feedback path of the state variable Z_0 to prevent the formation of essential hazards that could affect the transition $S_a \rightarrow S_b$ under the condition $\overline{X} \cdot Y$.

NOTE.– In the case of an asynchronous state machine with a single input, the effect of an essential hazard is often manifested by the transformation of a transition into a sequence of three transitions.

An essential hazard can be identified by analyzing the asynchronous state machine, whose operation is described by the flow table shown in Table 3.10. This is a single input state machine whose state diagram is represented in Figure 3.23(a). The Karnaugh map shown in Figure 3.23(b) is constructed by encoding each of the states using a binary combination of the variables Y and Z. The state S_a is represented by 00, S_b by 01, S_c by 11 and S_d by 10.

When the machine is in the state S_a and the input X is set to 0, a change in the logic state of the input X should cause a transition to the state S_b. However, if during the transition from S_a to S_b, the effect of the state change of the variable Z is taken into account before that of the input X, the variable Y can take the logic state 1 and the machine will then move to the state S_c. And when the state change of the input X is finally taken into account, the state machine moves to the state S_d.

PS	NS $X = 0$	1
S_a	S_a	S_b
S_b	S_c	S_b
S_c	S_c	S_d
S_d	S_a	S_d

Table 3.10. *Flow table*

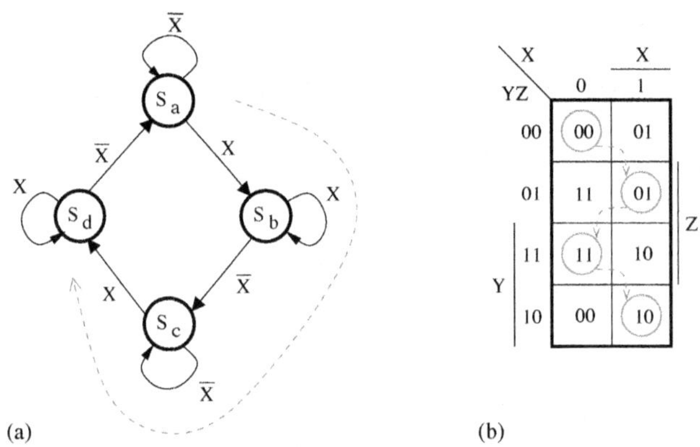

Figure 3.23. *a) State diagram; b) Karnaugh map illustrating an essential hazard*

Due to the essential hazard, a single input state machine that should go from one state to another, as in the case $S_a \rightarrow S_b$, instead carries out a sequence of transitions to settle in a different state from the expected one, namely $S_a \rightarrow S_b \rightarrow S_c \rightarrow S_d$.

Figure 3.24(a) shows the logic circuit of the state machine based on the logic equations obtained from the flow table as follows:

$$Y^+ = X \cdot Y + \overline{X} \cdot Z + Y \cdot Z \quad [3.18]$$

and:

$$Z^+ = X \cdot \overline{Y} + X \cdot Z + \overline{Y} \cdot Z \quad [3.19]$$

The timing diagram shown in Figure 3.24(b) illustrates the effect of an essential hazard on the working of the state machine. Initially, the state of the state machine is characterized by $YZ = 00$. When the input X changes from the logic state 0 to 1, taking into account the propagation delay of the inverter connected to the signal X, the effect of the state change is propagated through the machine so that the state $YZ = 01$ is first reached. As the signal $\overline{X}$ has not yet switched from the logic level 1 to 0, the state of the signal $\overline{X} \cdot Z$ becomes 1 instead of being held at 0. This causes the variable Y to move to the logic state 1 and the machine takes the state $YZ = 11$. When the signal $\overline{X}$ then takes the logic state 1, the machine wrongly enters the state $YZ = 10$.

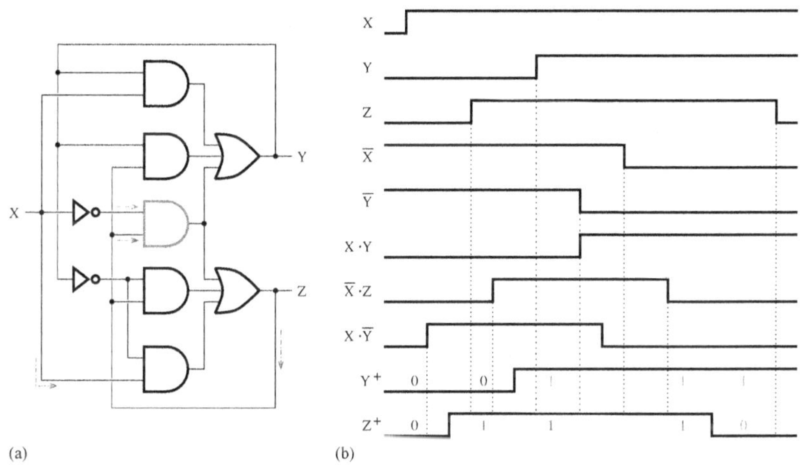

Figure 3.24. *a) Logic circuit of the state machine; b) timing diagram illustrating the effect of the essential hazard on the state machine operation*

The essential hazard can be eliminated by inserting delay elements in the feedback path of the state variable Z, as shown in Figure 3.25, to prevent incorrect transitions from taking place.

3.7.2.2. *d-trio hazard*

A d-trio hazard is caused by the propagation delay introduced on one of the two paths extending from one input to a logic gate. Due to the effect of a d-trio hazard, an asynchronous state machine first transits to an erroneous transient state before settling in the desired state. A d-trio hazard often results in a delayed transition between the initial and the final states, but it can also cause an undesired state change of an output.

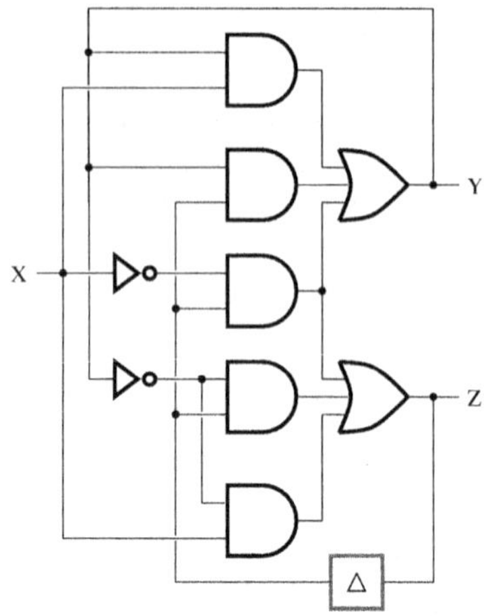

Figure 3.25. *Logic circuit of the state machine with delay elements*

Consider the logic circuit shown in Figure 3.26, where the $\overline{R}$ signal is active-low and is used to drive the state machine to the initial state. For operation in the fundamental mode, the logic equations of the state variables can be written as follows:

$$Y^+ = \overline{R}(\overline{X} \cdot Y + \overline{X} \cdot Z) \quad [3.20]$$

and:

$$Z^+ = \overline{R}(X + Z) \quad [3.21]$$

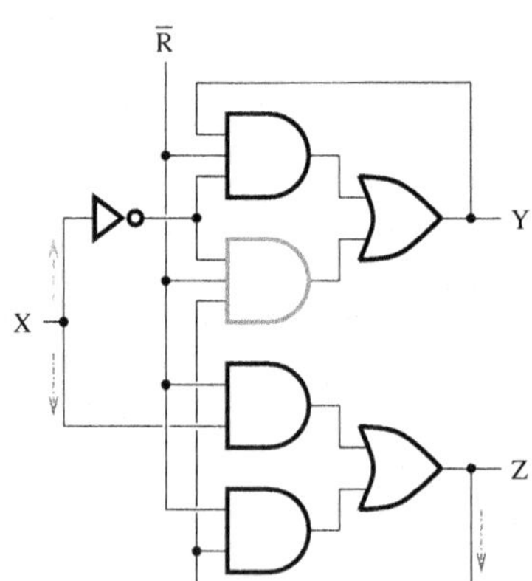

Figure 3.26. *Logic circuit of an asynchronous state machine*

During a normal operation, $\overline{R} = 1$, and the transition table can be constructed as shown in Table 3.11. By assigning binary codes 00, 01, 11 and 10 to the states S_a, S_b, S_c and S_d, respectively, the flow table can be represented as shown in Table 3.12, where each stable state is encircled.

PS	NS	
YZ	Y^+Z^+	
	$X = 0$	1
00	(00)	01
01	11	(01)
10	(10)	01
11	(11)	01

Table 3.11. *Transition table*

PS	NS	
	$X = 0$	1
S_a	(S_a)	S_b
S_b	S_c	(S_b)
S_d	(S_d)	S_b
S_c	(S_c)	S_b

Table 3.12. *Flow table*

In the presence of the d-trio hazard, the path followed by the state machine is illustrated in the state diagram in Figure 3.27(a), and also in the Karnaugh map in Figure 3.27(b). A change in the logic state for the input X can follow a direct path and also an indirect path via Z to reach the AND gate connected to $\overline{X}$ and Z. From the state S_a, setting the input X to 1 results in the state variable Z being set at 1. When, to determine the state variable Y, the effects of these modifications are taken into account in the order of their occurrence, the machine goes to the state S_b. If, as a result of the propagation delay caused on the direct path by X, the effect of the logic state change in Z is propagated and reaches the aforementioned AND gate before that of the input X, the state variable Y takes the logic state 1 and the machine moves to the state S_c. When the effect of the state change of X is then taken into account, the logic state of Y changes again and the machine returns to the state S_b.

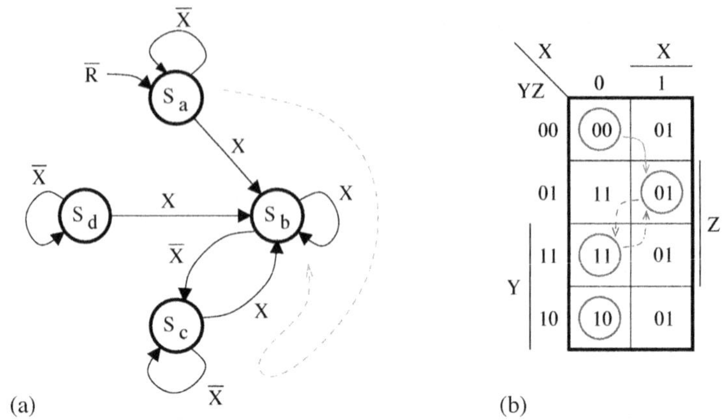

Figure 3.27. *a) State diagram; b) Karnaugh map illustrating the path taken by the state machine due to the d-trio hazard*

Due to the effect of the d-trio hazard, the state machine that should move from one state to another, as shown by the transition $S_a \rightarrow S_b$, actually undergoes undesirable transitions before stabilizing itself in the target state, namely $S_a \rightarrow S_b \rightarrow S_c \rightarrow S_b$ or $S_a \rightarrow S_b \leftrightarrows S_c$. One solution to prevent the formation of d-trio hazards consists of adding delay elements along the feedback path for the variable Z.

3.7.2.3. *Essential and d-trio hazard detection*

An asynchronous state machine operating in the fundamental mode can also be affected by essential hazards or d-trio hazards. A section of the state diagram for such a state machine is reproduced in Figure 3.28.

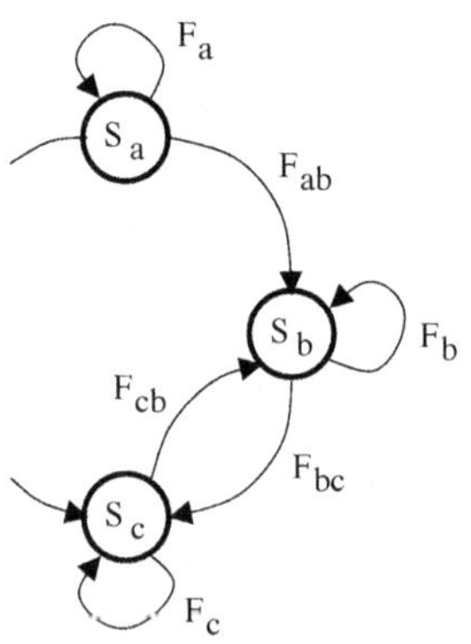

Figure 3.28. *Section of a state diagram*

Considering only state machines implemented as two-level logic circuits, the formation of an essential hazard or a d-trio hazard is only possible if a propagation delay of sufficient value is introduced in the direct path of the initiator input and at least the following requirements are satisfied:

– the branching condition, F_{ab}, must be contained[1] in the holding condition F_b;

– the branching condition F_{bc} must be contained in the holding condition F_a;

– only a change in the logic state of the hazard initiator input is allowed in the branching conditions F_{ab} and F_{bc} and all other inputs must remain constant.

The branching condition F_{ab} is not contained in the branching condition F_{cb} in the case of an essential hazard, while the condition F_{ab} is contained in the condition F_{cb} in the case of a d-trio hazard.

3.7.3. *Design of asynchronous state machines*

The synthesis of an asynchronous state machine can involve the following steps:

1) develop functional and temporal specifications for the circuit;

2) obtain the flow table:

i) derive the primitive flow table;

ii) verify that the timing constraints are satisfied and that the primitive flow table does not contain oscillatory cycles;

iii) minimize the primitive flow table to obtain the reduced flow table;

3) encode the states: the states must be encoded such that each transition from one state to another requires a change in only one variable. Otherwise, the flow table must be enlarged by inserting additional states so that the transitions only occur between logically adjacent states, and all critical race conditions must be searched and eliminated;

4) derive the logic equations for the next states and the outputs:

i) construct the appropriate Karnaugh maps using the transition table;

ii) if necessary, add redundant terms to the minimal expressions obtained to eliminate the static hazards;

1 Example: The logic expression $A \cdot B$ is contained in $A + \overline{B}$, while the expression $\overline{A} \cdot B$ is not contained in $A + \overline{B}$. Thus, every time the expression $A \cdot B$ is true, $A + \overline{B}$ is also true, but the converse may not be true.

iii) identify the essential and d-trio hazards that can only be eliminated by re-examining the problem or by inserting delay elements (for example two inverters in series) along the feedback path;

5) implement the logic circuit.

3.8. Application examples of asynchronous state machines

Asynchronous state machines are used as basic components in several data processing, communication and data verification applications. However, the examples considered here are characterized by a certain level of simplicity.

3.8.1. *Pulse synchronizer*

Implement a pulse synchronizer with one input for the clock signal, C, one control input, P, and one output, Z.

When the clock signal first changes to the logic state 1, and if the control signal is set to 1, a pulse, whose width is equal to a half period of the clock signal, is generated at the output. In the case where the clock signal moves to the logic state 1 when the control signal is set to 1 and then reset to 0 before the clock signal is again set to 1, the output remains at the logic state 0. If the clock signal goes to the logic state 1 when the control signal is set to 0, the logic state 0 is maintained at the output.

Two successive edges of the control signal are separated at least by a half period of the clock signal, and a single pulse is generated at the output whenever the control signal is set to 1.

The symbol and an example of a timing diagram for the pulse synchronizer are shown in Figures 3.29(a) and (b), respectively.

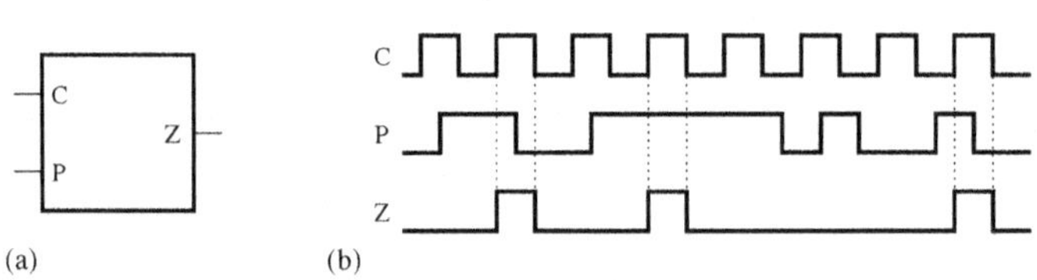

Figure 3.29. *a) Symbol and b) timing diagram of a pulse synchronizer*

The initial state diagram is constructed according to specifications, as shown in Figure 3.30, by listing the different possible transitions. The initial description of the pulse synchronizer uses eight states. Each combination formed by the two inputs and the output is used to define the condition for the transition toward a given state and the holding condition in this state. The initial flow table is represented in Table 3.13.

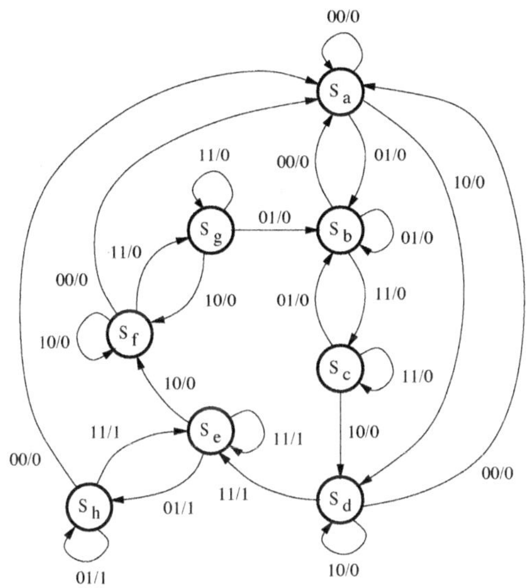

Figure 3.30. *Initial state diagram of the pulse synchronizer*

PS	NS				Output Z
	$PC = 00$	01	10	11	
S_a	S_a	S_b	S_d	–	0
S_b	S_a	S_b	–	S_c	0
S_c	–	S_b	S_d	S_c	0
S_d	S_a	–	S_d	S_e	0
S_e	–	S_h	S_f	S_e	1
S_f	S_a	–	S_f	S_g	0
S_g	–	S_b	S_f	S_g	0
S_h	S_a	S_h	–	S_e	1

Table 3.13. *Initial flow table of the pulse synchronizer*

Applying the simplification procedure for incompletely specified state machines, we can determine the following compatibility classes: (S_a, S_b, S_c), (S_d), (S_e, S_h), and (S_f, S_g). The reduced flow table shown in Table 3.14 is obtained by making the following assumptions:

$$S_0 = S_a = S_b = S_c \qquad [3.22]$$

$$S_1 = S_d \qquad [3.23]$$

$$S_2 = S_e = S_h \qquad [3.24]$$

and:

$$S_3 = S_f = S_g \qquad [3.25]$$

PS	NS				Output Z
	$PC = 00$	01	10	11	
S_0	(S_0)	(S_0)	S_1	(S_0)	0
S_1	S_0	–	(S_1)	S_2	0
S_2	S_0	(S_2)	S_3	(S_2)	1
S_3	S_0	S_0	(S_3)	(S_3)	0

Table 3.14. *Reduced flow table of the pulse synchronizer*

Natural binary code, or Gray code, can be chosen to represent the states of the pulse synchronizer.

Using the natural binary code, we can obtain the Karnaugh map represented in Figure 3.31(a), where the oriented arcs indicate the transitions that can be affected by the race conditions, one of which is non-critical while the other two are critical. The non-critical race condition does not really hinder the operation as the same final state is reached regardless of the path taken. On the other hand, the critical race conditions that can cause a malfunction of the circuit are eliminated by using Gray code. Hence, the Karnaugh map shown in Figure 3.31(b) exhibits only one non-critical race condition.

The Karnaugh maps for the variables X^+ and Y^+, as illustrated in Figures 3.32(a) and (b), are derived from the flow table, assuming that the states are represented using Gray code. Thus:

$$X^+ = C \cdot Y + P \cdot X \qquad [3.26]$$

and:

$$Y^+ = C \cdot Y + P \cdot \overline{C} \cdot \overline{X} + P \cdot \overline{X} \cdot Y \qquad [3.27]$$

where a redundant term is added to the minimal expression of Y^+ to prevent a circuit malfunction due to the hazard caused by unbalanced propagation delays. The logic equation for the output, Z, is given by:

$$Z = X \cdot Y \qquad [3.28]$$

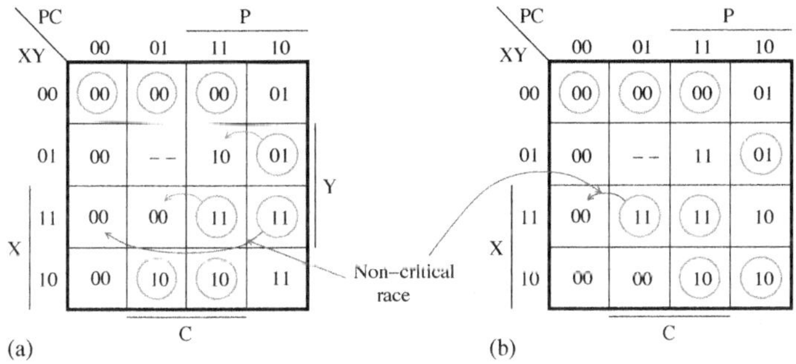

Figure 3.31. *Karnaugh maps for two different state encodings*

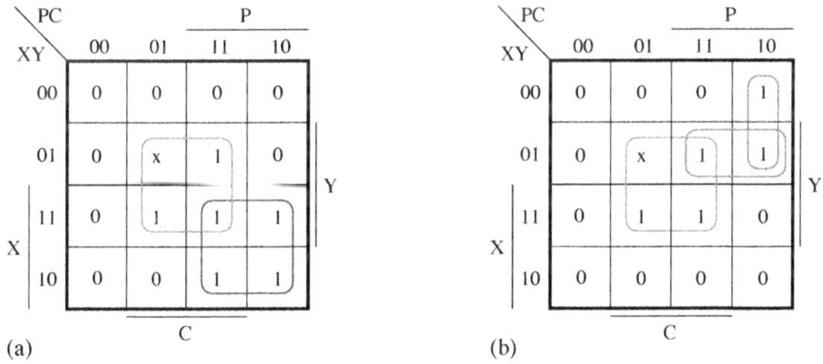

Figure 3.32. *Karnaugh maps: a) X^+; b) Y^+*

Figure 3.33 presents the logic circuit of the pulse synchronizer.

3.8.2. *Asynchronous counter*

Implement an asynchronous state machine that can count the pulses of an input signal. This is a modulo 4 counter having one input and two outputs.

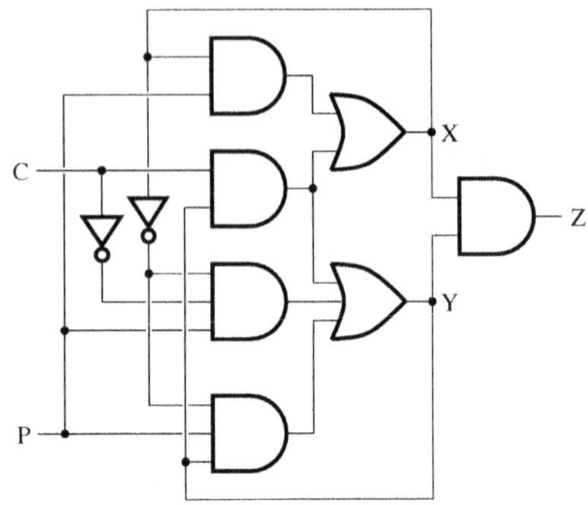

Figure 3.33. *Logic circuit of the pulse synchronizer*

To count the pulses, the asynchronous state machine must change its state each time a transition of the input signal is detected. Figure 3.34 presents the state diagram of the counter, where eight states are required to represent the transition of each of the four consecutive pulses of the input signal.

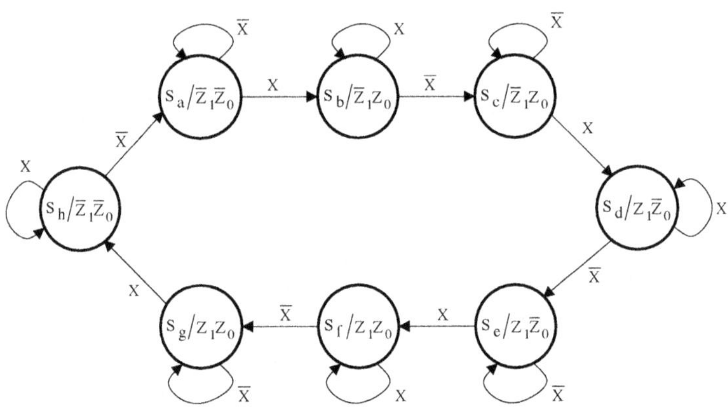

Figure 3.34. *State diagram of the modulo 4 counter*

Initially, the counter is in the state S_a and the input signal X is set to 0. It is held in one state as long as the signal X is at the same logic state, and changes state each time there is a transition of the signal X. The outputs Z_1 and Z_0 are identical for two states, because a pulse begins with a transition from 0 to 1 and ends with a transition from 1 to 0.

The primitive flow table and the transition table of the counter are represented in Tables 3.15 and 3.15, respectively. To eliminate any race condition, the binary codes are assigned to the states such that each transition only requires the modification of one of the state variables: Y_2, Y_1 or Y_0. The outputs, Z_1 and Z_0, give a binary representation of the number associated with a given pulse. The Karnaugh maps, as shown in Figure 3.35, are obtained from the transition table and can be used to determine the logic equations for the state variables and the outputs, as follows:

$$Y_2^+ = X \cdot Y_2 + Y_2 \cdot Y_0 + \overline{X} \cdot Y_1 \cdot \overline{Y_0} + Y_2 \cdot Y_1 \quad [3.29]$$

$$Y_1^+ = X \cdot Y_1 + Y_1 \cdot \overline{Y_0} + \overline{X} \cdot \overline{Y_2} \cdot Y_0 + \overline{Y_2} \cdot Y_1 \quad [3.30]$$

$$Y_0^+ = \overline{X} \cdot Y_0 + X \cdot Y_2 \cdot Y_1 + X \cdot \overline{Y_2} \cdot \overline{Y_1} + \overline{Y_2} \cdot \overline{Y_1} \cdot Y_0 + Y_2 \cdot Y_1 \cdot Y_0 \quad [3.31]$$

$$Z_1^+ = Y_2 \cdot Y_0 + Y_1 \cdot \overline{Y_0} + Y_2 \cdot Y_1 \quad [3.32]$$

and:

$$Z_0^+ = Y_0 \quad [3.33]$$

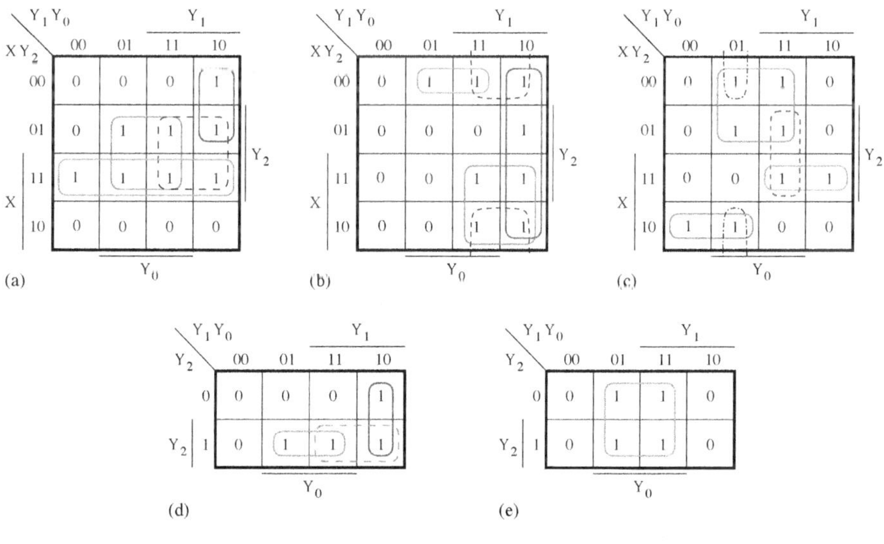

Figure 3.35. *Karnaugh maps: a) Y_2^+; b) Y_1^+; c) Y_0^+; d) Z_1; e) Z_0*

It should be noted that the equations for the state variables and for one of the outputs contain redundant terms to compensate for the effect of hazards on the counter operation.

PS	NS $X = 0$	NS 1	Outputs $Z_1 Z_0$
S_a	(S_a)	S_b	0 0
S_b	S_c	(S_b)	0 1
S_c	(S_c)	S_d	0 1
S_d	S_e	(S_d)	1 0
S_e	(S_e)	S_f	1 0
S_f	S_g	(S_f)	1 1
S_g	(S_g)	S_h	1 1
S_h	S_a	(S_h)	0 0

Table 3.15. *Primitive flow table of the modulo 4 counter*

PS $Y_2Y_1Y_0$	NS $Y_2^+Y_1^+Y_0^+$ $X = 0$	NS 1	Outputs $Z_1 Z_0$
000	(000)	001	0 0
001	011	(001)	0 1
011	(011)	010	0 1
010	110	(010)	1 0
110	(110)	111	1 0
111	101	(111)	1 1
101	(101)	100	1 1
100	000	(100)	0 0

Table 3.16. *Transition table of the modulo 4 counter*

The logic circuit of the asynchronous modulo 4 counter is represented in Figure 3.36. The $\overline{CLR}$ signal is active-low and is used to reset the counter.

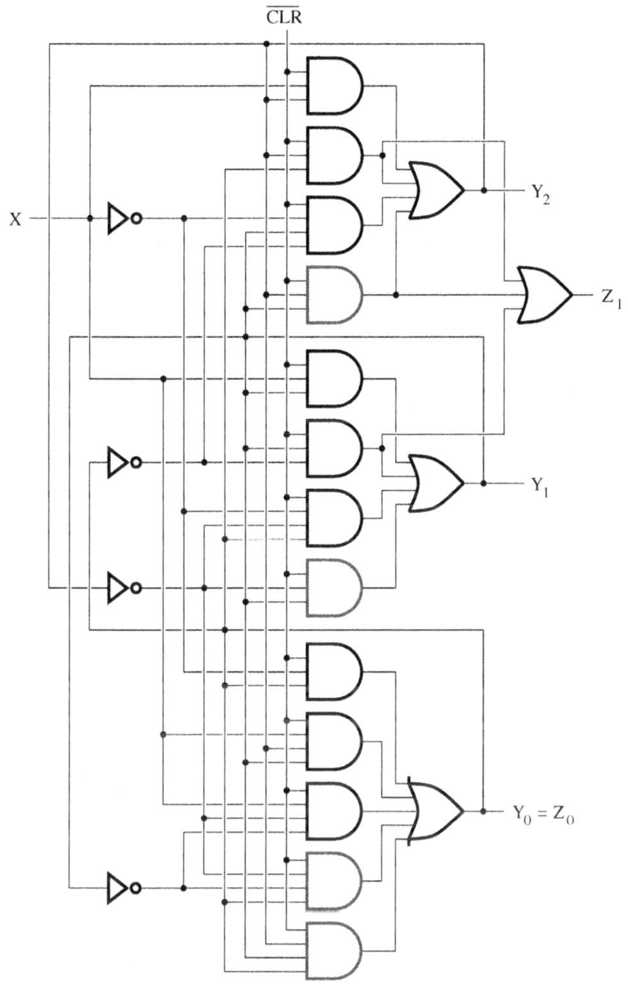

Figure 3.36. *Logic circuit of the counter*

3.9. Implementation of asynchronous machines using SR latches or C-elements

An asynchronous state machine can be implemented by combining logic gates and SR latches or C-elements. The excitation table of the SR latch or the C-element is used in conjunction with a transition table to construct the Karnaugh maps required for the determination of the logic equations for the inputs of the SR latches or C-elements and the output logic equations. In the case of SR latches, the logic equations for the inputs S_i and R_i must be such that $S_i \cdot R_i = 0$.

The advantage of SR latches or C-elements results from the fact that the generation of next states is not affected by static hazards.

Two types of sequential components can be used to implement asynchronous state machines: the SR latch, which operate in the fundamental mode, and the C-element, which is used to synchronize signals and can operate in modes other than the fundamental mode. Figure 3.37 presents the symbols of a C-element, an SR latch and a complementary C-element that has one active-high input and one active-low input.

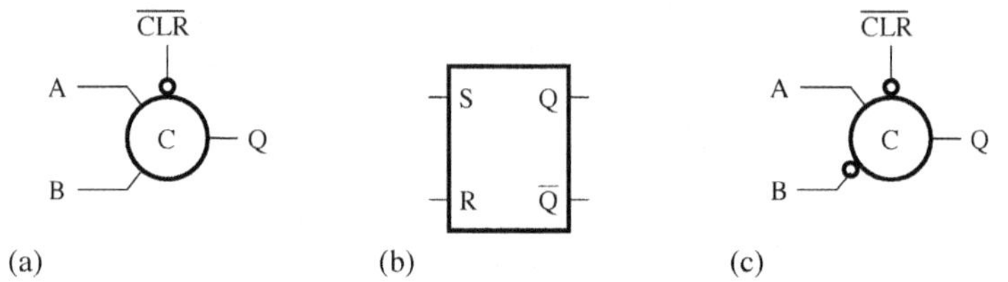

Figure 3.37. *a) Symbols of C-element, b) an SR latch and c) a complementary C-element*

Each sequential component can be characterized by an excitation table, as shown in Table 3.17 for a C-element, an SR latch and a complementary C-element. Note that there is a similarity between the SR latch and the complementary C-element.

$Q \rightarrow Q^+$	C-element A	C-element B	SR latch S	SR latch R	Complementary C-element A	Complementary C-element B
$0 \rightarrow 0$	0	x	0	x	0	x
	x	0			x	1
$0 \rightarrow 1$	1	1	1	0	1	0
$1 \rightarrow 0$	0	0	0	1	0	1
$1 \rightarrow 1$	1	x			1	x
	x	1	x	0	x	0

Table 3.17. *Comparison of excitation tables for the C-element, SR latch and complementary C-element*

To implement a state machine based on the logic equations for the inputs S_i and R_i, the inputs R_i must be logically complemented if the basic components to be used are C-elements instead of SR latches.

An arbiter is to be implemented for the serial communication system shown in Figure 3.38, where the memory is considered as a resource shared by circuits 1 and 2.

Only a single circuit can access the memory at one time. In the case of multiple requests, the arbiter circuit establishes an order of priority to allow access to the memory.

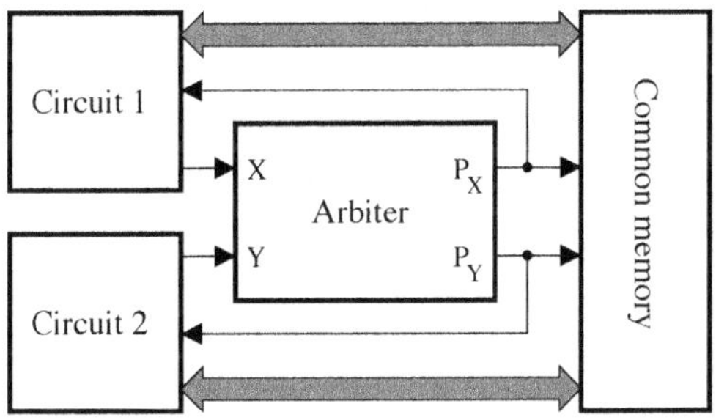

Figure 3.38. *Communication system*

Circuits 1 and 2 initiate communication by setting the signals X and Y, respectively, to 1. When the memory is available, the arbiter can grant access to circuit 1 by setting the signal P_X to 1, or to circuit 2 by setting the signal P_Y to 1. The signals X and Y are then reset to 0 by circuits 1 and 2, respectively, at the end of each transmission. That is, the arbiter can reset one of the signals, P_X or P_Y, to 0.

The flow table of the arbiter is illustrated in Table 3.18, where the initial state is represented by S_a. The binary codes 00, 01, 10 and 11 are assigned to the states S_a, S_b, S_c and S_d, respectively.

PS	NS				Outputs, $P_X P_Y$
	$XY = 00$	01	10	11	
S_a	(S_a)	S_b	S_c	–	0 0
S_b	S_a	(S_b)	S_d	(S_b)	0 1
S_c	S_a	S_d	(S_c)	(S_c)	1 0
S_d	–	S_b	S_c	–	1 0

Table 3.18. *Flow table*

Access to the memory is granted, on a priority basis, either to circuit 1 when the arbiter is in the state S_b, or to circuit 2 when the arbiter circuit is in the state S_c. As the inputs X and Y are assumed to not change state simultaneously, the transition from $XY = 00$ to $XY = 11$ is not possible from the state S_a, where the requests start to be received. The state S_d is introduced to prevent the establishment of a race condition between the states S_b and S_c.

The Karnaugh maps shown in Figure 3.39 can be used to determine the logic equations for the latch inputs S_A, R_A, S_B and R_B, and the state machine outputs P_X and P_Y. Thus:

$$S_0 = X \cdot \overline{Y} \qquad [3.34]$$

$$R_0 = \overline{X} \cdot Y \qquad [3.35]$$

$$S_1 = \overline{X} \cdot Y \qquad [3.36]$$

$$R_1 = \overline{X} \cdot \overline{Y} + X \cdot A \qquad [3.37]$$

$$P_X = A \qquad [3.38]$$

and:

$$P_Y = \overline{A} \cdot B \qquad [3.39]$$

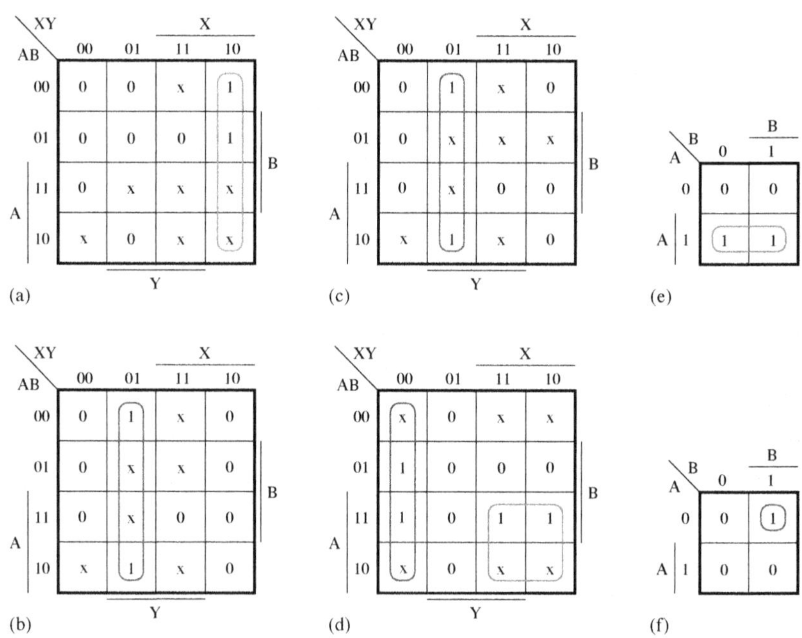

Figure 3.39. *Karnaugh maps: a) S_0; b) R_0; c) S_1; d) R_1; e) P_X; (f) P_Y*

Figure 3.40 shows the logic circuit of the arbiter, whose initialization (or transition to the state S_a) can be triggered by the active-low signal, $\overline{IS}$.

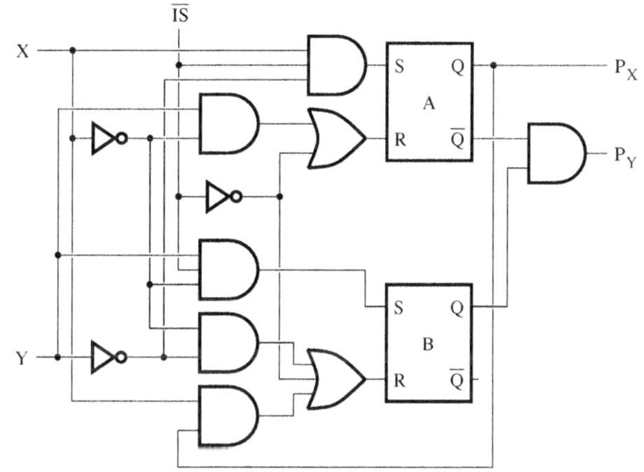

Figure 3.40. *Logic circuit of the arbiter based on SR latches*

The arbiter circuit can also be implemented using C-elements, as illustrated in Figure 3.41.

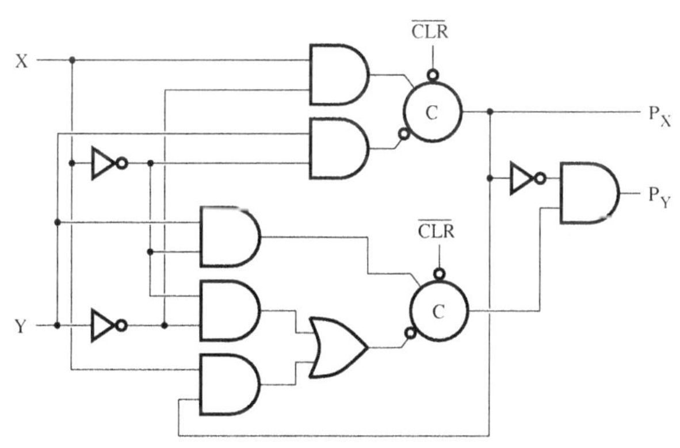

Figure 3.41. *Logic circuit of the arbiter based on C-elements*

3.10. Asynchronous state machine operating in pulse mode

Asynchronous state machines that can operate in pulse mode are designed using flip-flops activated by data signals and assuming that the input signals are pulses which do not overlap. They offer the advantage of being insensitive to the imperfections that may affect the operation of state machines in the fundamental mode. However, operation that does not depend on a clock signal is only possible at the price of a constraint imposed on the input signals of the state machine.

The implementation of state machines operating in pulse mode is based on the use of flip-flops in switching mode. Some possible configurations are illustrated in Figure 3.42. Because input pulses must have an adequate width and be separated enough to initiate transition from one state to another, it is preferable to use flip-flops triggered by the falling edge when using active-high signals or flip-flops triggered by the rising edge when using active-low signals.

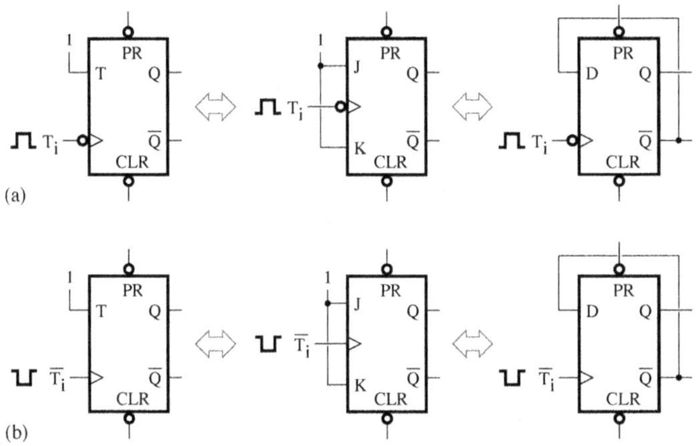

Figure 3.42. *Flip-flops activated by a) the falling edge or b) the rising edge of a data signal*

Determining the input logic equations for the flip-flops and for the outputs of the state machine from the transition table requires the use of an excitation table, such as that shown in Table 3.19, where the transitions $0 \rightarrow 1$ and $1 \rightarrow 0$ correspond to switching.

EXAMPLE 3.1.– sequence detector.

Implement a sequence detector whose output, Z, is set to 1 to indicate the detection of three consecutive Y pulses following on any number of X pulses, where X and Y represent the input signals.

The operation of a $X - Y - Y - Y$ sequence detector is described by the state diagram shown in Figure 3.43. The detector is a state machine which has four states and its flow table is represented in Table 3.20, where the undefined states must be interpreted taking into account the specification. Assigning the binary codes 00, 01, 11, and 10 to the states S_a, S_b, S_c, and S_d, respectively, and using the excitation table for a flip-flop operating in switching mode (or a T flip-flop), the Karnaugh maps can be constructed as shown in figures 3.44a, 3.44b, and 3.44c to determine the logic equations for the inputs T_A and T_B, and the output Z. It should be noted that the logic state 0 is assigned to each of the cells associated with the column $XY = 00$ and the cell $ABXY = 0001$, while a don't-care state is attributed to each cell related to the column $XY = 11$. The resulting logic equations can be written as follows:

$$T_A = A \cdot X + A \cdot \overline{B} \cdot Y + \overline{A} \cdot B \cdot Y \quad [3.40]$$

$$T_B = \overline{A} \cdot X + \overline{B} \cdot X + A \cdot B \cdot Y \quad [3.41]$$

and:

$$Z = A \cdot \overline{B} \cdot Y \quad [3.42]$$

Q	→	Q^+	T_i
0	→	0	0
0	→	1	1
1	→	0	1
1	→	1	0

Table 3.19. *Excitation table of a flip-flop configured to operate in pulse mode*

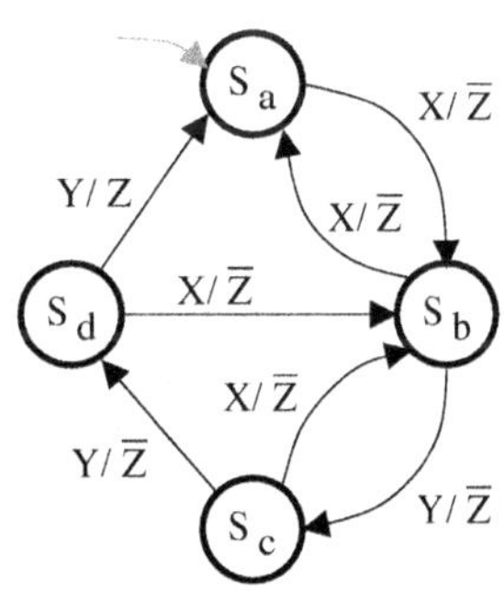

Figure 3.43. *State diagram of the sequence detector*

Figure 3.45 shows the logic circuit of the sequence detector.

PS	NS				Output, Z			
	$XY = 00$	01	10	11	$XY = 00$	01	10	11
S_a	-	-	S_b	-	-	-	0	-
S_b	-	S_c	S_a	-	-	0	0	-
S_c	-	S_d	S_b	-	-	0	0	-
S_d	-	S_a	S_b	-	-	1	0	-

Table 3.20. *Flow table of the sequence detector*

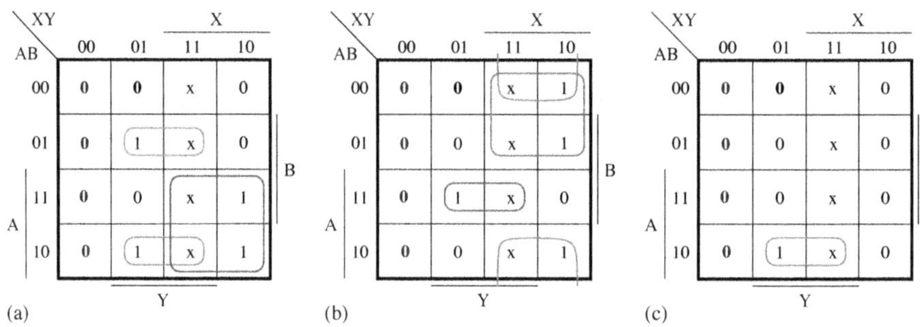

Figure 3.44. *Karnaugh maps: a) T_A; b) T_B; c) Z*

NOTE.– In the case of a state machine operating in pulse mode, the state diagram and the flow table have some specific characteristics:

Each branching condition always corresponds to the logic state of a single variable or an OR logic function of single variables. It is possible to represent these variables only in the non-complemented form when the signals are active-high or in the complemented form when the signals are active-low. No unconditional branching is allowed.

Only the flip-flop output toggling that characterize the operation in pulse mode allows the transition from one state to another for each combination of inputs. Thus, the holding condition in a given state must not appear in the state diagram or flow table.

The sum law is not verified but the mutual-exclusion requirement is satisfied due to the fact that input signal pulses must not overlap.

Any code can be adopted to represent the states of a state machine operating in pulse mode, even though binary encoding is used wherever possible to minimize the different logic equations.

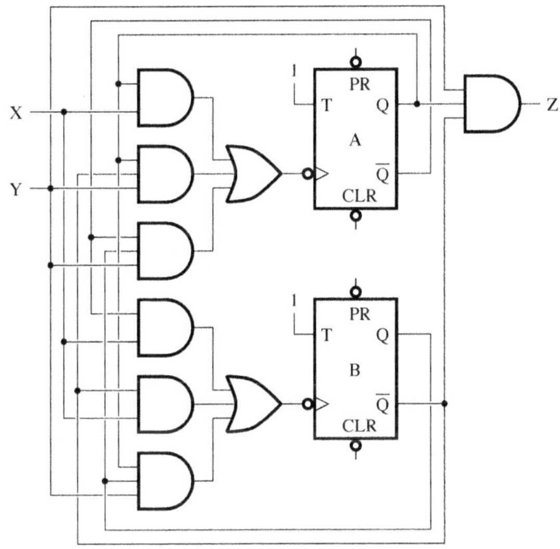

Figure 3.45. *Logic circuit of the sequence detector*

EXAMPLE 3.2.– Digital lock controller.

Consider the implementation of a digital lock controller whose state diagram is shown in Figure 3.46. The lock is opened by setting the output Z to 1 when the input combination $X - X - Y - X$ is detected. The flow table can be represented as shown in Table 3.21. It can be used to derive the Karnaugh maps shown in Figure 3.47 when the binary codes 00, 01, 11, and 10 are assigned to the states S_a, S_b, S_c, and S_d, respectively, and the excitation table for a flip-flop operating in the switching mode is used.

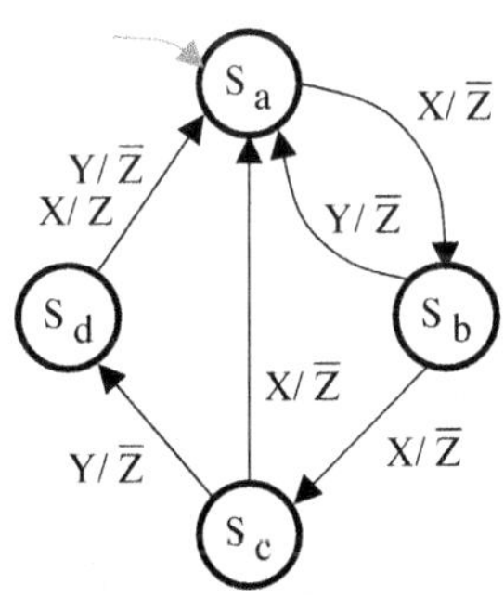

Figure 3.46. *State diagram of the controller for a digital lock*

PS	NS				Output, Z			
	$XY=00$	01	10	11	$XY=00$	01	10	11
S_a	–	–	S_b	–	–	–	0	–
S_b	–	S_a	S_c	–	–	0	0	–
S_c	–	S_d	S_a	–	–	0	0	–
S_d	–	S_a	S_a	–	–	0	1	–

Table 3.21. *Flow table of the controller for a digital lock*

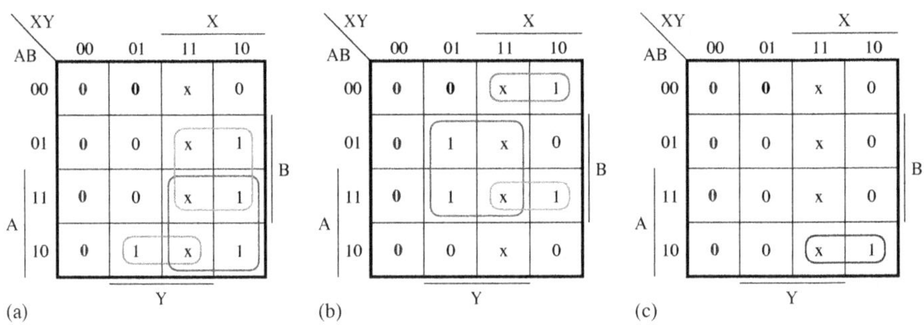

Figure 3.47. *Karnaugh maps: (a) T_A ; (b) T_B ; (c) Z*

The logic equations for the inputs of flip-flops and the state machine output can be written as follows:

$$T_A = A \cdot X + B \cdot X + A \cdot \overline{B} \cdot Y \qquad [3.43]$$

$$T_B = B \cdot Y + A \cdot B \cdot X + \overline{A} \cdot \overline{B} \cdot X \qquad [3.44]$$

and:

$$Z = A \cdot \overline{B} \cdot X \qquad [3.45]$$

Figure 3.48 presents the logic circuit of the controller for a digital lock.

3.11. Asynchronous state machine operating in burst mode

Asynchronous state machines operating in burst mode find applications in the implementation of controllers. They are characterized by the fact that the simultaneous change in logic state of multiple inputs is allowed and can be described using a Mealy model.

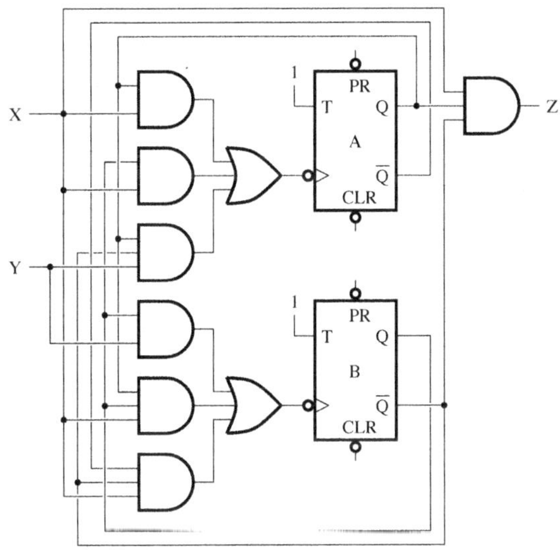

Figure 3.48. *Logic circuit of the controller for a digital lock*

To operate in burst mode, the state machine is assumed to remain in a state until the necessary change or transition is applied to each input of the given set of inputs. The order in which the inputs undergo the changes does not matter. As soon as the last change has taken place at the inputs, the machine initiates the transitions for a set of outputs, if any, and moves to the next state.

Table 3.22 shows the flow table of a state machine that has two inputs and two outputs and that can operate in burst mode.

PS	NS				Outputs, KL			
	$XY = 00$	01	10	11	$XY = 00$	01	10	11
' A	(A)	–	–	B	00	–	–	10
B	C	–	–	(B)	11	–	–	10
C	(C)	D	–	–	11	10	–	–
D	A	(D)	–	–	00	10	–	–

Table 3.22. *Flow table*

In practice, a state machine operating in burst mode can be represented by a state diagram where each transition from one state to another is associated with a label of the form I/O, where I denotes a set of transitions to be applied to the inputs and is also called the *input burst*; O corresponds to the set of transitions for the output signals and is also called the *output burst*. In this case, each signal appearing in a burst is annotated with the sign + or − to indicate whether it is supposed to undergo a rising or falling transition.

The specifications of a state machine operating in burst mode must satisfy two requirements:

– an input burst cannot be empty. Hence, when there is no change at the inputs the machine remains at the same state;

– the input burst for a transition from a given state cannot be contained in any other burst associated with a transition leaving that state. This property ensures that, at any time, the state machine can unambiguously perform a transition or stay in the present state.

Furthermore, to simplify the synthesis of hazard-free circuits, the same combination of logic states must be assigned to inputs as well as outputs for all transitions leading to the same state.

3.12. Exercises

EXERCISE 3.1.– (Implementation of a Rotation-direction Discriminator Based on a Programmable Circuit).

A rotation-direction discriminator as illustrated in Figure 3.49 is to be designed. The recognition of the direction of the rotation of a motor is realized by decoding the two signals, X and Y, produced by sensors. Each sensor gives, on each quarter turn of the motor shaft, a signal that can take the logic level 0 or 1 depending on whether the part touched is white or gray.

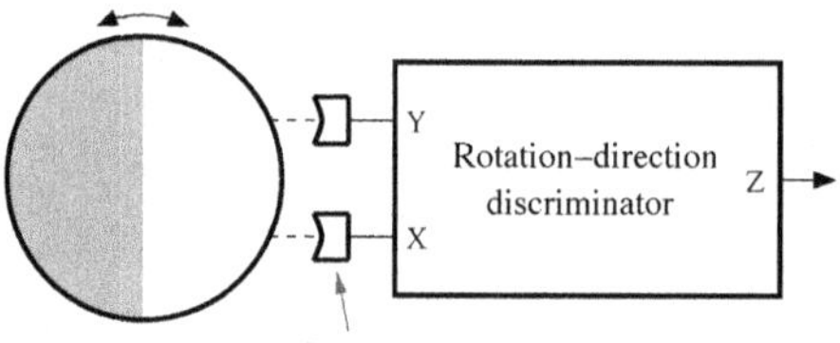

Figure 3.49. *Rotation-direction discriminator*

The output signal, Z, is set to 0 when the motor turns clockwise and to 1 when rotating counterclockwise.

– construct the primitive flow table;

– minimize the number of states and draw up the reduced flow table;

– encode the states and determine the logic equations for the synthesis of the discriminator using D flip-flops.

Represent the logic circuit of the discriminator using FPGA with four-input lookup table.

Assume that the signals X and Y do not vary simultaneously and that the frequency of the clock signal is higher than the frequency of both the X and Y signals.

EXERCISE 3.2.– (Bus Arbiter).

Let us consider a bus arbiter that allocates the bus to one of two components, A or B, with priority initially assigned to component A when there are two concurrent requests. In the block diagram in Figure 3.50, the bus arbiter provides the control signals for both three-state buffers connected to a common data bus.

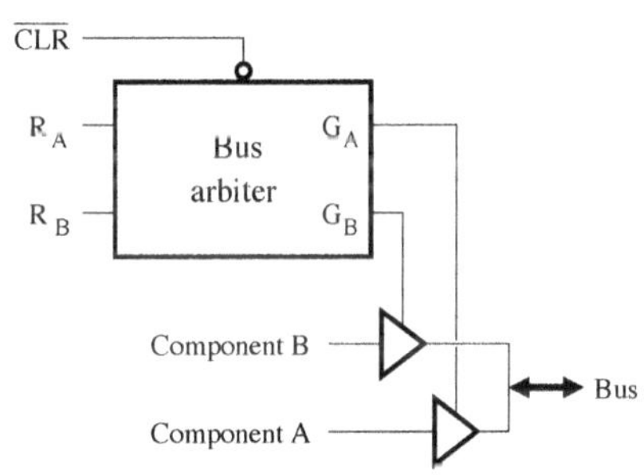

Figure 3.50. *Bus arbiter*

The bus arbiter must be implemented as a Moore model-based asynchronous state machine that operates as follows. When the two inputs R_A and R_B are set to 0, the machine returns to the initial state or is held in the initial state. For the inputs R_A and R_B that take the binary combination 10 or 11, the machine goes to the state where the bus is allocated to the component A and is held in this state as long as the input R_A remains at 1. On the contrary, when the combination 01 is assigned to the inputs R_A and R_B, the machine returns to the initial state where the bus is allocated to the component B and remains in this state as long as the input R_B remains at 1. The transition between the two states, where the bus is allocated to one of both components, occurs by assigning either the combination 01 or the combination 10 to the inputs R_A and R_B.

– construct the state diagram for the state machine;

– using Gray code to represent the states, determine the logic equations for the implementation of the machine when it operates in fundamental mode;

– represent the logic circuit of the state machine.

EXERCISE 3.3.– (T Flip-Flop).

A T flip-flop is described by the flow table shown in Table 3.23, where the input and output are denoted by T and Q, respectively. The initial state is represented by A.

PS	NS		Output, Q
	$T = 0$	1	
A	(A)	B	0
B	C	(B)	0
C	(C)	D	1
D	A	(D)	1

Table 3.23. *Flow table*

Implement this flip-flop using only logic gates, and from either SR latches or C-elements. Assume that this flip-flop operates in the fundamental mode and has an active-low reset input, and the initial state is denoted by A and is represented by the binary code 00.

EXERCISE 3.4.– (Multiple-Row State Assignment).

Consider the asynchronous machine whose flow table is represented in Table 3.24.

PS	NS				Output			
	$XY = 00$	01	10	11	$XY = 00$	01	10	11
A	(A)	C	(A)	B	0	0	0	1
B	A	C	A	(B)	0	1	0	1
C	A	(C)	A	B	0	0	-	0

Table 3.24. *Flow table of an asynchronous state machine*

i) Construct the transition table and determine the critical race condition when the states are encoded as follows: A (00), B (01) and C (10);

ii) Show that a multiple-row state assignment obtained by adding a state to the flow table helps to eliminate critical race conditions.

EXERCISE 3.5.– (Oscillatory Cycle).

Determine the imperfections that can affect the operation in the fundamental mode of the asynchronous machine whose state diagram is shown in Figure 3.51, where X and Y denote the input signals.

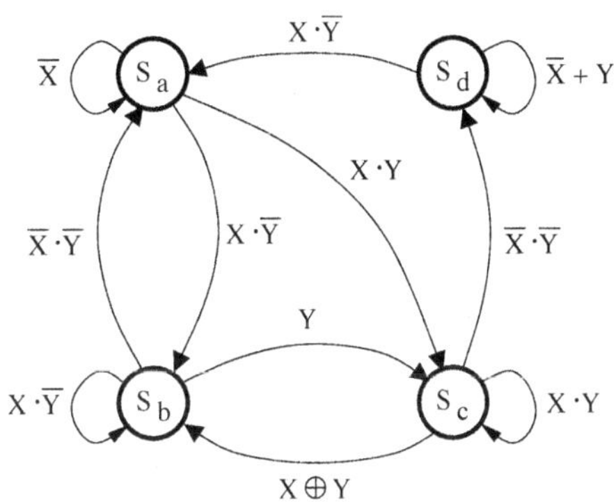

Figure 3.51. *State diagram*

Modify the state diagram for this state machine so that it can operate based on the flow table shown in Table 3.25.

PS	NS			
	$XY = 00$	01	10	11
S_a	(S_a)	(S_a)	S_b	S_c
S_b	S_a	(S_b)	(S_b)	S_c
S_c	S_d	S_b	S_b	(S_c)
S_d	(S_d)	(S_d)	S_a	(S_d)

Table 3.25. *Flow table*

Can we encode the states for this state machine using only two state variables?

EXERCISE 3.6.– (Cycle and Essential Hazard).

Identify the cycles that appear in the operation of the state machine described by the state diagram shown in Figure 3.52, where X and Y represent the inputs and Z is the output.

Modify this state diagram so that the state machine can operate according to the flow table shown in Table 3.26.

Can we eliminate all the critical race conditions by assigning a two-bit binary code to each state?

With the addition of the intermediate states, S_e, S_f, S_g and S_h, the flow table is modified as shown in Table 3.27. Determine the logic equations for the state variables and for the output assuming that the states are represented as follows: S_a (000), S_b (001), S_c (010), S_d (110), S_e (010), S_f (100) and S_g (111).

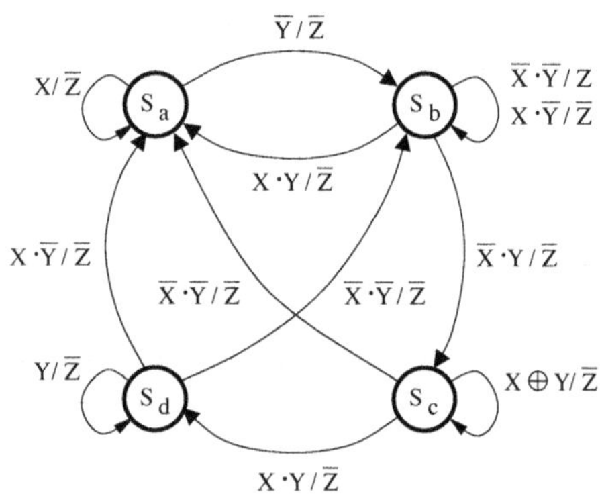

Figure 3.52. *State diagram*

PS	NS				Output, Z
	$XY = 00$	01	10	11	
S_a	S_b	S_c	(S_a)	(S_a)	0
S_b	(S_b)	S_c	(S_b)	S_a	$\overline{X}$
S_c	S_b	(S_c)	(S_c)	S_d	0
S_d	S_b	(S_d)	S_a	(S_d)	0

Table 3.26. *Flow table*

Are there any essential hazards that can affect the operation of this state machine?

Represent the logic circuit for this state machine.

EXERCISE 3.7.– (Essential and d-Trio Hazards).

The state machine that is described by the state diagram shown in Figure 3.53 has two inputs and one output and operates in the fundamental mode. The states S_a, S_b, S_c and S_d are represented by the binary codes 00, 01, 11 and 10, respectively.

– Construct the flow table for this state machine;

– derive the logic equations that are useful for the implementation of this state machine using logic gates;

– determine the imperfections that can affect the operation of this state machine;

– represent the logic circuit for this state machine.

PS	NS				Output, Z
	$XY = 00$	01	10	11	
S_a	S_b	S_c	(S_a)	(S_a)	0
S_b	(S_b)	S_c	(S_b)	S_a	$\overline{X}$
S_c	S_b	(S_c)	(S_c)	S_d	0
S_d	S_b	(S_d)	S_a	(S_d)	0
S_e	S_b	S_c	S_c	S_d	0
S_f	S_b	S_d	S_a	S_a	0
S_g	S_b	S_d	S_b	S_a	0
S_h	S_b	S_d	S_b	S_d	0

Table 3.27. *Flow table*

EXERCISE 3.8.– (Pulse Selector).

The state diagram shown in Figure 3.54 describes the operation of a pulse selector. Depending on the level of the signal applied to the input D, the input C is either transferred toward the output H (high level) or the output L (low level).

Assigning the binary codes 00, 01, 11 and 10 to the states S_a, S_b, S_c and S_d, respectively, determine the logic equations for the implementation of this selector based on SR latches.

Represent the logic circuit for this selector assuming that the signals are active-high and the state machine must be able to be initialized in the state S_a.

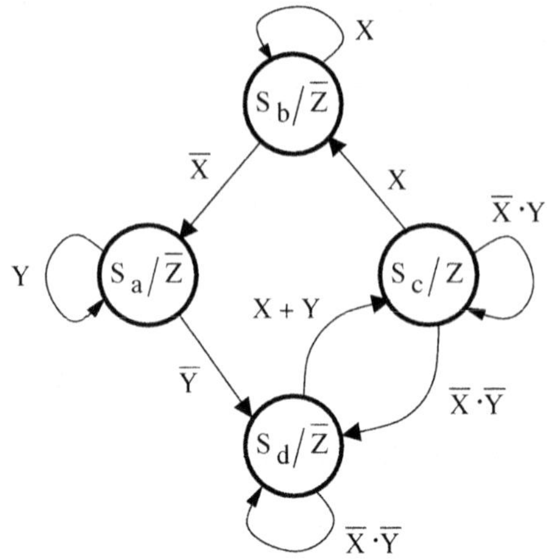

Figure 3.53. *State diagram*

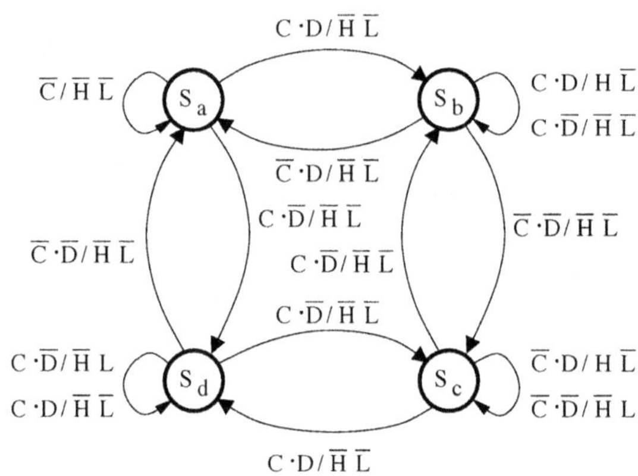

Figure 3.54. *State diagram*

EXERCISE 3.9.– (Implementation of a Circuit Operating in Pulse Mode).

The asynchronous state machine described by the state diagram in Figure 3.55 is to be implemented as a circuit operating in the pulse mode.

– Draw up the flow table for this state machine;

– representing the states S_a, S_b, S_c and S_d using the binary codes 00, 01, 11 and 10, respectively, construct the Karnaugh maps that can be used to determine the logic equations for the flip-flop inputs and the state machine output;

– represent the logic circuit for the state machine using flip-flops triggered by the falling edge of the signal;

– what is the logic function for this machine?

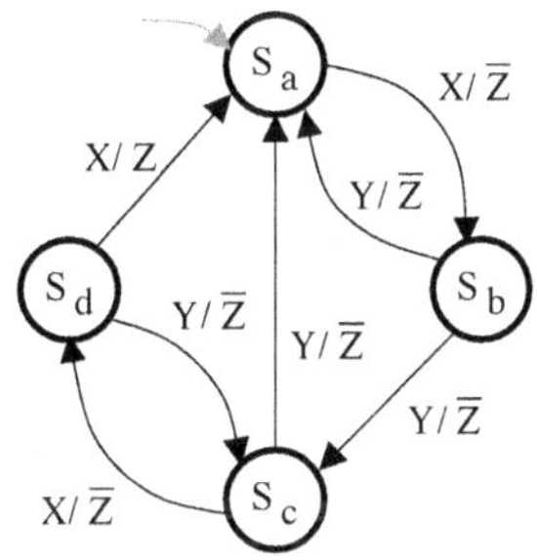

Figure 3.55. *State diagram*

EXERCISE 3.10.– (Analysis of a Circuit Operating in the Pulse Mode).

The logic circuit shown in Figure 3.56 is that of an asynchronous state machine designed to operate in the pulse mode. It is initialized by setting the $\overline{PR}$ input of the flip-flop A and the $\overline{CLR}$ signal of the flip-flop B to 0. The inputs are denoted by X and Y, and the two outputs by K and L.

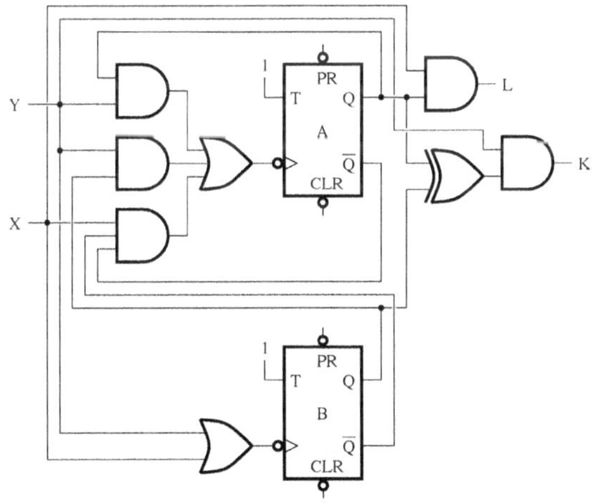

Figure 3.56. *Logic circuit*

– Determine the logic equations for the flip-flop inputs, T_A and T_B, and the state machine outputs, K and L;

– draw up the flow table for this state machine;

– construct the state diagram for this state machine;

– what is the logic function of this state machine?

3.13. Solutions

SOLUTION 3.1.– (Rotation-Direction Discriminator).

Each sensor can take the logic state 0 or 1. There are, thus, four possible input combinations: 00, 01, 10 and 11. The rotation-direction discriminator can, therefore, be described as a Mealy state machine whose state assignments are given in Table 3.28.

	Clockwise rotation				Anticlockwise rotation			
XY	00	01	11	10	00	10	11	01
PS	S_0	S_1	S_2	S_3	S_4	S_5	S_6	S_7

Table 3.28. *State assignments*

PS	NS				Output Z			
	$XY = 00$	01	11	10	$XY = 00$	01	11	10
S_0	(S_0)	S_1	–	S_4	0	0	–	1
S_1	S_7	(S_1)	S_2	–	1	0	0	–
S_2	–	S_6	(S_2)	S_3	–	1	0	0
S_3	S_0	–	S_5	(S_3)	0	–	1	0
S_4	S_0	–	S_5	(S_4)	0	–	1	1
S_5	–	S_6	(S_5)	S_3	–	1	1	0
S_6	S_7	(S_6)	S_2	–	1	1	0	–
S_7	(S_7)	S_1	–	S_4	1	0	–	1

Table 3.29. *Primitive flow table of the discriminator*

PS	NS				Output Z			
	$XY = 00$	01	11	10	$XY = 00$	01	11	10
S_a	(S_a)	S_b	S_d	(S_a)	0	0	1	1
S_b	(S_b)	(S_b)	S_c	S_a	1	0	0	1
S_c	S_b	(S_c)	(S_c)	S_d	1	1	0	0
S_d	S_a	S_c	(S_d)	(S_d)	0	1	1	0

Table 3.30. *Reduced flow table of the discriminator*

The flow table with a single stable state per row (or primitive flow table) is shown in Table 3.29.

The pairs of compatible states can be identified by analyzing the flow table. The reduced flow table shown in Table 3.30 is constructed by merging the states as follows:

$$S_0 \equiv S_4 \equiv S_a,\ S_1 \equiv S_7 \equiv S_b,\ S_2 \equiv S_6 \equiv S_c, \text{ and } S_3 \equiv S_5 \equiv S_d$$

To reduce the size of the combinational circuit, adjacent codes are attributed first to the states leading to the same next state for a given input combination. Choosing to represent the states with two variables, A and B, the binary codes 00, 01, 11 and 10, can be assigned to the states S_a, S_b, S_c and S_d, respectively.

Figure 3.57 presents the Karnaugh maps for A^+, B^+ and Z, as derived from the flow table. The different logic equations are given by:

$$A^+ = A \cdot X + A \cdot Y + X \cdot Y \quad [3.46]$$

$$B^+ = B \cdot \overline{X} + B \cdot Y + \overline{X} \cdot Y \quad [3.47]$$

and:

$$Z = \overline{A} \cdot \overline{B} \cdot X + \overline{A} \cdot B \cdot \overline{Y} + A \cdot B \cdot \overline{X} + A \cdot \overline{B} \cdot Y \quad [3.48]$$

The rotation-direction discriminator can be implemented using FPGA with four-input LUT and D flip-flops as shown in Figure 3.58.

SOLUTION 3.2.– (Bus Arbiter).

The operation of the bus arbiter can be described by state diagram shown in Figure 3.59.

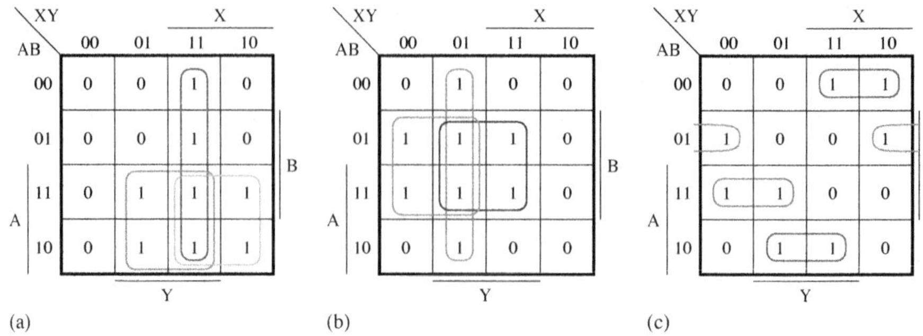

Figure 3.57. *Karnaugh maps: a) A^+; b) B^+; c) Z*

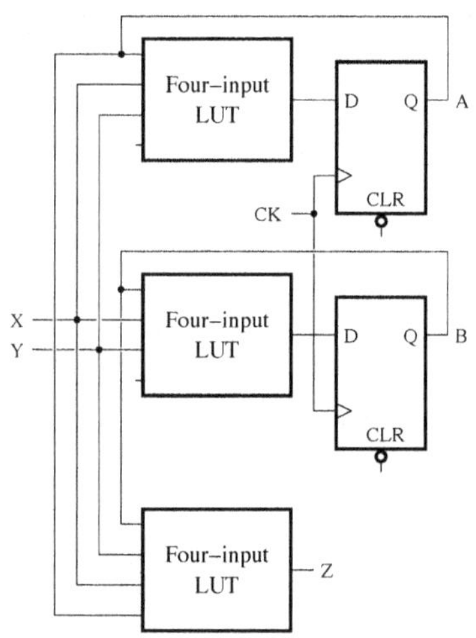

Figure 3.58. *Logic circuit of the discriminator*

Using Gray code to represent states, the transition table is constructed as shown in Table 3.31. The transitions between the unused state S_3 and other states have been defined in order to reduce the effects of possible critical race conditions on the operation of the state machine.

For the state machine operation in the fundamental mode, we can obtain the Karnaugh maps represented in Figures 3.60 and 3.61. The logic equations for the next states can be written as follows:

$$X^+ = \overline{R_A} \cdot R_B + R_B \cdot X \cdot Y \qquad [3.49]$$

and:

$$Y^+ = R_A + R_B \qquad [3.50]$$

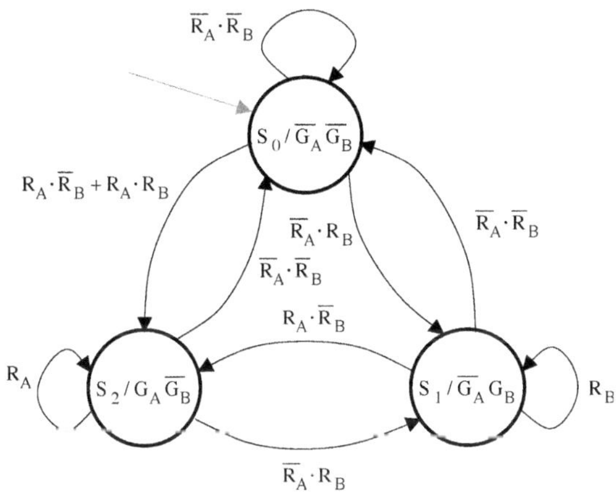

Figure 3.59. *State diagram*

PS	XY	NS X^+Y^+ $R_AR_B = 00$	01	10	11	Outputs G_AG_B	
S_0	00	(00)	11	01	01	0	0
S_1	01	00	11	(01)	(01)	1	0
S_2	11	00	(11)	01	(11)	0	1
S_3	10	00	11	01	01	0	0

Table 3.31. *Transition table*

The outputs are not directly dependent on the input signals, R_A and R_B. We obtain, based on the transition table, the following logic equations:

$$G_A = \overline{X} \cdot Y \qquad [3.51]$$

and:

$$G_B = X \cdot Y \qquad [3.52]$$

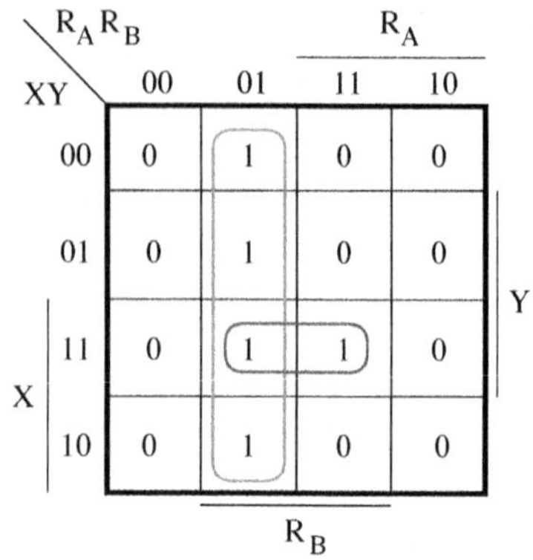

Figure 3.60. *Function X^+*
$X^+ = \overline{R_A} \cdot R_B + R_B \cdot X \cdot Y$

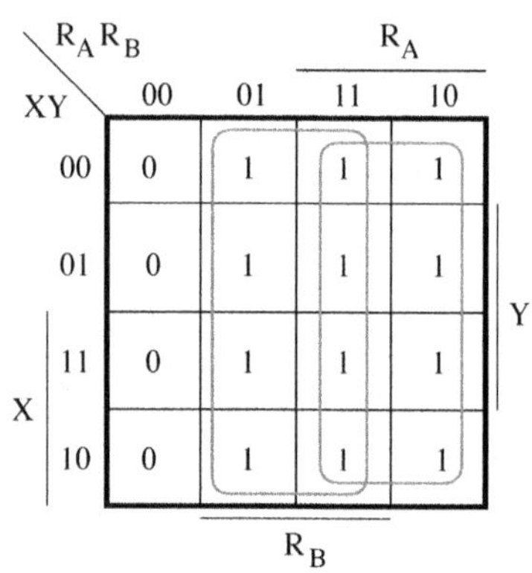

Figure 3.61. *Function Y^+*
$Y^+ = R_A + R_B$

Figure 3.62 presents the implementation of the bus arbiter. The reset signal, $\overline{CLR}$, is active-low.

SOLUTION 3.3.– (T Flip-Flop).

Assigning the binary codes 00, 01, 10 and 11 to the states A, B, C and D, respectively, the Karnaugh maps can be constructed from the flow table, as shown in

Figure 3.63. The logic equations for the state variables and the state machine output are given by:

$$Y_1^+ = Y_1 \cdot T + Y_0 \cdot \overline{T} + Y_1 \cdot Y_0 \qquad [3.53]$$

$$Y_0^+ = \overline{Y_1} \cdot T + Y_0 \cdot \overline{T} + \overline{Y_1} \cdot Y_0 \qquad [3.54]$$

and:

$$Q = Y_1 \qquad [3.55]$$

where each of the redundant terms $Y_1 \cdot Y_0$ and $\overline{Y_1} \cdot Y_0$ is added to compensate for the effect of static hazards.

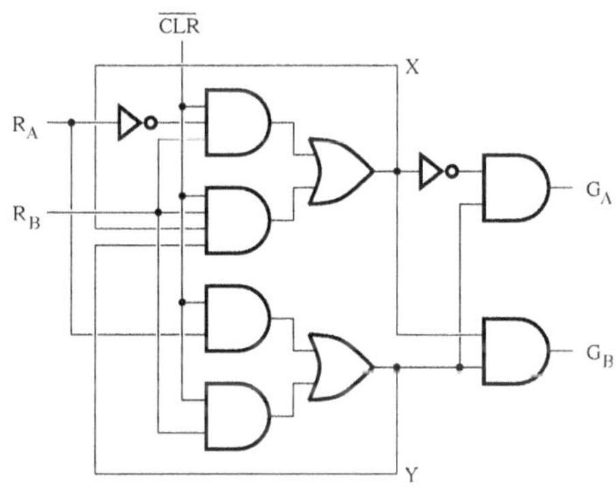

Figure 3.62. *Bus arbiter*

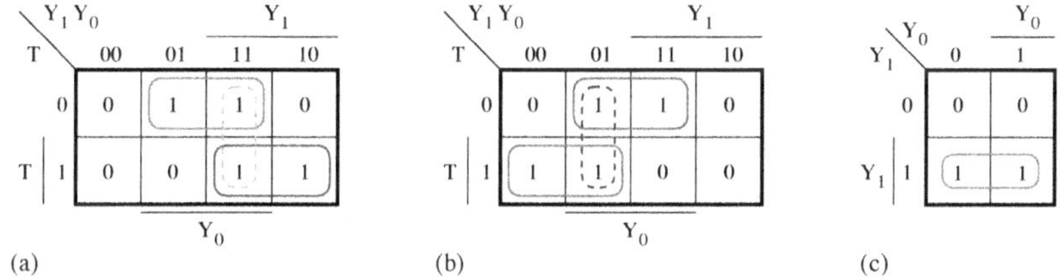

Figure 3.63. *Karnaugh maps: a) Y_1^+; b) Y_0^+; c) Q*

Figure 3.64 shows the logic circuit of the T flip-flop, where $\overline{IS}$ denotes the reset signal.

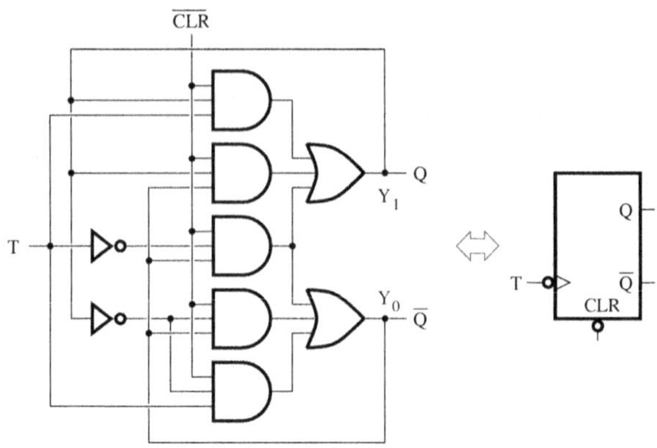

Figure 3.64. *Logic circuit*

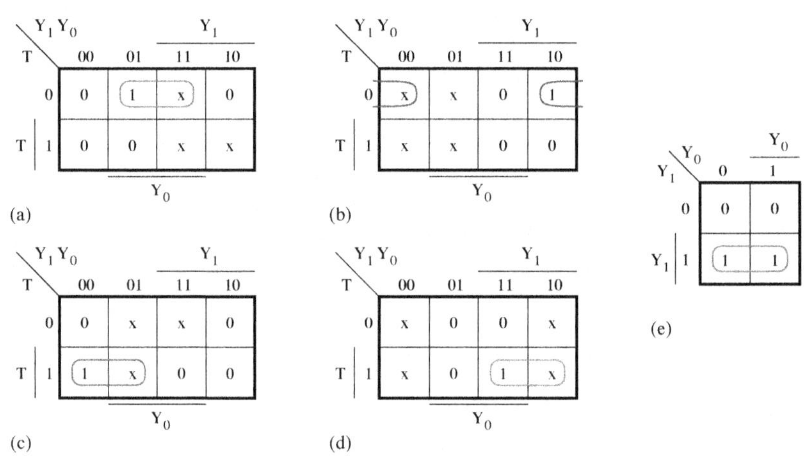

Figure 3.65. *Karnaugh maps: a) S_1; b) R_1; c) S_0; d) R_0; e) Q*

In the case of the *T* flip-flop implementation based on SR latches, the Karnaugh maps are constructed as shown in Figure 3.65, using the state table of the *T* flip-flop and the excitation table for the SR latch. We thus have:

$$S_1 = Y_0 \cdot \overline{T} \qquad [3.56]$$

$$R_1 = \overline{Y_0} \cdot \overline{T} \qquad [3.57]$$

$$S_0 = \overline{Y_1} \cdot T \qquad [3.58]$$

$$R_0 = Y_1 \cdot T \qquad [3.59]$$

and:

$$Q = Y_1 \qquad [3.60]$$

The logic circuit of the *T* flip-flop is represented in Figure 3.66, where $\overline{IS}$ is the reset signal.

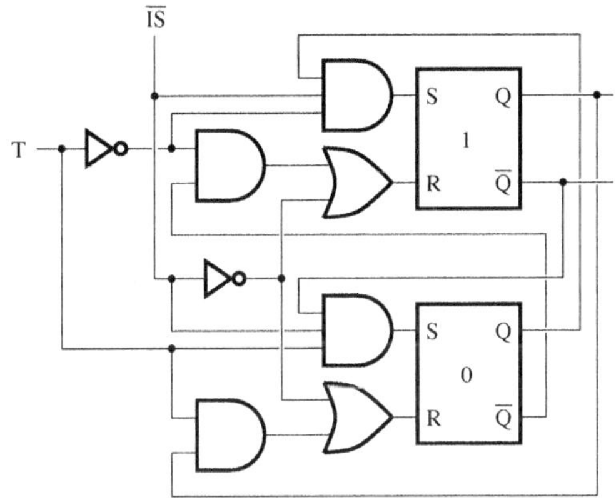

Figure 3.66. *Logic circuit based on SR latches*

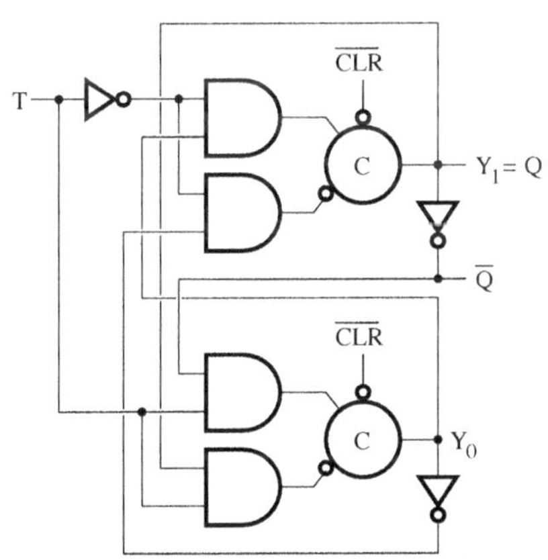

Figure 3.67. *Logic circuit based on C-elements*

Figure 3.67 presents the *T* flip-flop logic circuit based on C-elements.

SOLUTION 3.4.– (Shared-Row State Assignment).

i) The analysis of the flow table shows that the transition from state B to state C is affected by a critical race condition, because it requires the simultaneous change in two state variables.

Other encoding possibilities with two state variables also lead to a possible critical race condition for at least one of the transitions.

Figure 3.68 shows the state diagram of the state machine.

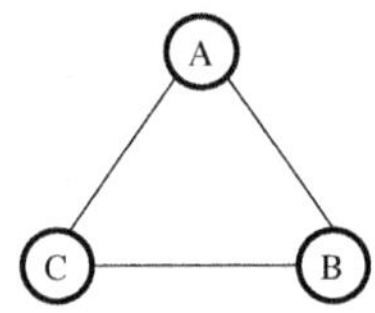

Figure 3.68. *Transition diagram*

ii) Adding a state, D, between states B and C, helps to eliminate the critical race condition. The extended flow table is represented in Table 3.32, where the states are encoded as follows: A (00), B (01), C (10) and D (11).

PS	NS				Output			
	$XY = 00$	01	10	11	$XY = 00$	01	10	11
A	(A)	C	(A)	B	0	0	0	1
B	A	C	A	(B)	0	0	0	1
C	A	(C)	A	B	0	0	1	0
D	–	C	–	B	–	1	–	0

Table 3.32. *Extended flow table*

SOLUTION 3.5.– (Oscillatory Cycle).

The analysis of the state diagram shows the existence of an oscillatory cycle between the states S_b and S_c. Under the condition Y, the state machine moves from the state S_b to the state S_c, while the condition $A \oplus B$ cause a transition from the state S_c to the state S_b. The relationship, $(A \oplus B) \cdot B = \overline{A} \cdot B$, reveals the existence of the condition $\overline{A} \cdot B$ that, by allowing the transition from the state S_b to the state S_c and the transition from the state S_c to the state S_b, allowed the machine to enter an oscillatory cycle.

Based on the flow table, the state diagram of the state machine can be represented as shown in Figure 3.69. The critical race condition is eliminated because $(A \oplus B) \cdot (A \cdot B) = 0$.

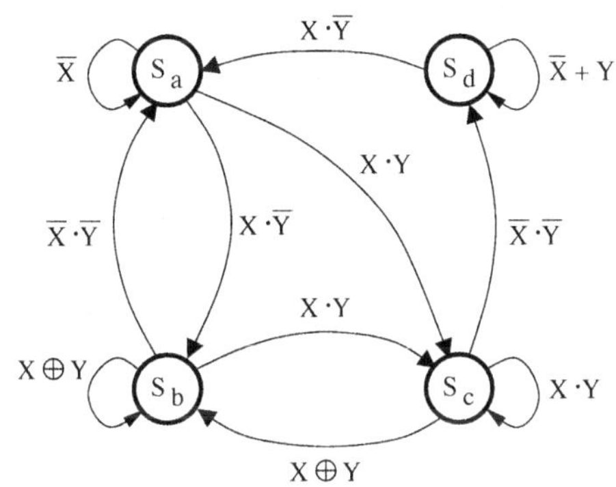

Figure 3.69. *State diagram*

The states of this state machine cannot be encoded using only two state variables without running the risk of creating critical race conditions.

SOLUTION 3.6.– (Cycle and Essential Hazard).

The condition $\overline{X} \cdot \overline{Y}$ allows the transition sequence $S_c \to S_a \to S_b$, while the condition $\overline{X} \cdot Y$ causes the transitions $S_a \to S_b \to S_c$. In the operation of the state machine as described by the state diagram, there are two cycles.

The state diagram modified according to the flow table is represented in Figure 3.70. It can be seen that, in comparison with the initial state diagram, the cycles were eliminated because of the modifications.

To eliminate the critical race conditions, the binary code assigned to each state of the state machine must have at least three variables.

Representing each state by its binary code, we can draw up the table shown in Table 3.33, which can be used to construct the Karnaugh maps required for the determination of the logic equations for the state variables. From the Karnaugh maps with entered variables in Figure 3.71, the logic equations for the state variables and the state machine output can be obtained, as follows:

$$Z_2^+ = \overline{Z_1} \cdot \overline{X} \cdot Y + \overline{Z_2} \cdot X \cdot Y + Z_2 \cdot Z_1 \cdot Y \quad [3.61]$$

$$Z_1^+ = \overline{X} \cdot Y + Z_1 \cdot Y + \overline{Z_2} \cdot Z_1 \cdot X \quad [3.62]$$

$$Z_0^+ = \overline{X} \cdot \overline{Y} + \overline{Z_2} \cdot \overline{X} + Z_0 \cdot \overline{Y} + \overline{Z_2} \cdot Z_1 \cdot \overline{Y} \quad [3.63]$$

and:

$$Z = \overline{Z_2} \cdot \overline{Z_1} \cdot Z_0 \cdot \overline{X} \quad [3.64]$$

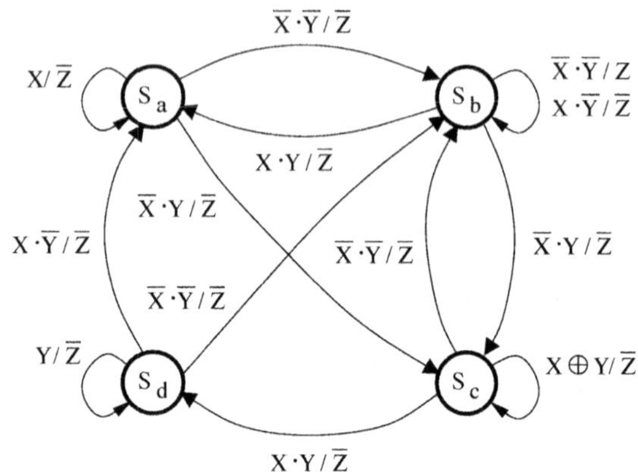

Figure 3.70. *State diagram*

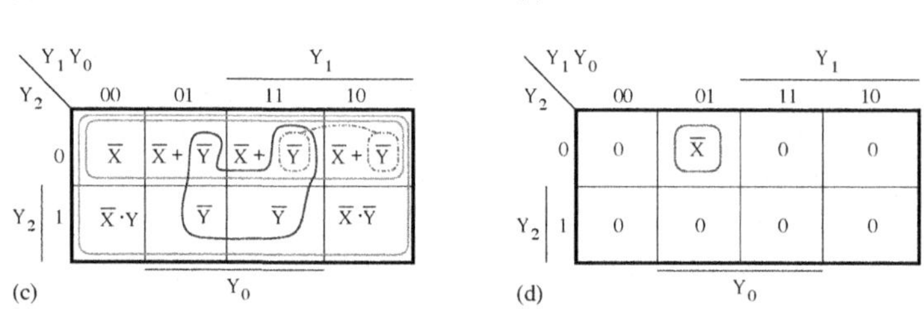

Figure 3.71. *Karnaugh maps: a) Z_2^+; b) Z_1^+; c) Z_0^+; d) Z*

The term $Z_2 \cdot Z_1 \cdot Y$ helps to eliminate the effect of any static hazard associated with the presence of X and $\overline{X}$ when expressing Z_2^+.

PS $Z_2Z_1Z_0$	Inputs	NS $Z_2^+Z_1^+Z_0^+$	PS $Z_2Z_1Z_0$	Inputs	NS $Z_2^+Z_1^+Z_0^+$
000	$\overline{X} \cdot \overline{Y}$	001	010	$\overline{X} \cdot \overline{Y}$	001
	$\overline{X} \cdot Y$	011		$\overline{X} \cdot Y$	011
	$X \cdot \overline{Y}$	000		$X \cdot \overline{Y}$	011
	$X \cdot Y$	000		$X \cdot Y$	110
001	$\overline{X} \cdot \overline{Y}$	001	100	$\overline{X} \cdot \overline{Y}$	001
	$\overline{X} \cdot Y$	011		$\overline{X} \cdot Y$	110
	$X \cdot \overline{Y}$	001		$X \cdot \overline{Y}$	000
	$X \cdot Y$	000		$X \cdot Y$	000
011	$\overline{X} \cdot \overline{Y}$	001	101	$\overline{X} \cdot \overline{Y}$	001
	$\overline{X} \cdot Y$	011		$\overline{X} \cdot Y$	110
	$X \cdot \overline{Y}$	011		$X \cdot \overline{Y}$	001
	$X \cdot Y$	110		$X \cdot Y$	000
110	$\overline{X} \cdot \overline{Y}$	001	111	$\overline{X} \cdot \overline{Y}$	001
	$\overline{X} \cdot Y$	110		$\overline{X} \cdot Y$	110
	$X \cdot \overline{Y}$	000		$X \cdot \overline{Y}$	001
	$X \cdot Y$	110		$X \cdot Y$	110

Table 3.33. *Table showing present states/next states*

Because of the difference in propagation delays associated with the paths taken by X and Z_1 to reach the AND gate implementing the function $Z_1 \cdot X \cdot Y$, the transition $S_a \rightarrow S_c$, under the condition $\overline{X} \cdot Y$ can be affected by an essential hazard.

Assuming that the machine is initially in the state S_a and the two inputs X and Y are set to 1, a transition from 1 to 0 at the input X brings the variables Z_1 and Z_0 to the logic state 1, thus causing a transition of the state machine to the state S_c.

When the effect of the state change in Z_1 is taken into account by the above-mentioned AND gate before that of X, the variable Z_2 goes to the logic state 1 (instead of staying at 0). The state variable Z_0 is then reset to 0, while the state of Z_1 remains unchanged. And finally, the state machine settles in the state S_d.

Under the effect of the essential hazard, the transition $S_a \rightarrow S_c$ is transformed into a sequence of transitions $S_a \rightarrow S_c \rightarrow S_d$ or $000 \rightarrow 011 \rightarrow 110$.

The logic circuit of the machine is represented in Figure 3.72, where delay elements are inserted in the feedback path for Z_1 to prevent the formation of any essential hazard.

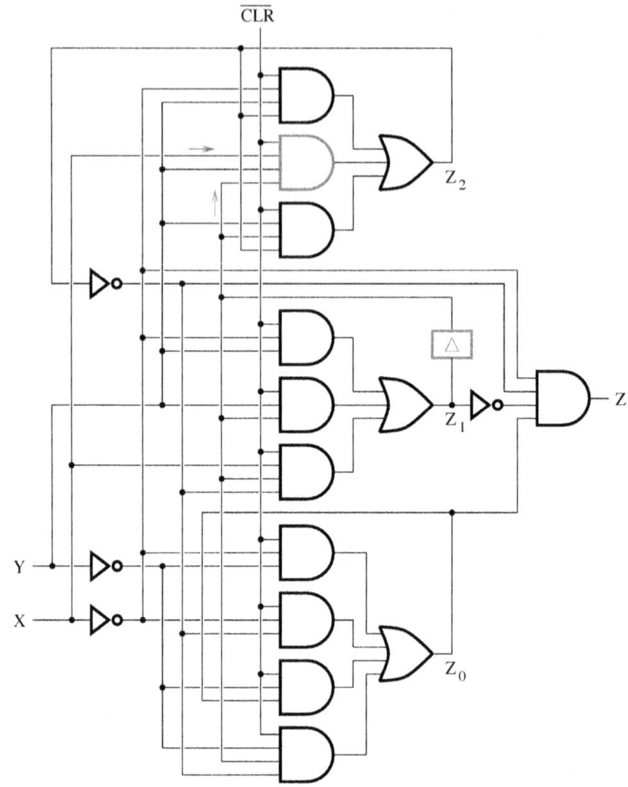

Figure 3.72. *Logic circuit*

The propagation delay for the inverter, the AND gate and the OR gate are τ_i, τ_{AND} and τ_{OR}, respectively. Designating the propagation delays associated with the direct and indirect paths starting from the input X by $\triangle t_e$ and $\tau_i + \tau_{AND} + \tau_{OR}$, respectively, the propagation delay introduced by the delay elements must be such that $\triangle t_e < \tau_i + \tau_{AND} + \tau_{OR} + \triangle t_c$.

SOLUTION 3.7.

–Essential and d-trio hazards:

The operation of the state machine can be described by the flow table shown in Table 3.34.

PS	NS				Output, Z
	$XY = 00$	01	10	11	
S_a	S_d	(S_a)	S_d	(S_a)	0
S_b	S_a	S_a	(S_b)	(S_b)	0
S_c	S_d	(S_c)	S_b	S_b	1
S_d	(S_d)	S_c	S_c	S_c	0

Table 3.34. *Flow table*

Representing each state by the corresponding binary code, the flow table can be used to construct the Karnaugh maps shown in Figure 3.73. The logic equations for the next states and the state machine output can then be written as:

$$Z_1^+ = \overline{Z_0} \cdot \overline{Y} + Z_1 \cdot \overline{X} + Z_1 \cdot \overline{Z_0} \qquad [3.65]$$

$$Z_0^+ = Z_0 \cdot X + Z_1 \cdot X + Z_1 \cdot Y \qquad [3.66]$$

and:

$$Z = Z_1 \cdot Z_0 \qquad [3.67]$$

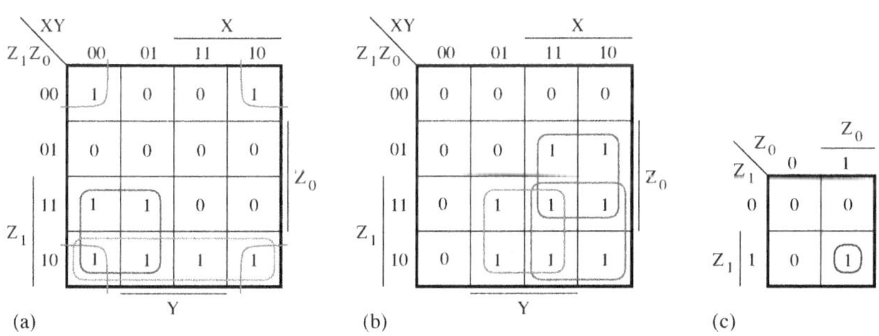

Figure 3.73. *Karnaugh maps: a) Z_1^+; b) Z_0^+; c) Z*

– Essential hazard:

For the transition $S_c \rightarrow S_b$ or $11 \rightarrow 01$ that takes place under the condition X, the change from 0 to 1 of the input X causes the transition from 1 to 0 of the state variable Z_1 and the state machine can move to the state S_b.

If the propagation delay introduced on the input path X is such that the state change in Z_1 is taken into account by the AND gates while the input X is still at the logic state 0, the logic state of each of the inputs of the lowermost OR gate will become 0, allowing the variable Z_0 to reset to 0 and the state machine to erroneously enter the state S_a.

Under the effect of an essential hazard, the machine which should make the transition $S_c \rightarrow S_b$, or $11 \rightarrow 01$, rather undergoes the sequence of transitions $S_c \rightarrow S_b \rightarrow S_a$ or $11 \rightarrow 01 \rightarrow 00$.

– d-trio hazard:

Analyzing the operation of the state machine, we can identify a d-trio hazard that can affect the transition $S_a \rightarrow S_d$, or $00 \rightarrow 10$, under the condition $\overline{Y}$.

The transition from 1 to 0 of the input Y, from the state S_a or 00, allows the state variable Z_1 to take the logic state 1 and the state machine to move to the state S_d. Because of the propagation delay introduced in the input path Y, the effect of the transition of the variable Z_1 reaches the lowermost AND gate before that of the input Y. The response of this AND gate is first determined by the logic state 1 of the variable Z_1 and the input Y, resulting in the state variable Z_0 being set to 1 and the transition of the state machine to the state S_c. When the effect of the transition from 1 to 0 of the input Y is afterward taken into account by the AND gate, the state variable Z_0 is reset to 0 and the state machine goes to the state S_d.

Thus, the transition $S_a \rightarrow S_d$, o'r $00 \rightarrow 10$, becomes a sequence of transitions, $S_a \rightarrow S_d \rightarrow S_c \rightarrow S_d$, or $00 \rightarrow 10 \rightarrow 11 \rightarrow 10$, due to the effect of the d-trio hazard.

The state diagram shown in Figure 3.74(a) illustrates the path between the source state and the final state for each of the essential and d-trio hazards. Figure 3.74(b) presents the logic circuit of the state machine with delay elements inserted on the feedback path to prevent the formation of an essential or d-trio hazard.

Because of component imperfections (or parasitic capacitances), the path connecting the input X to the AND gate is characterized by the propagation delay Δt_e.

The essential hazard is caused by the race between the input X and the state variable Z_1 to reach the lowermost OR gate.

Let τ_i, τ_{AND} and τ_{OR} be the propagation delays of the inverter, the AND gate and the OR gate, respectively. The propagation delay on the direct path leading to the lowermost OR gate is $\Delta t_e + \tau_{AND}$, while the propagation delay on the feedback path starting from the OR gate takes the form, $\tau_i + 2\tau_{AND} + \tau_{OR}$. The formation of the essential hazard is only possible if $\Delta t_e + \tau_{AND} > \tau_i + 2\tau_{AND} + \tau_{OR}$, that is:

$$\Delta t_e > \tau_i + \tau_{AND} + \tau_{OR} \qquad [3.68]$$

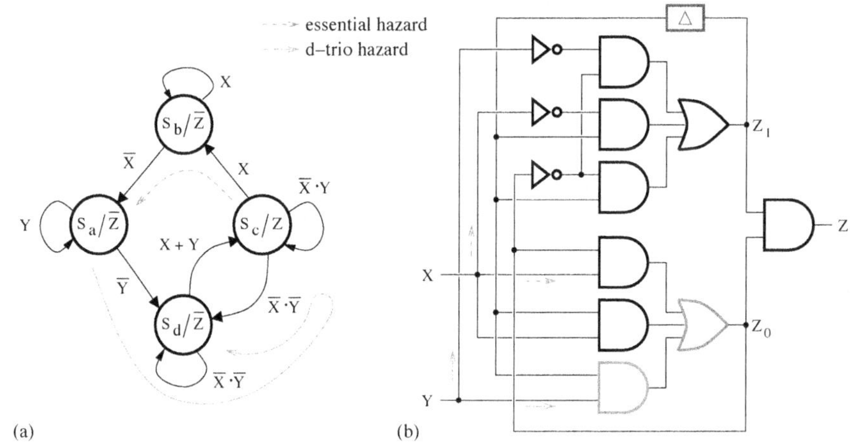

Figure 3.74. *a) State diagram; b) logic circuit*

One solution to prevent the formation of an essential hazard consists of adding delay elements along the feedback pathway so that:

$$\Delta t_e < \tau_i + \tau_{AND} + \tau_{OR} + \Delta t_c \qquad [3.69]$$

where the propagation delay of the delay elements is designated by Δt_c.

The d-trio hazard is due to the difference between the propagation delays of the paths taken by the input Y, Δt_d and the state variable Z_1, $\tau_i + \tau_{AND} + \tau_{OR}$, to reach the lowermost AND gate. It will affect the operation of the state machine if the next relationship is satisfied:

$$\Delta t_d > \tau_i + \tau_{AND} + \tau_{OR} \qquad [3.70]$$

Adding delay elements along the feedback path is effective in compensating for the effect of d-trio hazards if the following condition is verified:

$$\Delta t_d < \tau_i + \tau_{AND} + \tau_{OR} + \Delta t_c \qquad [3.71]$$

where the propagation delay of the delay elements is designated by Δt_c.

SOLUTION 3.8.– (Pulse Selector).

The flow table shown in Table 3.35 can be constructed based on the state diagram.

PS	NS				Outputs	
					H	L
	$XY = 00$	01	10	11		
S_a	S_a	S_a	S_d	S_b	0	0
S_b	S_c	S_a	S_b	S_b	D	0
S_c	S_c	S_c	S_b	S_d	D	$\overline{D}$
S_d	S_a	S_c	S_d	S_d	0	$\overline{D}$

Table 3.35. *Flow table*

Representing the states S_a, S_b, S_c and S_d by the binary codes 00, 01, 11 and 10, respectively, the flow table of the state machine and the excitation table for the SR latch can be used to construct the Karnaugh maps in Figure 3.75.

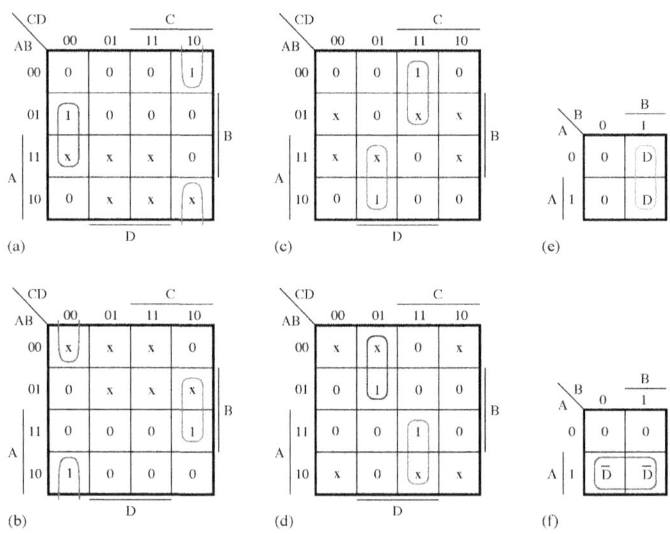

Figure 3.75. *Karnaugh maps: a) S_1; b) R_1; c) S_0; d) R_0; e) H; f) L*

The input equations for the SR latches and the state machine outputs, H and L, can be written as follows:

$$S_1 = B \cdot \overline{C} \cdot \overline{D} + \overline{B} \cdot C \cdot \overline{D} \quad [3.72]$$

$$R_1 = B \cdot C \cdot \overline{D} + \overline{B} \cdot \overline{C} \cdot \overline{D} \quad [3.73]$$

$$S_0 = A \cdot \overline{C} \cdot D + \overline{A} \cdot C \cdot D \quad [3.74]$$

$$R_0 = A \cdot C \cdot D + \overline{A} \cdot \overline{C} \cdot D \quad [3.75]$$

$$H = B \cdot D \quad [3.76]$$

and:

$$L = A \cdot \overline{D} \tag{3.77}$$

Figure 3.76 shows the logic circuit of the pulse selector with the reset signal being represented by $\overline{IS}$.

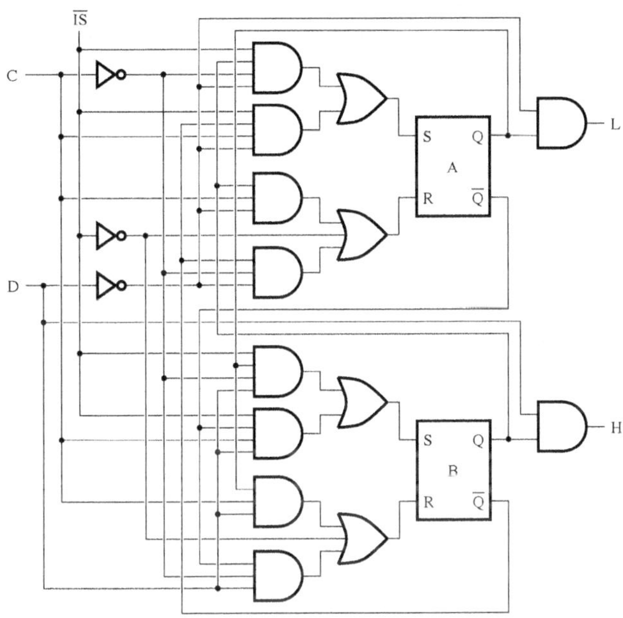

Figure 3.76. *Logic circuit*

SOLUTION 3.9.– (Implementation of a Circuit Operating in the Pulse Mode).

The flow table may be constructed as shown in Table 3.36 based on the state diagram.

PS	NS				Output, Z			
	$XY = 00$	01	10	11	$XY = 00$	01	10	11
S_a	–	–	S_b	–	–	–	0	–
S_b	–	S_c	S_a	–	–	0	0	–
S_c	–	S_a	S_d	–	–	0	0	–
S_d	–	S_c	S_a	–	–	0	1	–

Table 3.36. *Flow table*

Assigning the binary codes 00, 01, 11 and 10 to the states S_a, S_b, S_c and S_d, respectively, and using the excitation table of the T flip-flop, the Karnaugh maps can be constructed as shown in Figures 3.77(a)(c) in order to determine the logic equations for the inputs T_A and T_B and the output Z. That is:

$$T_A = B \cdot Y + A \cdot \overline{B} \cdot X \qquad [3.78]$$

$$T_B = A \cdot Y + B \cdot X + \overline{A} \cdot X \qquad [3.79]$$

and:

$$Z = A \cdot \overline{B} \cdot X \qquad [3.80]$$

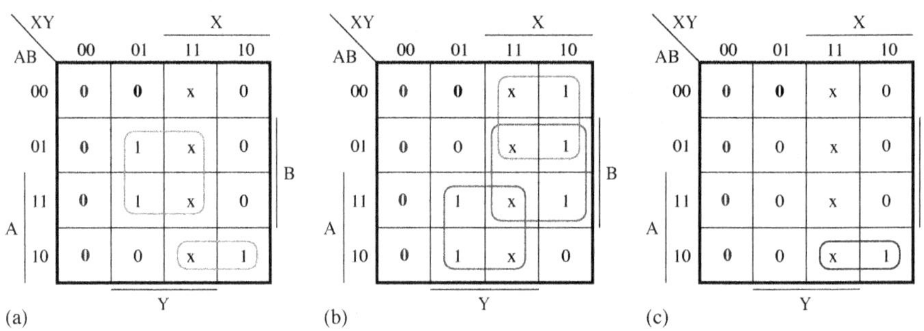

Figure 3.77. *Karnaugh maps: a) T_A; b) T_B; c) Z*

Figure 3.78 presents the logic circuit of the asynchronous state machine.

This is a $X - Y - X - X$ sequence detector.

SOLUTION 3.10.– (Analysis of a Circuit Operating in Pulse Mode).

The following logic equations can be obtained by analyzing the logic circuit:

$$T_A = \overline{A} \cdot \overline{B} \cdot X + A \cdot Y + B \cdot Y \qquad [3.81]$$

$$T_B = X + Y \qquad [3.82]$$

$$K = (A \oplus B)Y \qquad [3.83]$$

and:

$$L = A \cdot X \qquad [3.84]$$

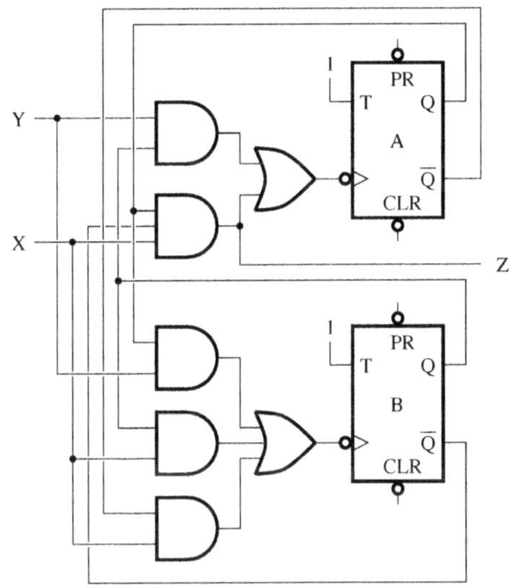

Figure 3.78. *Logic circuit*

Using the characteristic equation of the T flip-flop, which is given by:

$$Q^+ = \overline{T} \cdot Q + T \cdot \overline{Q} \quad [3.85]$$

we can obtain:

$$A^+ = \overline{T_A} \cdot A + T_A \cdot \overline{A} \quad [3.86]$$

$$= A \cdot \overline{Y} + \overline{A} \cdot \overline{B} \cdot X + \overline{A} \cdot B \cdot X \quad [3.87]$$

and:

$$B^+ = \overline{T_B} \cdot B + T_B \cdot \overline{B} \quad [3.88]$$

$$= \overline{B} \cdot X + \overline{B} \cdot Y + B \cdot \overline{X} \cdot \overline{Y} \quad [3.89]$$

Taking into account the characteristics inherent to circuits that operate in the pulse mode and using the expressions for A^+ and B^+, the transition table can be constructed as shown in Table 3.37.

In the case where the binary codes 00, 01, 11 and 10 represent the states S_a, S_b, S_c and S_d, respectively, the flow table can be constructed as shown in Table 3.38.

Figure 3.79 presents the state diagram obtained from the flow table.

PS, AB	NS, A^+B^+				Outputs, KL			
	$XY = 00$	01	10	11	$XY = 00$	01	10	11
00	–	01	01	–	–	00	00	–
01	–	10	10	–	–	10	00	–
11	–	00	10	–	–	00	00	–
10	–	01	11	–	–	10	01	–

Table 3.37. *Transition table*

PS	NS				Outputs, KL			
	$XY = 00$	01	10	11	$XY = 00$	01	10	11
S_a	–	S_b	S_b	–	–	00	00	–
S_b	–	S_d	S_d	–	–	10	00	–
S_c	–	S_a	S_d	–	–	00	00	–
S_d	–	S_b	S_c	–	–	10	01	–

Table 3.38. *Flow table*

The operation of the state machine from the initial state S_d corresponds to one of the four following cases:

	Case 1		Case 2		Case 3				Case 4			
Inputs:	Y	Y	Y	X	X	Y	–	X	X	Y	–	Y
Outputs KL:	10	01	10	00	01	00	00	00	01	00	00	01

where each hyphen may be replaced by X or Y.

This machine can identify which of the inputs, X or Y, is at the beginning and end of a binary sequence.

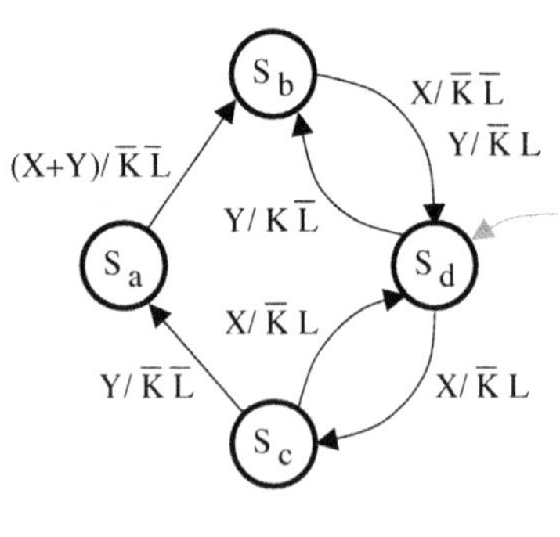

Figure 3.79. *State diagram*

Appendix

Overview of VHDL Language

A.1. Introduction

VHSIC Hardware Description Language (VHDL, where VHSIC denotes very high speed integrated circuits) is a hardware description language used to represent the behavior and architecture of a digital system. VHDL is characterized by the fact that it allows for the easy expression of the parallelism that is inherent to a circuit.

One of the goals of VHDL is to facilitate the development of digital circuits. Thus, the specifications of a system, described in VHDL, can be verified by simulation well before using a synthesis tool for transcription in the form of logic gates or programmable circuit (PROM, PAL, PLA, PLD, FPGA).

In VHDL, the description of any component has two aspects:

– the interface with the external world is described in the section named `ENTITY`;

– the function or structure to be implemented is described in the section named `ARCHITECTURE`.

In general, the description of a function or a structure is based on the use of concurrent instructions. However, in cases that might be too complex to be described in concurrent instructions, an algorithmic description, called `PROCESS`, can be chosen for convenience. The instructions used within a process are no longer concurrent but, rather, sequential. A `PROCESS`-type declaration provides a behavioral description but not a structural one.

A.2. Principles of VHDL

VHDL is used to model a digital system as an assembly of entities that can be described on one of the following three levels:

– structural;

– behavioral;

– dataflow.

Unlike the structural description, the behavioral description is not directly related to the basic elements of a system, but it is based on the use of sequential algorithms. Dataflow description uses concurrent instructions to assign the corresponding values to different signals.

A.2.1. *Names*

In VHDL, a name is used to identify the following elements: ENTITY, ARCHITECTURE, PACKAGE, PACKAGE BODY and CONFIGURATION. It must be composed of a letter followed by any number of letters or numbers without any spaces. An underscore can be used within a name, but not at the beginning or end. Additionally, two consecutive underscores are not allowed.

It should be noted that VHDL does not differentiate between upper case and lower case characters.

Certain names or identifiers are used as keywords in VHDL. These are *reserved* words:

```
ABS ACCESS AFTER ALIAS ALL AND ARCHITECTURE ARRAY ASSERT
ATTRIBUTE BEGIN BLOCK BODY BUFFER BUS CASE COMPONENT
CONFIGURATION CONSTANT DISCONNECT DOWNTO ELSE ELSIF END ENTITY
EXIT FILE FOR FUNCTION GENERATE GENERIC GROUP GUARDED IF IMPURE
IN INERTIAL INOUT IS LABEL LIBRARY LINKAGE LITERAL LOOP MAP MOD
NAND NEW NEXT NOR NOT NULL OF ON OPEN OR OTHERS OUT PACKAGE PORT
POSTPONED PROCEDURE PROCESS PURE RANGE RECORD REGISTER REJECT REM
REPORT RETURN ROL ROR SELECT SEVERITY SIGNAL SHARED SLA SLL SRA
SRL SUBTYPE THEN TO TRANSPORT TYPE UNAFFECTED UNITS UNTIL USE
VARIABLE WAIT WHEN WHILE WITH XNOR XOR
```

A.2.2. *Comments*

A comment line begins with two hyphens and is ignored by the VHDL compiler.

EXAMPLE A.1.– (Comment Line).

```
-- This is a comment
-- Description of a state machine
```

A.2.3. *Library and packages*

A library is a collection of precompiled design entities. It must be declared at the beginning of a VHDL file with the instruction `LIBRARY`.

VHDL supports two predefined logic libraries: the default work library, `WORK`, where the compiled descriptions are stored, and the library containing definitions on the types and basic functions, `STD`.

A `PACKAGE` is used to group declarations and descriptions of types, sub-types, components, constants or subprogrammes to store them in the library.

The `USE` instruction is used to declare a package before its use.

A.2.4. *Ports*

The `PORT` declaration for an entity gives the definition for the input and output pin of the component. The direction in which a pin works is specified using one of the following modes: `IN` (input), `OUT` (output), `INOUT` (input/output) and `BUFFER` (an output, but which can also be connected to a feedback path returning inside the entity).

A.2.5. *Signal and variable*

A signal transports information between the input nodes, output nodes and internal nodes. A value is allocated to a signal using the operator $<=$ and the change is effective from the next iteration of the simulation (with a delay, delta).

A signal declared in a `PACKAGE` is global; in an `ENTITY`, it is common to all architectures, and in an `ARCHITECTURE`, it is local.

A variable is used to store intermediate results. It is only employed in a process and is, thus, always local. The variable's value can be modified using an $:=$ operator. Unlike a signal, a variable takes its new value as soon as it is allocated.

A.2.6. *Data types and objects*

The type determines the values that can be assigned to a pin or that can be taken by a signal or a variable. In VHDL, any object that explicitly or implicitly belongs to a type and any data (signal, variable, etc.) must be declared before use by indicating its type.

Existing types are as follows:

– scalar: integer, real, enumeration (characterized by an enumeration of all possible values), physical (which makes it possible to represent values of physical quantities);

– composites: tables, recordings (defined as a collection of named elements (or fields) whose values may be of different types);

– pointers;

– files.

VHDL has a certain number of predefined types and subtypes. Their declarations are included in the STANDARD package of the STD library.

To use these types, the following directives must be included in the source code:

```
LIBRARY ieee;
USE ieee.std_logic_1164.ALL;
USE ieee.numeric_std.ALL;
```

It should be noted that it is not possible to mix objects of different types in an expression without specifying conversion functions. Development environments generally provide packages that contain type conversion functions. To use these functions, we can just access the std_ logic_arith package of the ieee library.

```
USE ieee. std_logic_arith.ALL;
```

Additionally, a user can define new types (or subtypes) using basic types or types predefined by all compilers.

A.2.7. *Attributes*

An attribute is a characteristic associated with a type or an object that can be evaluated either at the time of compilation or during the simulation. Each attribute is referenced by its name, which comprises a prefix, an apostrophe, and the attribute itself. The prefix must correspond to a type, subtype, table or block.

There are two classes of attributes: predefined attributes and attributes declared and specified by the user.

EXAMPLE A.2.– Among the attributes, we have:

– a'LEFT is the leftmost subscript of a vector or table, a;

– a'RANGE[(n)] refers to the range of dimension n of a;

– a'LENGTH[(n)] refers to the number of elements of the n-th index of a;

– s'STABLE[(t)] is a Boolean signal that is true if s has not changed value in the time interval t (optional);

– s'EVENT represents a Boolean function that is true if an event occurs on a signal in the current simulation cycle.

The following VHDL code can be used to obtain a circuit with a maximum delay of 10 on all the output ports using the attribute max_delay:

```
ENTITY example IS
   PORT (a, b: IN BIT;
            c: IN BIT_VECTOR (1 TO 3);
         x, y: OUT BIT;
            z: OUT BIT_VECTOR (1 TO 3));
   ATTRIBUTE max_delay OF x, y, z: SIGNAL IS 10.0;
END example;
```

A.2.8. *Entity and architecture*

A digital system is described in a VHDL file as an entity or several interconnected entities. Each entity is modeled by an ENTITY declaration and an ARCHITECTURE section, which is made up of entities, processes and interconnected components operating concurrently.

EXAMPLE A.3.– (Entity and architecture).

```
LIBRARY ieee;
USE ieee.std_logic_1164.ALL;

ENTITY andgate IS
  PORT (a: IN STD_LOGIC;
        b: IN STD_LOGIC;
        f: OUT STD_LOGIC);
END andgate;

ARCHITECTURE andgate_beh OF andgate IS
BEGIN
  PROCESS(a, b)
  BEGIN
    f <= a AND b;
```

```
  END PROCESS;
END andgate_beh;
```

This is a behavioral description for a two-input AND gate. The sensitivity list for the process includes the inputs a and b, such that whenever one or the other undergoes a change, the process is executed again resulting in the generation of an updated output.

A three-state buffer can be described in VHDL as follows:

```
LIBRARY ieee;
USE ieee.std_logic_1164.ALL;

ENTITY tristate IS
    PORT ( x: IN STD_LOGIC;
          en: IN STD_LOGIC;
           y: OUT STD_LOGIC);
END tristate;

ARCHITECTURE tristate_beh OF tristate IS
BEGIN
    y <= x WHEN (en = '1') ELSE 'Z';
END tristate_beh;
```

A.3. Concurrent instructions

In a logic circuit, each gate or logic operator may be considered as a concurrent structure. A change applied to several concurrent structures affects all of them simultaneously.

A concurrent instruction is used to assign the value of a Boolean expression or a constant to a gate or signal. It is useful for the description of a Boolean equation. As the logic operators NOT, AND, OR, and XOR have the same priority in VHDL, the order of priority must be explicitly established by parentheses.

EXAMPLE A.4.– A full adder can be described in VHDL as follows:

```
LIBRARY ieee;
USE ieee.std_logic_1164.ALL;

ENTITY FullAdder IS
  PORT (a, b, ci: IN STD_LOGIC;
        c0, s   : OUT STD_LOGIC);
```

```
END FullAdder;

ARCHITECTURE FullAdder_beh OF FullAdder IS
BEGIN
  PROCESS(a, b, ci)
  BEGIN
    -- Concurrent statement
    c0 <= (a AND b) OR (b AND ci) OR (a AND ci);
    s <= a XOR b XOR ci;
  END PROCESS;
END FullAdder_beh;
```

This is dataflow type architecture.

A.3.1. *Concurrent instructions with selective assignment*

A concurrent instruction with selective assignment is used to assign different values to a port or signal depending on the values taken by a selection signal.

EXAMPLE A.5.– A 4 : 1 multiplexer can be described using concurrent instructions with selective assignment as follows:

```
LIBRARY ieee;
USE ieee.std_logic_1164.ALL;

ENTITY Mux4_1 IS
  PORT (x0, x1, x2, x3: IN STD_LOGIC;
        s: IN STD_LOGIC_VECTOR(1 downto 0);
        y: OUT STD_LOGIC);
END Mux4_1;

ARCHITECTURE Mux4_1beh OF Mux4_1 IS
BEGIN
  WITH s SELECT
    y <= x0 WHEN "00",
         x1 WHEN "01",
         x2 WHEN "10",
         x3 WHEN "11";
END Mux4_1beh;
```

A.3.2. *Concurrent instructions with conditional assignment*

A concurrent instruction with selective assignment is used to modify the state of a port or a signal based on the result of a test or condition.

EXAMPLE A.6.– The following VHDL description corresponds to that of an 8 : 3 priority encoder:

```
LIBRARY ieee;
USE ieee.std_logic_1164.ALL;

ENTITY PriorityEncoder83 IS
  PORT (din  : IN STD_LOGIC_VECTOR(7 downto 0);
        dout : OUT STD_LOGIC_VECTOR(2 downto 0)
        );
END PriorityEncoder83;
mux2_1beh
ARCHITECTURE PriorityEncoder83_beh OF PriorityEncoder83 IS
BEGIN
  dout <= "111" WHEN din(7)='1' ELSE
          "110" WHEN din(6)='1' ELSE
          "101" WHEN din(5)='1' ELSE
          "100" WHEN din(4)='1' ELSE
          "011" WHEN din(3)='1' ELSE
          "010" WHEN din(2)='1' ELSE
          "001" WHEN din(1)='1' ELSE
          "000";
END PriorityEncoder83_beh;
```

A.4. Components

The VHDL representation of a component consists of a user interface (or ENTITY) and a description of the function implemented (or ARCHITECTURE).

The structural description of a more complex component can be implemented by interconnecting other components. It is composed of three sections:

– declarations of components models used as defined in the entities of these components;

– declarations of the internal signals that will interconnect the components;

– the instantiation of each component, which consists of specifying the parameters (GENERIC MAP) and the internal wiring or the input and output connections (PORT MAP).

EXAMPLE A.7.– A four-bit adder can be described using the following VHDL code:

```
LIBRARY ieee;
USE ieee.std_logic_1164.ALL;

ENTITY Adder_4bit IS
   PORT(
         carryi: IN STD_LOGIC;
         x, y  : IN STD_LOGIC_VECTOR(3 DOWNTO 0);
         carry0: OUT STD_LOGIC;
         sum   : OUT STD_LOGIC_VECTOR(3 DOWNTO 0)
         );
END Adder_4bit;

ARCHITECTURE Adder_4bit_arc OF Adder_4bit IS
   -- Internal carry
   SIGNAL c: STD_LOGIC_VECTOR (3 DOWNTO 1);
   COMPONENT FullAdder   -- Defined in FullAdder.vhd
   PORT (a, b, ci: IN STD_LOGIC;
         c0, s   : OUT STD_LOGIC
          );
   END COMPONENT;
BEGIN       -- Four full adder components interconnected
  FA1: FullAdder PORT MAP (x(1),y(1),carryi,c(1),sum(1));
  FA2: FullAdder PORT MAP (x(2),y(2),c(1),c(2),sum(2));
  FA3: FullAdder PORT MAP (x(3),y(3),c(2),c(3),sum(3));
  FA4: FullAdder PORT MAP (x(4),y(4),c(3),carry0,sum(4));
END Adder_4bit_arc;
```

Each instance of a component is a unique copy of this component with a name and a list of ports (or inputs and outputs). The `PORT MAP` instruction is used to describe the wiring of the different ports of a component:

– the ports can be associated by position:

```
gate1:  and PORT MAP (a, b, f);
```

– the ports can be associated by denomination:

```
gate2:  and PORT MAP (s1 => a, s2 => b, en => f);
```

– a port can be open:

```
muxcir:  mux21 PORT MAP (i1, open, s);
```

A value must be assigned by default to an unconnected (or open) input.

A.4.1. *Generics*

Generics, or the GENERIC instruction, is used to describe parameterized blocks (or entities), which can then be instantiated with the values of parameters to implement a component. It is declared at the entity level and not the architecture level. In the case of generic parameters, it is possible to specify the values by default, using the := operator.

EXAMPLE A.8.– The following VHDL description is that of a 2 : 1 multiplexer whose generic parameters are the length of words, (tb+1), and the propagation delay, (pd):

```
LIBRARY IEEE;
USE IEEE.std_logic_1164.ALL;

ENTITY mux2_1g IS
     GENERIC(tb : natural := 15;      -- top bit
             pd : time := 100 ps);   -- propagation delay
      PORT(din0 : IN  STD_LOGIC_VECTOR (tb DOWNTO 0);
           din1 : IN  STD_LOGIC_VECTOR (tb DOWNTO 0);
            sel : IN  STD_LOGIC;
           dout : OUT STD_LOGIC_VECTOR (tb DOWNTO 0));
END ENTITY mux2_1g;

ARCHITECTURE mux2_1beh OF mux2_1g IS
BEGIN   -- no process needed with concurrent statements
  dout <= din1 WHEN sel='1' OR sel='H'
               ELSE din0 AFTER pd;
END mux2_1beh;
```

A.4.2. *The GENERATE Instruction*

The GENERATE instruction allows for an iterative or conditional elaboration of a group of concurrent instructions, thus providing a more compact description.

EXAMPLE A.9.– Another description of the four-bit adder can be obtained using the GENERATE instruction as follows:

```
LIBRARY ieee;
USE ieee.std_logic_1164.ALL;

ENTITY Adder_4bit IS
   PORT(
         carryi: in STD_LOGIC;
```

```
        x, y  : in STD_LOGIC_VECTOR(3 DOWNTO 0);
        carry0: out STD_LOGIC;
        sum   : out STD_LOGIC_VECTOR(3 DOWNTO 0)
        );
END Adder_4bit;

ARCHITECTURE Adder_4bit_arc OF Adder_4bit IS
   SIGNAL c: STD_LOGIC_VECTOR (4 DOWNTO 0); -- Carry bits
   COMPONENT FullAdder   -- Defined in FullAdder.vhd
   PORT (a, b, ci: in STD_LOGIC;
         c0, s   : out STD_LOGIC
          );
   END COMPONENT
BEGIN
  c(0) <= carryi;
  -- Instantiate four full adders
  Adders:   -- Note that a label is required here
  FOR i IN 1 TO 4 GENERATE
     FA: FullAdder
        PORT MAP(x(i), x(i), c(i-1), c(i), sum(i));
  END GENERATE;
  carry0 <= c(4);
END Adder_4bit_arc
```

In general, the For instruction is used to repeat the execution of the same sequence of instructions several times. All that is required is designating, in the syntax, the parameter that will serve as a counter (or down counter), its initial value and final values.

The WHILE instruction provides another approach for executing the same sequence of instructions multiple times. The sequence of instructions is executed as long as a condition is satisfied. However, the WHILE instruction is primarily used to describe the models for simulation and implementation of test benches.

A.4.3. *Process*

A process (PROCESS) behaves, from an external point of view, as a concurrent instruction, even though it consists of sequential instructions just like the classical control structures of programming languages (if-else, case-when, for/while), which offer the possibility of implementing more complex logic functions.

The execution of a process only takes place when there is a change in the logic state of one or more signals, whose names are defined in the sensitivity list at the process declaration step.

A.5. Sequential structures

VHDL supports the sequential structures in which the order of the instructions affects the result of the execution.

A sequential structure must always be placed in a process. Even though it can be used to describe a sequential circuit as well as a combinational circuit, it is especially used to describe sequential circuits such as flip-flops, registers, counters and finite state machines.

A.5.1. *The IF instruction*

The IF instruction is used to implement conditional loops. It allows for the execution of a series of operations provided a condition is satisfied. Several IF loops can be nested within each other.

When an IF instruction is followed by several instructions, it is essential to terminate the series of instructions by the expression END IF. If, however, there is only one instruction, the END IF expression is not required.

The IF ... THEN ... ELSE expression is used to execute another series of instructions when the condition is not satisfied.

The IF ... THEN ... ELSEIF ... ELSE instruction is used to sequence a series of instructions, while avoiding nested IF instructions.

EXAMPLE A.10.– A D flip-flop activated by the high level of the En signal can be described in VHDL as follows:

```
LIBRARY ieee;
USE ieee.std_logic_1164.ALL;

ENTITY dlatch IS
     PORT (d, en : IN STD_LOGIC;
           q, qb : OUT STD_LOGIC);
END dlatch;

ARCHITECTURE dlatch_beh OF dlatch IS
BEGIN
  PROCESS (d, en)
    BEGIN
      IF en='1' THEN
         q <= d;
```

```
        qb <= not d;
      END IF;
    END PROCESS;
END dlatch_beh;
```

The following VHDL description corresponds to that of a D flip-flop triggered by the rising edge of a clock signal and having one asynchronous reset input:

```
LIBRARY ieee;
USE ieee.std_logic_1164.ALL;

ENTITY dff_clear IS
     PORT (clk, d, clr : in STD_LOGIC;
           q, qb       : out STD_LOGIC);
END dff_clear;

ARCHITECTURE dff_beh OF dff_clear IS
BEGIN
  dff_process: PROCESS (clk, clr)
     BEGIN
           IF (clr='0') THEN
                q <= '0';
                qb <= '1';
           ELSIF RISING_EDGE(clk) THEN
                q <= d;
                qb <= not d;
           END IF;
     END PROCESS;
END dff_beh;
```

The detection of one of the edges of the clock signal can be implemented using the expression `RISING_EDGE(clk)` (or `clk'EVENT and clk = '1'`) for the rising edge and the expression `FALLING_EDGE(clk)` (or `clk'EVENT and clk = '0'`) for the falling edge.

An n-bit right-shift register can be described in VHDL as follows:

```
LIBRARY ieee;
USE ieee.std_logic_1164.ALL;

ENTITY shift_reg IS
GENERIC(n: integer:=4);
PORT( clk, din, rst: IN STD_LOGIC;
```

```
      dout          : OUT STD_LOGIC);
END shift_reg;

ARCHITECTURE behavioral OF shift_reg IS
SIGNAL reg : STD_LOGIC_VECTOR (n-1 DOWNTO 0);
BEGIN

  PROCESS(clk)
  BEGIN
    IF RISING_EDGE(clk) THEN
      IF(rst='1') THEN
        -- n bit synchronous reset
        reg(n-1 DOWNTO 0) <= (others => '0');
      ELSE
        -- Shift right
        reg(n-1 DOWNTO 0) <= din & reg(n-2 DOWNTO 0);
      END IF;
    END IF;
  END PROCESS;

  dout <= reg(n-1);

END behavioral;
```

The following VHDL description corresponds to that of an n-bit synchronous binary counter/down counter:

```
LIBRARY ieee;
USE ieee.std_logic_1164.ALL;
USE ieee.std_logic_unsigned.ALL;

ENTITY up_counter_en IS
   GENERIC (
            bit_width : INTEGER := 4; -- Structure
            out_delay : TIME := 3 ns  -- Behavior
           );
   PORT (
      reset_n : IN STD_LOGIC;
      clk : IN STD_LOGIC;
      en : IN STD_LOGIc;
      down_count: IN STD_LOGIC;
      count : OUT STD_LOGIC_VECTOR(bit_width-1 downto 0)
         );
END up_counter_en;
```

```
ARCHITECTURE behav OF up_counter_en IS
SIGNAL count_s : std_logic_vector(BIT_WIDTH-1 downto 0);
BEGIN

  PROCESS (clk, reset_n)
  BEGIN
    IF(reset_n='0') THEN
      count_s <= (others => '0'); -- Asynchronous reset
    ELSIF RISING_EDGE(clk) THEN
      IF (en='1') THEN
         IF (down_count='0') THEN
            count_s <= count_s + 1;
         ELSE
            count_s <= count_s - 1;
         END IF;
      ELSE
         count_s <= count_s;
      END IF;
    END IF;
  END PROCESS;

    count <= count_s AFTER out_delay;

END behav;
```

The WAIT instruction can be used to model asynchronous circuits or systems.

The following VHDL code describes an asynchronous circuit with two inputs, x and y, and one output, z. If a transition from 0 to 1 at the input x is immediately followed by a transition from 1 to 0 at the other input, y, then the output z takes the logic state 1. The output z remains at this state, 1, until the input x is reset to 0 or the input y is set to 1. Hence:

```
LIBRARY ieee;
USE ieee.std_logic_1164.ALL;

ENTITY edge_detector IS
  PORT (x, y: IN BIT;
        z   : OUT BIT);
END edge_detector;

ARCHITECTURE asynchronous OF edge_detector IS
BEGIN
```

```
  PROCESS
  BEGIN
    WAIT UNTIL x='1';
    WAIT ON x, y;
    IF y='0' AND NOT y'STABLE THEN
       z <= '1';
       WAIT ON x, y;
       z <= '0';
    END IF;
  END PROCESS;
END asynchronous;
```

The IF declaration is used to check whether the change taking place at the input y, after the rising edge at the input x, is the falling edge.

The following VHDL description gives the structural description of a modulo 10 synchronous counter with a reset signal and an enable signal. The counter is composed of a sequential section (four flip-flops) and a combinational section. Two counters can be cascaded using the rco signal. That is:

```
LIBRARY ieee;
USE ieee.std_logic_1164.ALL;
USE ieee.std_logic_unsigned.ALL;
USE ENTITY WORK.dff

ENTITY mod10cnt IS
  PORT (clk, clr, en : IN BIT;
              count : OUT BIT_VECTOR(3 DOWNTO 0);
                rco : OUT BIT);

ARCHITECTURE mod10cnt_arc OF mod10cnt IS
  COMPONENT dff
       PORT (clk, d, clr : IN BIT;
                   q, qb : OUT BIT);
  END COMPONENT;

  SIGNAL d, q, qb : BIT_VECTOR(3 DOWNTO 0);

BEGIN

  seq_sec : FOR i IN 3 DOWNTO 0 GENERATE
            b : dff PORT MAP (clk, d(i), clr, q(i), qb(i));
            END GENERATE;
```

```
  d(3) <= (en AND q(2) AND q(1) AND q(0)) OR (q(3) AND
                                           ((NOT en) OR qb(0)));
  d(2) <= (en AND qb(2) AND q(1) AND q(0)) OR
           (q(2) AND ((NOT en) OR qb(1) OR qb(0)));
  d(1) <= (en AND qb(3) AND qb(1) AND q(0)) OR (q(1) AND
                                            ((NOT en) OR qb(0)));
  d(0) <= en XOR q(0);

  count <= q;
  rco <= q(3) AND q(0);

END mod10cnt_arc;
```

A.5.2. *CASE instruction*

The CASE instruction is used to execute one among several instruction sequences depending on the value of the same expression.

In the CASE instruction, an expression is evaluated and its value is compared to that of each of the possible choices and the instructions associated with the corresponding WHEN clause are executed.

The following restrictions are placed on the possible choices:

– it is not acceptable to have two choices if they overlap;

– all the eventual values of the tested expression must be part of the set of choices, unless the NO OTHERS clause is used as the last choice.

EXAMPLE A.11.– Using the CASE instruction, the VHDL description for a 4 : 1 multiplexer takes the following form:

```
LIBRARY IEEE;
USE IEEE.std_logic_1164.ALL;

ENTITY  Mux4_1 IS
   PORT (
         s0, s1, in0, in1, in2, in3: IN  STD_LOGIC;
         output                    : OUT STD_LOGIC
        );
END Mux4_1;

ARCHITECTURE Mux4_1beh OF Mux4_1 IS
BEGIN
```

```
  Mux: PROCESS(s0, s1, in0, in1, in2, in3)
  VARIABLE  sel: STD_LOGIC_VECTOR(1 DOWNTO 0);
  BEGIN
    sel := s1 & s0;    -- concatenate s1 and s0
    CASE sel IS
      WHEN  "00"  =>  output <= in0;
      WHEN  "01"  =>  output <= in1;
      WHEN  "10"  =>  output <= in2;
      WHEN  "11"  =>  output <= in3;
    END CASE;
  END PROCESS Mux;
END Mux4_1beh;
```

The CASE instruction is suitable for the description of a state table (or state diagram).

Consider the 1011 sequence detector that operates according to the state table shown in Table A.1.

PS	NS		Output Y	
	$X = 0$	1	$X = 0$	1
S_0	S_0	S_1	0	0
S_1	S_2	S_1	0	0
S_2	S_0	S_3	0	0
S_3	S_0	S_0	0	1

Table A.1. *State table of the 1011 sequence detector (Mealy model)*

This is a Mealy model based state machine whose VHDL description is as follows:

```
-- Mealy state machine: 1011 sequence detector
LIBRARY IEEE;
USE IEEE.std_logic_1164.ALL;

ENTITY DetectSq IS
       PORT (reset, clk, x : IN STD_LOGIC;
                         y : OUT STD_LOGIC);
END DetectSq;

ARCHITECTURE behavioral OF DetectSq IS
```

```
   TYPE state_type IS (s0, s1, s2, s3);
   SIGNAL presentS, nextS: state_type;

BEGIN
    PROCESS (reset, clk) -- Clocked (state) process
    BEGIN
      IF (reset=’0’) THEN
          presentS <= s0;
      ELSIF RISING_EDGE(clk) THEN
         CASE presentS IS
            WHEN s0 =>  IF x=’1’ THEN nextS <= s1;
                       ELSE nextS <= s0;
                       END IF;
            WHEN s1 =>  IF x=’0’ THEN nextS <= s2;
                       ELSE nextS <= s1;
                       END IF;
            WHEN s2 =>  IF x=’1’ THEN nextS <= s3;
                       ELSE nextS <= s0;
                       END IF;
            WHEN s3 =>  nextS <= s0;
         END CASE;
      END IF;
    END PROCESS;

    PROCESS (nextS)          -- Combinational process
    BEGIN
       presentS <= nextS;
       CASE presentS IS
          WHEN s0 => y <= ’0’;
          WHEN s1 => y <= ’0’;
          WHEN s2 => y <= ’0’;
          WHEN s3 => IF x=’1’ THEN y <= ’1’;
                    ELSE y <= ’0’;
                    END IF;
       END CASE;
   END PROCESS;

END behavioral;
```

A.6. Testbench

To simulate a VHDL module, another VHDL code, called *testbench*, must be written. A testbench can read the test signals from a file and apply them to the model under test. The output signals are then recovered for analysis.

A process resulting in the generation of signals does not have a sensitivity list.

The execution of a cyclic process restarts each time that it reaches the declaration `END PROCESS`, unless it has been interrupted earlier by an unconditional wait instruction.

The `WAIT` instruction indefinitely suspends the process from the time it is executed by the simulator.

EXAMPLE A.12.– The operation of the AND logic gate model can be verified using the following VHDL testbench:

```
-- And gate testbench
LIBRARY IEEE;
USE IEEE.std_logic_1164.ALL;
USE WORK.andgate;

ENTITY andgate_tb IS
END andgate_tb;

ARCHITECTURE tb OF andgate_tb IS

   COMPONENT andgate IS
   PORT(a, b : IN STD_LOGIC;
           f : OUT STD_LOGIC);
   END COMPONENT;

   SIGNAL a, b, f: STD_LOGIC;

BEGIN
       -- Create a test instance of the and gate
   andut: andgate PORT MAP(a => a, b => b, f => f);
        -- Now define a process to apply some stimulus
        -- over time
   PROCESS
      CONSTANT period: TIME := 40 ns;
   BEGIN
      a <= '0';
```

```
      b <= '0';
      WAIT FOR period;
      ASSERT (f='1')
      REPORT "Test failed!" severity error;
      a <= '0';
      b <= '1';
      WAIT FOR period;
      ASSERT (f='1')
      REPORT "Test failed!" severity ERROR;
      a <= '1';
      b <= '0';
      WAIT FOR period;
      ASSERT (f='1')
      REPORT "Test failed!" severity ERROR;
      a <= '1';
      b <= '1';
      WAIT FOR period;
      ASSERT (f='0')
      REPORT "Test failed!" severity ERROR;

      WAIT;   -- stop running

   END PROCESS;
END tb;

CONFIGURATION cfg_tb OF andgate_tb IS
   FOR tb
   END FOR;
END cfg_tb;
```

A CONFIGURATION declaration is used to connect an entity to a specific architecture for synthesis or simulation.

The following VHDL testbench can be used to carry the functional simulation of a 1011 sequence detector. The sequence of bits applied at the input of the detector is 1101110101. Thus:

```
LIBRARY ieee;
USE ieee.std_logic_1164.ALL;
USE WORK.DetectSq;

ENTITY tb_DetectSq IS
END tb_DetectSq;
```

```
ARCHITECTURE tb OF tb_DetectSq IS

   COMPONENT DetectSq
         PORT (reset, clk, x : IN STD_LOGIC;
                           y : OUT STD_LOGIC);
   END COMPONENT;

   SIGNAL reset, clk, x: STD_LOGIC:= '0';
   SIGNAL y: STD_LOGIC;

   CONSTANT clk_period : time := 10 ns;

BEGIN

    -- Instantiate the unit under test (UUT)
   uut: ENTITY DetectSq PORT MAP (
                            reset => reset,
                            clk   => clk,
                            x     => x,
                            y     => y );

   -- Clock process definitions
   clk_process : PROCESS
   BEGIN
        clk <= '0';
        WAIT FOR clk_period/2;
        clk <= '1';
        WAIT FOR clk_period/2;
   END PROCESS;

   -- Stimulus process : Apply the bits in the sequence
   -- one by one.
   stim_proc: process
   BEGIN
        x <= '1';              --1
      WAIT FOR clk_period;
        x <= '1';              --11
      WAIT FOR clk_period;
        x <= '0';              --110
      WAIT FOR clk_period;
        x <= '1';              --1101
      WAIT FOR clk_period;
        x <= '1';              --11011
```

```
      WAIT FOR clk_period;
        x <= '1';               --110111
      WAIT FOR clk_period;
        x <= '0';               --1101110
      WAIT FOR clk_period;
        x <= '1';               --11011101
      WAIT FOR clk_period;
        x <= '0';               --110111010
      WAIT FOR clk_period;
        x <= '1';               --1101110101
      WAIT FOR clk_period;
      WAIT;
   END PROCESS;

END;

CONFIGURATION cfg_tb_DetectSq OF tb_DetectSq IS
    FOR tb
    END for;
END cfg_tb_DetectSq;
```

Bibliography

[BRO 08] BROWN S., VRANESIC Z., *Fundamentals of Digital Logic with VHDL Design*, 3rd ed., McGraw-Hill Education, NY, 2008.

[CLE 00] CLEMENTS A., *The Principles of Computer Hardware*, 3rd ed., Oxford University Press, 2000.

[COM 95] COMER D. J., *Digital Logic and State Machine Design*, 3rd ed., Oxford University Press, NY, 1995.

[DUE 01] DUECK R. K., *Digital Design with CPLD Applications and VHDL*, Delmar Thomson Learning, NY, 2001.

[GIV 03] GIVONE D., *Digital Principles and Design*, McGraw-Hill, NY, 2003.

[HAY 93] HAYES J. P., *Introduction to Digital Logic Design*, Addison-Wesley, MA, 1993.

[HAY 98] HAYES J. P., *Computer Architecture and Organization*, McGraw-Hill, NY, 1998.

[KAT 05] KATZ R. H., BORRIELO G., *Contemporary Logic Design*, 2nd ed., Prentice Hall, NJ, 2005.

[MAN 01] MANO M. M., *Digital Design*, 3rd ed., Prentice Hall, NJ, 2001.

[MAR 10] MARCOVITZ A. B., *Introduction to Logic Design*, 3rd ed., McGraw-Hill Education, NY, 2010.

[NDJ 11] NDJOUNTCHE T., *CMOS Analog Integrated Circuits: High-Speed and Power-Efficient Design*, CRC Press, FL, 2011.

[ROT 04] ROTH Jr. C. H., *Fundamental of Logic Design*, 5th ed., Brooks/Cole – Thomson Learning, Belmont, CA, 2004.

[SAN 02] SANDIGE R. S., *Digital Design Essentials*, Prentice Hall, NJ, 2002.

[TIN 00] TINDER R. F., *Engineering Digital Design*, Academic Press, CA, 2000.

[TOC 03] TOCCI R. J., Ambrosio F. J., *Microprocessors and Microcomputers*, 6th ed., Prentice Hall, NJ, 2003.

[WAK 00] WAKERLY J. F., *Digital Design Principles and Practices*, 3rd ed., Prentice Hall, NJ, 2000.

[WIL 98] WILKINSON B., *The Essence of Digital Design*, Prentice Hall Europe, UK, 1998.

[YAR 97] YARBROUGH J. M., *Digital Logic – Applications and Design*, West Publishing Company, MN, 1997.

Index

F, G

H, I, J

L, M

O, P

R, S

Other titles from

iSTE

in

Electronics Engineering

2016

BAUDRAND Henri, TITAOUINE Mohammed, RAVEU Nathalie
The Wave Concept in Electromagnetism and Circuits: Theory and Applications

FANET Hervé
Ultra Low Power Electronics and Adiabatic Solutions

NDJOUNTCHE Tertulien
Digital Electronics 1: Combinational Logic Circuits
Digital Electronics 2: Sequential and Arithmetic Logic Circuits

2015

DURAFFOURG Laurent, ARCAMONE Julien
Nanoelectromechanical Systems

2014

APPRIOU Alain
Uncertainty Theories and Multisensor Data Fusion

CONSONNI Vincent, FEUILLET Guy
Wide Band Gap Semiconductor Nanowires 1: Low-Dimensionality Effects and Growth
Wide Band Gap Semiconductor Nanowires 2: Heterostructures and Optoelectronic Devices

GAUTIER Jean-Luc
Design of Microwave Active Devices

LACAZE Pierre Camille, LACROIX Jean-Christophe
Non-volatile Memories

TEMPLIER François
OLED Microdisplays: Technology and Applications

THOMAS Jean-Hugh, YAAKOUBI Nourdin
New Sensors and Processing Chain

2013

COSTA François, GAUTIER Cyrille, LABOURE Eric, REVOL Bertrand
Electromagnetic Compatibility in Power Electronics

KORDON Fabrice, HUGUES Jérôme, CANALS Agusti, DOHET Alain
Embedded Systems: Analysis and Modeling with SysML, UML and AADL

LE TIEC Yannick
Chemistry in Microelectronics

2012

BECHERRAWY Tamer
Electromagnetism: Maxwell Equations, Wave Propagation and Emission

LALAUZE René
Chemical Sensors and Biosensors

LE MENN Marc
Instrumentation and Metrology in Oceanography

SAGUET Pierre
Numerical Analysis in Electromagnetics: The TLM Method

2011

ALGANI Catherine, RUMELHARD Christian, BILLABERT Anne-Laure
Microwaves Photonic Links: Components and Circuits

BAUDRANT Annie
Silicon Technologies: Ion Implantation and Thermal Treatment

DEFAY Emmanuel
Integration of Ferroelectric and Piezoelectric Thin Films: Concepts ans Applications for Microsystems

DEFAY Emmanuel
Ferroelectric Dielectrics Integrated on Silicon

BESNIER Philippe, DÉMOULIN Bernard
Electromagnetic Reverberation Chambers

LANDIS Stefan
Nano-lithography

2010

LANDIS Stefan
Lithography

PIETTE Bernard
VHF / UHF Filters and Multicouplers

2009

DE SALVO Barbara
Silicon Non-volatile Memories / Paths of Innovation

DECOSTER Didier, HARARI Joseph
Optoelectronic Sensors

FABRY Pierre, FOULETIER Jacques
Chemical and Biological Microsensors / Applications in Fluid Media

GAUTIER Jacques
Physics and Operation of Silicon Devices in Integrated Circuits

MOLITON André
Solid-State Physics for Electronics

PERRET Robert
Power Electronics Semiconductor Devices

SAGUET Pierre
Passive RF Integrated Circuits

2008

CHARRUAU Stéphane
Electromagnetism and Interconnections

2007

RIPKA Pavel, TIPEK Alois
Modern Sensors Handbook

Lightning Source UK Ltd.
Milton Keynes UK
UKOW05n2054271116
288645UK00001B/3/P

9 781848 219861